Integrated Molecular and Cellular Biophysics

Valerica Raicu • Aurel Popescu

Integrated Molecular and Cellular Biophysics

 Springer

Valerica Raicu
University of Wisconsin
Milwaukee, WI
USA

Aurel Popescu
University of Bucharest
Bucharest-Magurele
Romania

ISBN 978-1-4020-8267-2   e-ISBN 978-1-4020-8268-9

Library of Congress Control Number: 2008921929

All Rights Reserved
© 2008 Springer Science + Business Media B.V.
No part of this work may be reproduced, stored in a retrieval system, or transmitted in any form or by any means, electronic, mechanical, photocopying, microfilming, recording or otherwise, without written permission from the Publisher, with the exception of any material supplied specifically for the purpose of being entered and executed on a computer system, for exclusive use by the purchaser of the work.

Printed on acid-free paper

9 8 7 6 5 4 3 2 1

springer.com

Contents

Preface ix

Introduction 1

1 The Molecular Basis of Life 7
   1.1 Molecular Interactions 7
      1.1.1 Interactions Between Polar Entities 8
      1.1.2 van der Waals Interactions 10
   1.2 Water and Polar Interactions 18
      1.2.1 Physical Properties of Water 19
      1.2.2 Overview of the Importance of Water for the Living Matter State 28
      1.2.3 The pH 29
   1.3 Hydrophobic "Interactions" and Molecular Self-Association 30
      1.3.1 The Hydrophobic Effect 30
      1.3.2 Amphiphilic Molecules 31
   References 37

2 The Composition and Architecture of the Cell 39
   2.1 The Cell: An Overview 39
      2.1.1 Eukaryotic Cells 40
      2.1.2 Cellular Self-Reproduction 43
      2.1.3 Cellular Metabolism 46
   2.2 Proteins 47
      2.2.1 Protein Structure 47
      2.2.2 Protein Folding 57
   2.3 Deoxyribonucleic Acid (DNA): The Cell's Legislative Power 60
      2.3.1 DNA Structure 60
      2.3.2 DNA Replication (Gene Autoreproduction) 66
   2.4 Determination of Molecular Structure 68
   References 70

3 Cell Membrane: Structure and Physical Properties ... 73
- 3.1 Membrane Structure ... 73
  - 3.1.1 Chemical Composition of the Plasma Membrane ... 74
  - 3.1.2 Spatial Architecture of the Plasma Membrane ... 75
- 3.2 Surface Charges ... 78
  - 3.2.1 Origin of the Surface Charges ... 78
  - 3.2.2 Electrical Double Layer ... 79
  - 3.2.3 Gouy-Chapman-Stern Theory of the Electrical Double Layer ... 79
- 3.3 Static Electrical Properties of Planar Membranes ... 88
  - 3.3.1 Electrical Parameters as Complex Quantities ... 88
  - 3.3.2 Dielectric Relaxation of a Dielectric Multi-Layer ... 90
  - 3.3.3 Dielectric Properties of Random Suspensions of Particles with Particular Relevance to Biological Cells ... 95
- References ... 98

4 Substance Transport Across Membranes ... 101
- 4.1 Brief Overview ... 101
- 4.2 Diffusion in Biological Systems ... 103
  - 4.2.1 Fick's Laws of Diffusion ... 103
  - 4.2.2 Simple Diffusion Through Membranes ... 105
  - 4.2.3 Determination of Membrane Permeability from Membrane Potential Energy Profile ... 107
- 4.3 Osmosis and Osmotic Pressure ... 110
  - 4.3.1 van't Hoff's Laws ... 111
  - 4.3.2 Deviations from van't Hoff Laws ... 113
  - 4.3.3 Osmotic Pressure of Biological Liquids ... 114
  - 4.3.4 The Cellular "Osmotic Pressure Menace" ... 115
- 4.4 Facilitated Transport ... 116
  - 4.4.1 Channel-Mediated Transport ... 116
  - 4.4.2 Carrier-Mediated Transport ... 116
  - 4.4.3 Main Characteristics of Facilitated Transport ... 120
- 4.5 Active Ion Transport ... 120
- References ... 122

5 Reaction, Diffusion and Dimensionality ... 123
- 5.1 Equilibrium and the Law of Mass Action ... 123
  - 5.1.1 Molecular Association ... 123
  - 5.1.2 Determination of Affinity Constant by Equilibrium Dialysis ... 125
  - 5.1.3 Competitive Binding ... 127
  - 5.1.4 Allosteric Activation and Inhibition of Binding ... 128
- 5.2 Introduction to Fractals ... 131
  - 5.2.1 "...The Measure of Everything" ... 132
  - 5.2.2 Examples of Fractals of Biological Interest ... 136
  - 5.2.3 Practical Considerations and Limitations of the Above Theory ... 138

Contents vii

    5.3   Fractal Diffusion and the Law of Mass Action 140
              5.3.1   Diffusion on Fractal Lattices 141
              5.3.2   The Carrier-Mediated Transport of Glucose Revisited 143
    References 145

6   Electrophysiology and Excitability 147
    6.1   Electric Charges and the Transmembrane Potential............... 147
              6.1.1   Models for Calculating the Transmembrane Potential 149
    6.2   Excitable Membranes 155
              6.2.1   Electrotonic Versus Action Potentials 155
              6.2.2   The Voltage-Clamp Technique and Transmembrane
                        Ionic Currents 160
              6.2.3   The Hodgkin-Huxley Model of Action Potential 164
    References 171

7   Structure and Function of Molecular Machines.................... 173
    7.1   Channels and Pores 173
              7.1.1   Electrical Behavior of Individual Ion Channels 174
              7.1.2   Structural Characterization by X-Ray Crystallography 177
    7.2   X-Ray Investigations of Channels and Pores 181
    7.3   Ion Pumps 185
              7.3.1   The Na-K-ATPase 185
              7.3.2   Other Ionic Pumps 188
    7.4   Light Absorption in Photosynthesis 189
              7.4.1   Brief Overview of the Mechanism of Photosynthesis....... 189
              7.4.2   Thermodynamics of Light Absorption 190
    References 192

8   Protein-Protein Interactions 195
    8.1   Probing Protein Association In Vivo 195
              8.1.1   Elementary Theory of Fluorescence and FRET 197
              8.1.2   FRET-Based Determinations of Interaction Stoichiometry .. 201
    8.2   Structural Studies of Protein–Protein Interactions 209
              8.2.1   Principles of Nuclear Magnetic Resonance (NMR) 209
              8.2.2   Chemical Shift and the NMR Spectroscopy 211
              8.2.3   NMR Studies of Protein-Protein Interactions 214
    References 216

Answers to Selected Quizzes 219

Appendix 227

Index 231

Preface

Biophysics represents perhaps one of the best examples of interdisciplinary research areas, where concepts and methods from disciplines such as physics, biology, biochemistry, colloid chemistry, and physiology are integrated. It is by no means a new field of study and has actually been around, initially as quantitative physiology and partly as colloid science, for over a hundred years. For a long time, biophysics has been taught and practiced as a research discipline mostly in medical schools and life sciences departments, and excellent biophysics textbooks have been published that are targeted at a biologically literate audience.

With a few exceptions, it is only relatively recently that biophysics has started to be recognized as a physical science and integrated into physics departments' curricula, sometimes under the new name of biological physics. In this period of crystallization and possible redefinition of biophysics, there still exists some uncertainty as to what biophysics might actually represent. A particular tendency among physicists is to associate biophysics research with the development of powerful new techniques that should eventually be used not by physicists to study physical processes in living matter, but by biologists in their biological investigations. There is value in that judgment, and excellent books have been published that introduce the interested reader to the use of physical principles for the development of new methods of investigation in life sciences. The tremendous progress registered in the development of photonics tools could serve as an excellent example in this regard. However, that is definitely not the whole story! Although the authors of this book have received their undergraduate and graduate education in Physics, and have gained a deep appreciation of the power of classical physical concepts as applied to other sciences, they believe that biophysics is not "merely" applied physics. There is also a component to biophysics that may well qualify as fundamental physics. We will illustrate this idea by drawing a parallel to astrophysics. It is perhaps widely understood that astrophysics is not fully equivalent to, let us say, classical mechanics applied to the study of the motion of planets, neither can it be reduced to the development of optical instruments for observation of such motions of planets, which are, admittedly, all indispensable tools in astrophysics research. Take for instance the existence of the microwave background, the black holes, the still puzzling rotational curves of galaxies which seem to suggest the existence of dark matter, or the even more puzzling accelerated expansion of the universe, which may imply the existence of exotic forms of dark energy: these phenomena can all qualify as fundamentally new physics or at least they did at the time of their discovery. Returning to biophysics, it ought to be said that, while previously unforeseen extensions of the classical physics (such as thermodynamics, electrodynamics and quantum mechanics) can go a long way into explaining the intricate behavior of biological systems, interesting new physics can be expected to be discovered, for example in the study of biological networks, or in the study of (protein) dynamics at mesoscale.

In our opinion, the new biophysicist should start his or her biophysics education with a background in physics and mathematics. According to our experience, it is favorable to graft biological ideas and concepts onto a physical and mathematical backbone that is constructed and consolidated early in one's career. We would like to stress out that much of the well-known physics has been fruitfully applied to the field of biology over decades by extraordinarily qualified quantitative physiologists and physical chemists. The physicist becoming a biophysicist should be willing to forge new directions of exploration into the physical reality and be ready to learn relevant biology and biochemistry terminology, much the same way that an astrophysicist employs the language of astronomy besides employing physical concepts and tools.

The present book aims to introduce the graduate physics students and also the interested physicists at any stage in their postdoctoral career to topics in "classical" biophysics and also to some up-to-date topics in current biophysics. In keeping with the widely accepted practice in other sub-fields of physics, we try to present the material in a quantitative, mathematical way, and assume familiarity with most of the known physical laws and concepts. The advanced undergraduate physics student and the life scientists interested in biophysics are also invited to join in, by building on their knowledge of basic physics. For the benefit of our readers that might need to refresh their knowledge of life sciences, essential notions of biochemistry and biology are introduced, as necessary, throughout the book, and especially in much of chapter 2. Life scientists may choose to skip those sections altogether, and delve directly into the quantitative biophysics sections.

The authors combined have over five decades of teaching and research experience in biophysics, and the selection of the topics covered in this book is based on what appeared to them as more commonly studied topics in biophysics. However, given the vastness of the biophysics research, originating in the diversity of the living matter, no single book would be expected to do justice to all the interesting topics in biophysics. Therefore, any attempt to group a wide variety of biophysics subjects into a single book is inevitably biased by the authors' background and expertise. This textbook should be regarded merely as a starting point for the study of biophysics, which may be furthered by consulting the references provided, as well as the readers' own selection of biophysics literature.

The way towards biophysics research may be long and not always easy. However, we invite you, our reader, to take on this road without any hesitation, and to share with us the continual feeling of wonder and surprise at the amazing degree of coherence and beauty of the living matter organization and operation.

University of Wisconsin, Milwaukee, WI, USA	Valerica Raicu
University of Bucharest, Bucharest-Magurele, Romania	Aurel Popescu

Acknowledgements

The material covered in this textbook has been used in one of the authors' (VR) graduate level course on molecular and cellular biophysics in the Physics Department at the University of Wisconsin-Milwaukee (UWM). As such, this book has benefited from many useful suggestions made by students as well as postdoctoral fellows and faculty members attending the course. In addition, a large number of people have graciously accepted to read one or more chapters of the book and made invaluable suggestions: Dilano Saldin, Richard Sorbello, Vladislav Yakovlev and Russell Fung of the University of Wisconsin-Milwaukee, James Wells of the University of Toronto, Daniel Sem of Marquette University, and Gina Raicu. We owe special thanks to Marius Schmidt (UWM) for carefully reading all but two chapters of the book, for making several suggestions and for providing a figure for chapter 7.

A large number of colleagues, family members and friends have contributed several figures as well as portions of text, as follows: Russell Fung prepared Figs. 1.12 and 1.13; Dilano Saldin has contributed text to section 2.2.4; several artwork figures in chapters 2 and 4 were prepared by Ana-Maria, Laura and Adela Raicu, while some chemical formulas in chapter 2 were written by Jason Chandler; Constantin T. Craescu, from Institute Curie (France) kindly provided Figs. 8.10 and 8.11; figures 6.4, 7.1 and 7.2 were reproduced with permission by Jaakko Malmivuo and Oxford University Press from the online version of his book (coauthored by Robert Plonsey) entitled "Bioelectromagnetism;" several publishers and authors have generously provided permission for reproducing figures throughout the book.

V. Raicu is immensely grateful to his family, particularly to his wife, Gina, and his daughters, Laura, Ana-Maria and Adela, for their understanding and support during the weekends and evenings that he spent away from them writing the book.

A. Popescu thanks the University of Wisconsin-Milwaukee, WI, USA, and the International Centre for Theoretical Physics "Abdus Salam", Trieste, Italy, for financial support. V. Raicu has been supported by the UWM Research Growth Initiative.

Introduction

1 The Role and Scope of Biophysics Research

Physical and mathematical methods have been employed by medical researchers and physiologists in the study of living organisms for a good part of the last two centuries. By the beginning of the twentieth century, experimental and theoretical methods of investigation were already used by some physicists in the study of a broad range of problems in biology. Some notable researchers were the experimentalist Hugo Fricke, who was the first to estimate the thickness of the cell membrane from electrical measurements of impedance on red blood cell suspensions (Fricke, 1927), and Nicolas Rashevsky, a tireless promoter of a new field of research that he called Mathematical Biophysics (Rashevsky, 1938, 1948, 1961). Rashevsky's efforts to systematize a large body of knowledge and theoretically formalize a range of biological, physiological and even social problems has had a significant impact on the emergence of biophysics as a legitimate field of investigation among life sciences (Cull, 2007). Other highlights in biophysics include: contributions to genetics by Delbrück (a physicist who attracted other physicists to biology), formulation of the Hodgkin-Huxley model of action potential propagation, determination of myoglobin structure by Kendrew and hemoglobin structure by Perutz, secondary structure determination of proteins by Pauling and Corey and of DNA by Watson and Crick, and, more recently, determination of crystal structures of ion channels by MacKinnon, all of whom have been awarded the Nobel Prize for their discoveries.

The formal recognition of biophysics as a self-consistent field of research has been granted perhaps by the First International Biophysics Congress, which took place in Stockholm in 1961. Over the past few decades, several definitions of biophysics have been advanced (Kellenberger, 1986; Mascarenhas, 1994). For instance, the inside cover of the *European Biophysics Journal* currently states that "*biophysics ... is defined as the study of biological phenomena by using physical methods and concepts*". It should be only added to that definition that "*biophysics is not a discipline proper like genetics, biochemistry and molecular biology, but is expected to promote interdisciplinary bridging*" (Kellenberger, 1986).

Biophysicists investigate a very diverse array of problems focussing on various levels of organization of the living matter, from molecules to cells and multicellular organisms. One of the challenges in biophysics currently is to understand how linear molecules such as proteins fold quickly and precisely into their functional structures. Once the physical processes involved in protein folding will be completely understood, it will become possible to create algorithms that predict a protein's three-dimensional structure starting from its primary structure (i.e., the sequence of amino acids). This could allow one to design more specific drugs necessary for highly efficient and cheaper therapy. But perhaps the greatest challenge nowadays is to understand the emergent behavior of the living world, for instance, how cognitive properties such as perception and memory emerge from biomolecular and cellular activity in the brain.

The authors of this book believe that, following the amazing progress in our understanding of biological matter over the past few decades, further important discoveries could be made by using the quantitative approach of physics in biology. After successful research in deciphering the *human genome* has been essentially completed, biophysical concepts and techniques may be used in attaining another notable objective: understanding the relationship between structure and function at the scale of the entire *proteome*, which is represented by the totality of proteins and their interactions in cells. Other areas of investigation in biology in which the skills of biophysicists could be most productively involved are (Varmus, 1999): development of new methods for studying properties of single macromolecules as well as complexes of large molecules, such as proteins and nucleic acids; interpretation of the complex process of development and differentiation of cells, for example, how initially similar cells will evolve into muscle tissue, while others will become brain tissue (or other type of tissue); development of a "radical physical explanation" for complex cellular functions.

It may be anticipated that, through its daring purpose and generous scope, biophysics will emerge as a *sine qua non* component of the fundamental and applied research in the new millennium. Biophysics, in synergy with other life and physical sciences, will likely help achieve a more profound understanding of the laws governing life process on Earth and maybe elsewhere in the Universe.

2 The Subfields of Biophysics

At the time of writing of this book, the field of biophysics is already too broad to be described as a single problem or theme, and there actually exist several subfields of biophysics. One could classify the main subfields of biophysics by using as criteria either the level of organization of the living matter or the subfields of physics from which biophysics mainly borrows its methods and approaches.

Based on the first criterion, one distinguishes:

1. *Quantum biophysics*, investigating the behavior of living matter at sub-molecular and molecular level. This overlaps quite substantially with *quantum biochemistry*, but its affiliation to biophysics seems to be naturally justified.

2. *Molecular and supramolecular biophysics* (sometimes called *chemical biophysics*) dealing, for example, with charge and energy transfer between biomolecules and their complexes.
3. *Cellular biophysics*, investigating single cell properties and processes (e.g., cellular excitation, membrane mechanics, membrane structure and function, etc.).
4. *Biophysics of complex systems* (e.g., neural networks, gene networks, and sensory systems).

Using the second criterion, one could speak of:

(a) *Biomechanics* with its sub-domains, *bioacoustics, biorheology*, and *hemodynamics*
(b) *Biothermodynamics* (e.g., bioenergetics of cellular respiration, muscle contraction and membrane transport)
(c) *Bioelectricity*, which investigates generation and evolution of membrane potentials, and electrical activity of different tissues and organs
(d) *Physiological optics*, investigating a wide palette of biophysical phenomena, starting with the primary physical interaction between light and visual pigments and continuing with the interpretation of the image by the visual cortex
(e) *Photobiophysics*, which uses physical and chemical approaches to investigate processes of photosynthesis and bioluminescence
(f) *Radiation biophysics*, which deals with the interaction of the living matter with ionizing and non-ionizing radiation
(g) *Theoretical and computational biophysics*, which attempts to conceive new modalities of analyzing and, especially, interpreting experimental data

The above sub-division of biophysics in specialized sub-fields has been done by analyzing the titles of numerous scientific journals, monographs, textbooks, and scientific conferences bearing such titles, or by looking at the subdivisions of such forums of communicating biophysics research. For instance, the *Biophysical Journal* includes among its main sections: *Photobiophysics, Cell biophysics* and *Theoretical and computational biophysics*.

3 About this Book

This book focuses on biophysical structures and processes rather than experimental techniques. Many excellent books and review papers exist, which deal with biophysical techniques, and our intent is not to duplicate any existing literature. Instead, we chose to describe experimental techniques only inasmuch as it makes possible meaningful discussion on the biophysics topics of interest. However, methods books and journal articles are cited throughout the book, which our readers are invited to refer to in order to broaden and deepen their knowledge on specific techniques and methods.

This book is organized into eight chapters that integrate basic biological and biochemical information with the necessary mathematics and physics. Basic knowledge of thermodynamics, statistical physics, electricity and magnetism, and standard quantum mechanics is assumed. The book begins with a description of the fundamental physical interactions occurring between biological molecules, and then introduces the physics and physical chemistry of phospholipid association into what constitutes the molecular matrix of the cell membrane: the phospholipid bilayer (chapter 1). A description of the general structure of a biological cell, covering macromolecular as well as organelle level, is provided in chapter 2, which also introduces aspects of DNA structure and replication as well as protein structure and folding. Global properties of the cell membrane, such as its passive electrical and electrokinetic properties are presented in chapter 3, while the classical treatment of substance transport across membrane (including passive diffusion and active transport) is discussed in chapter 4. Modern physical approaches to reactions and diffusion in biological systems are introduced in chapter 5, which emphasizes in particular some of the possible physical and biological effects induced by inhomogeneity and fractional (or fractal) dimensionality in biological systems. The active electrical behavior of membranes of excitable cells is described mathematically in chapter 6, starting from an interpretation of classical experimental data. The molecular basis of the action potential is also presented in some detail, while the exact nature of the structures of the molecules involved (channels and ion pumps) is revealed in chapter 7, which presents those molecular machines, as well as another interesting nanomachine: the light-harvesting antenna involved in conversion of light into chemical energy in the cell during the process of *photosynthesis*. The book ends with a description of structural and kinetic aspects of protein-protein interactions (chapter 8), which lie at the heart of the protein association into supramolecular systems with well-defined (although not always known) functional roles in the cell.

The material covered in this book has been used by V.R. in a course on Molecular, Cellular and System Biophysics that he teaches at the University of Wisconsin-Milwaukee. Topics in systems biophysics are not included in this book and are not taught in formal lectures by the instructor, although they are covered through projects on literature research conducted by students, which culminate with short talks given towards the end of the semester. Usually the students choose from a list of proposed topics, which may include: brain electrophysiology and neural networks; sensory systems (olfactory and taste, somatic and hearing, and visual); blood vessels and the principle of optimal design; red blood cell and rouleaux formation (kinetics, energetics, and physiological relevance); regulation of the blood pressure – the role of noise in the baroreflex system; optimum delivery of substance into volume; transport through self-similar networks; allometric scaling; other topics of special interest to the students that are already involved in research.

References

Cull, P. (2007) The mathematical biophysics of Nicolas Rashevsky, *BioSystems*, **88**: 178
Fricke, H. (1927) The electric capacity of suspensions of red corpuscles of a dog, *Phys. Rev.*, **26**: 682
Kellenberger, E. (1986) Role of the physicist in biology, *Europhysics News*, **17**: 1

Mascarenhas, S. (1994) What is biophysics? *College in Biophysics: Experimental and Theoretical Aspects of Biomolecules*, 26 September–14 October 1994, Trieste, Italy

Rashevsky, N. (1938) *Mathematical Biophysics: Physico-Mathematical Foundations of Biology*, University of Chicago Press, Chicago, IL

Rashevsky, N. (1948) *Mathematical Biophysics: Physico-Mathematical Foundations of Biology*, 2nd ed. University of Chicago Press, Chicago, IL

Rashevsky, N (1961) *Mathematical Principles in Biology and Their Application*, Charles C. Thomas, Springfield, IL

Varmus, H. (1999) The impact of physics on biology and medicine, *Physics World*, Special Issue, September 1999, pp. 27

Chapter 1
The Molecular Basis of Life

1.1 Molecular Interactions

Biological matter is comprised of small molecules (e.g., water), macromolecules (biopolymers), supramolecular assemblies or macromolecular complexes, which assemble into subcellular particles and cells, which in their turn form supracellular systems such as tissues and organs. One of the key ingredients of biological systems is represented by the complex, specific or non-specific, *intra-* and *inter-molecular* interactions, which determine the structure and, further, the biological functions associated to structure. Spatial packing or *folding* of macromolecules, such as proteins, nucleic acids or polysaccharides, is regulated by intramolecular interactions, which occur between segments of the same molecule. Equally important for the 3D structure of macromolecules are intermolecular interactions with solvent molecules (constituted by water molecules in biological systems), ions, and various small molecules from the cytosol (discussed in chapter 2). Intermolecular interactions between macromolecules are mostly conducive to formation of supramolecular structures such as protein complexes, which could be either transitory (e.g., binding of oxygen to hemoglobin) or permanent, relative to the lifespan of a cell or a superior organism (e.g., the tetrameric complex of hemoglobin).

There exists a wide variety of physical interactions relevant to the structure and function of biological systems: attractive or repulsive *electrostatic interactions* (e.g., charge-charge, charge-dipole, charge-multipole, dipole-multipole, etc.), attractive *electrodynamic interactions, hydrogen bonds*, and *hydrophobic "interactions."*

Observation: The distinction between electrostatic and electrodynamic interactions will become more apparent when we will describe them below.

Depending on the specific details, physical interactions could be either *long-range nonspecific* or *short-range specific* physical interactions. The long-range interactions are the first to control the repulsion or attraction between molecules, thereby determining whether *short-range biophysical interactions* come into play.

From a quantitative point of view, physical interactions can be described either as *forces* or as *energies* of interaction. Intermolecular interaction energy has generally an electrostatic origin, and is much lower than covalent bond energies, which involve electron sharing. For example, in liquid HCl, the intermolecular attractive energy (approximated by the energy required for evaporation) is 16 kJ/mol, while the energy of the covalent bond between H and Cl, calculated as the energy required for bond breaking, is 431 kJ/mol (Israelachvili, 1992).

The strength of intermolecular interactions determines the physical properties of a substance (e.g., melting and boiling temperatures, surface tension, etc.).

1.1.1 Interactions Between Polar Entities

A rigorous mathematical treatment of electrostatic interactions in biological macromolecules is complicated by a number of peculiarities, which we will list below.

1. Most biological macromolecules are poly-ionic mosaics, carrying both positive and negative charges, even at their *isoelectric* pH, at which the total charge is zero (see section 1.2.3 for a definition of pH). In other words, macromolecules possess space-distributed charges, which make the simple expression of Coulombian interaction energy inadequate to describe their interactions.
2. The charge distribution in a macromolecule is highly anisotropic, and subject continual changes induced by environmental changes (e.g., the intracellular pH).
3. Macromolecules are not rigid bodies but they undergo continuously conformation changes – called "breathing" movements. To make matters worse, during the interaction process itself, the interacting molecules do not assume fixed properties, but are rather "adapting" to one another (a process sometimes called "induced fit") through changes in their geometry and spatial charge distribution. Therefore, the strength of the interaction is also time dependent. To avoid intractable mathematical equations, an approximate time-average description of the interactions may be used.
4. The dielectric constant of the microenvironment of the macromolecules, which enters the mathematical expressions for electrical interactions, varies substantially from the bulk of the solvent to the surface of the macromolecule and farther into the macromolecule. For instance, the relative (to free space) permittivity of bulk aqueous solutions at room temperature, $\varepsilon_r \approx 80$, decreases dramatically near the surface of a macromolecule (to only a few dielectric units). In addition, many proteins are actually embedded in heterogeneous media; one such medium is the cell membrane, which, as we shall see, is composed of polar and non-polar layers with very different dielectric constants.

It becomes obvious that some assumptions and simplifications are necessary when modeling real biological systems, in order to obtain tractable mathematical expressions of macromolecular interactions. Therefore, one should be fully aware of the approximate character of the mathematical treatment of electrostatic interactions between biological macromolecules and between them and particles within the aqueous milieu (such as the cytosol, which will be discussed in chapter 2).

1.1 Molecular Interactions

1.1.1.1 Charge-Charge Interactions

Although interactions between only two point charges are seldom encountered in biological systems, we shall start our discussion by reminding the reader of the expression of their Coulombian energy, on which all other expressions of interaction energies are based, as we shall see. The electrostatic energy of interaction, W_{ES}, between two point charges, q_1 and q_2, separated by a distance, r, and embedded in a homogenous medium of relative permittivity, ε_r, is given (in IS units) by:

$$W_{ES} = \frac{1}{4\pi\varepsilon_0\varepsilon_r}\frac{q_1 q_2}{r} \approx 9 \times 10^9 \frac{q_1 q_2}{\varepsilon_r r} \qquad (1.1)$$

where $\varepsilon_0 = 8.854 \times 10^{-12}\,\text{F m}^{-1}$ represents the dielectric constant of free space. In biophysics, this formula is used for the description of the following type of interactions: ion-ion interactions, interactions between ions and charged chemical groups attached to a macromolecule, or between two (or more) charged chemical groups belonging either to different macromolecules or to the same macromolecule.

1.1.1.2 Interactions Between a Charge and a Permanent or Induced Dipole

Let us assume that a point charge, q, is situated at a distance r on the line that passes through the electric dipole, $\vec{p} = q_1 s\hat{n}$, formed by two point charges $+q_1$ and $-q_1$ separated by a distance $s \ll r$; $\hat{n}$ is a unit vector. The charge and the electric dipole are embedded in a medium of relative dielectric constant, ε_r (Fig. 1.1).

To calculate the interaction energy between the point charge, q, and the dipole, p, we need first to derive the equation for the electrical potential, ϕ_{dipole}, created by the dipole at the location of the charge q; ϕ_{dipole} is obtained by assuming the additivity of the potential created by each of the two charges constituting the dipole:

$$\phi_{dipole} = \frac{1}{4\pi\varepsilon_0\varepsilon_r}\left[\frac{q_1}{r} + \frac{-q_1}{|\vec{r}+s\hat{n}|}\right] \approx \frac{1}{4\pi\varepsilon_0\varepsilon_r}\frac{q_1 s\hat{n}\cdot\vec{r}}{r^3} = \frac{1}{4\pi\varepsilon_0\varepsilon_r}\frac{\vec{p}\cdot\vec{r}}{r^3} \qquad (1.2)$$

where we have used the approximation $\frac{1}{|\vec{r}+s\hat{n}|} = 1/r\left|1+s\frac{\hat{n}}{\vec{r}}\right| \approx \frac{1}{r} + s\hat{n}\cdot\hat{r}\frac{\partial}{\partial r}\left(\frac{1}{r}\right) = \frac{1}{r} - s\hat{n}\cdot\vec{r}\frac{1}{r^3}$.

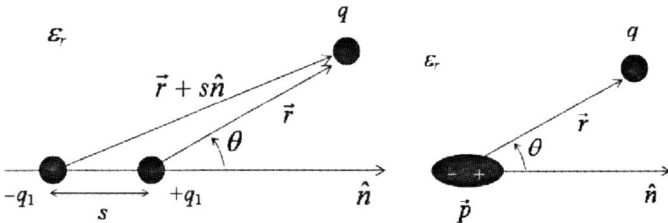

Fig. 1.1 Physical and geometrical parameters of the system formed by a point charge and an electric dipole. Note, in the case with $s \ll r$, the equivalence of the system to the right to the one to the left.

Therefore,

$$W_{ES} = q\phi_{dipole} = \frac{q}{4\pi\varepsilon_0\varepsilon_r}\frac{\vec{p}\cdot\vec{r}}{r^3} = \frac{q}{4\pi\varepsilon_0\varepsilon_r}\frac{p\cos\theta}{r^2} \qquad (1.3)$$

Note that, in contrast to the energy of interaction between two point charges, which decreases with the distance, r, the energy of interaction between a point charge and a dipole decreases more rapidly, that is, with the *square of the distance*.

In biophysical systems, charge-dipole interactions may occur, e.g., between water molecules (i.e., permanent dipoles), which have a very high dipole moment of about $1.8\,\mathrm{D}$ [$1\,\mathrm{D}$ (Debye) $= 3.336 \times 10^{-30}\,\mathrm{C\,m}$], and the charges attached to the macromolecule chains, or between water molecules and ions from the cytosol.

Even more important are the interactions between point charges and *induced dipoles* in electro-neutral but polarizable particles (e.g., large atoms and molecules). For instance, molecules with *conjugated double bonds* (i.e., with delocalized π electrons) are more polarizable, their electrons being displaced over long distances and thus presenting an induced dipole with a large moment, $p = qs$.

Observation: As described in standard chemistry textbooks, π electrons populate π orbitals. Two π orbitals are formed when two $2p_y$ orbitals belonging to different atoms are brought close together in a molecule. The electron densities are found above and below the bond axis.

In this case, the net attraction between the point charge and the induced dipole is stronger (being proportional to the separation, s, between charges) than between a point charge and small atoms/molecules.

1.1.2 van der Waals Interactions

Interatomic (intermolecular) attractions were first discovered experimentally by van der Waals, who was interested in the deviation of the behavior of real gases from the *ideal gas equation of state*. He discovered that the attraction strength varies *inversely proportional* to the *sixth power* of the distance between molecules. Currently nowadays, the term *van der Waals interactions comes to represent* three classes of interactions: (a) *orientation* or *Keesom interactions* (first described by Keesom in 1912) between permanent electric multipoles (i.e., dipoles, quadrupoles, octopoles, etc.); (b) *Debye induction interactions* (Debye, 1920) between *permanent* multipoles and *induced* multipoles; (c) *London dispersion interactions* (*London*, 1930) between *instantaneous* dipoles/multipoles and *instantaneous induced* dipoles/multipoles.

The most common van der Waals interactions are those between dipoles, which indeed obey the inverse sixth power dependence on separation (see below); in this book the term *van der Waals interactions* is used to denote all types of electrodynamic intermolecular interactions.

1.1.2.1 Keesom Interactions

Let us consider, for simplicity, two continuous distributions of electrical charges, with their centers at the origins O and O' of two Cartesian coordinate systems, $Oxyz(S_1)$ and $O'x'y'z'(S_2)$, and characterized, respectively, by: the volumes V and V', the net charges q and q', electrical charge densities $\rho = dq/dV$ and $\rho' = dq'/dV'$, and dipole moments $\vec{p}$ and $\vec{p}'$ (Fig. 1.2). The charge element $dq(\vec{r})$ of the volume element dV, at $\vec{r}$ in S_1 senses an electric potential, $\phi'_{S_1}(\vec{r})$, due to the charge distribution in S_2. We consider that the two systems of charges have already formed, and therefore the energy of formation of the two systems is not of concern here. Thus, the electrostatic energy of interaction between $dq(\vec{r})$ and $\phi'_{S_1}(\vec{r})$ is:

$$dW_{ES}(\vec{r}) = \phi'_{S_1}(\vec{r}) dq(\vec{r}) = \phi'_{S_1}(\vec{r}) \rho(\vec{r}) dV \tag{1.4}$$

which, integrated over the entire volume, V, gives:

$$W_{ES} = \int_V \phi'_{S_1}(\vec{r}) \rho(\vec{r}) dV \tag{1.5}$$

To obtain the interaction energy for the case of dipolar interactions, the electrostatic potential, $\phi'_{S_1}(\vec{r})$, is expanded in a Taylor series (Rein, 1975):

$$\phi'_{S_1}(\vec{r}) = \phi'_{S_1}|_{\vec{r}=0} + \frac{1}{1!} \vec{r} \cdot [\nabla_{\vec{r}} \phi'_{S_1}(\vec{r})]|_{\vec{r}=0} + \cdots \tag{1.6}$$

where higher-order terms are considered but not written, for brevity. Note that the 'prime' means that the potential is due to charge q', while subscript S_1 (or S_2) reads "relative to S_1 (or S_2)". $\nabla_{\vec{r}}$ is the gradient taken with respect to $\vec{r}$, while $\nabla_{\vec{r}'}$ is the gradient taken with respect to $\vec{r}'$. By substituting (1.6) into (1.5), we obtain:

$$W_{ES}(\vec{r}) = \phi'_{S_1}(\vec{r})|_{\vec{r}=0} \int_V \rho(\vec{r}) dV + \left[\nabla_{\vec{r}} \phi'_{S_1}(\vec{r})|_{\vec{r}=0} \right] \cdot \int_V \vec{r} \rho(\vec{r}) dV + \cdots \tag{1.7}$$

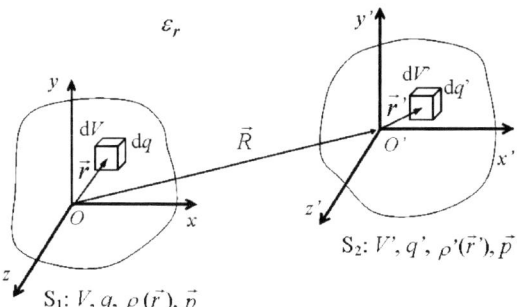

Fig. 1.2 Geometric and physical parameters of two electrical charge distributions with their "centers" situated distance R apart. The significance of the symbols is: V – volume, q – total charge, $\rho(\vec{r})$ – charge density, p – dipole moment; unprimed symbols refer to system S_1, while primed refer to S_2.

Since we have:

$$\int_V \rho(\vec{r})dV = q, \int_V \vec{r}\rho(\vec{r})dV = \vec{p}, \text{ and } \phi'_{S_1}(\vec{r})|_{\vec{r}=0} = \phi'_{S_2}(\vec{r}')|_{\vec{r}'=-\vec{R}} \quad (1.8)$$

equation (1.7) becomes:

$$W_{ES}(\vec{R}) = q\phi'_{S_2}(\vec{r}')|_{\vec{r}'=-\vec{R}} + \vec{p} \cdot \nabla_{\vec{r}'} \phi'_{S_2}(\vec{r}')|_{\vec{r}'=-\vec{R}} + \cdots \quad (1.9)$$

Next, recall the expression of the electrical potential produced by a charge distribution at an arbitrary distance $\vec{\xi}$ (see, e.g., Stratton, 1941),

$$\phi_S(\vec{\xi}) = \frac{1}{4\pi\varepsilon_0\varepsilon_r}\left[\frac{q}{\xi} + \frac{\vec{p}\cdot\vec{\xi}}{\xi^3} + \cdots\right] \quad (1.10)$$

to obtain, for the case of q', $\vec{r}'$ and $\vec{p}'$ relative to the reference frame S_2:

$$\nabla_{\vec{r}'}\phi'_{S_2}(\vec{r}') = \frac{1}{4\pi\varepsilon_0\varepsilon_r}\left[q'\nabla_{\vec{r}'}\left(\frac{1}{r'}\right) + \frac{1}{r'^3}\nabla_{\vec{r}'}(\vec{p}'\cdot\vec{r}') + (\vec{p}'\cdot\vec{r}')\nabla_{\vec{r}'}\left(\frac{1}{r'^3}\right)\cdots\right] \quad (1.11)$$

Further, by using equation (1.11) and the general identities:

$$\nabla\left(\frac{1}{\xi}\right) = -\frac{\vec{\xi}}{\xi^3}, \nabla\left(\frac{1}{\xi^3}\right) = -\frac{3\vec{\xi}}{\xi^5}, \text{ and } \nabla(\vec{p}\cdot\vec{\xi}) = \vec{p} \quad (1.12)$$

the final expression for the electrostatic energy of interaction is obtained from (1.9), as:

$$W_{ES}(\vec{R}) = \underbrace{\frac{qq'}{4\pi\varepsilon_0\varepsilon_r R}}_{\text{charge–charge}} \underbrace{-\frac{q\vec{p}'\cdot\vec{R}}{4\pi\varepsilon_0\varepsilon_r R^3} + \frac{q'\vec{p}\cdot\vec{R}}{4\pi\varepsilon_0\varepsilon_r R^3}}_{\text{charge–dipole}}$$

$$+ \underbrace{\frac{1}{4\pi\varepsilon_0\varepsilon_r R^3}\left[\vec{p}\cdot\vec{p}' - \frac{3(\vec{p}\cdot\vec{R})(\vec{p}'\cdot\vec{R})}{R^2}\right]}_{\text{dipole–dipole}} + \cdots \quad (1.13)$$

Quiz 1. Prove relations (1.12).

In equation (1.13), the Keesom interactions are represented by the last term,

$$W_{ES}^K(\vec{R}) = \frac{1}{4\pi\varepsilon_0\varepsilon_r R^3}\left[\vec{p}\cdot\vec{p}' - \frac{3(\vec{p}\cdot\vec{R})(\vec{p}'\cdot\vec{R})}{R^2}\right] \quad (1.14)$$

which reflects the instantaneous interaction between two dipoles.

Observation: By using the above approach, not only the energy of the dipole-dipole interaction was obtained (last term in the right-hand side of equation (1.13)) but also the

1.1 Molecular Interactions

already known energy of charge–charge interaction (the first term) and the expressions for charge–dipole interactions (the second term), which is of the same type as represented by equation (1.3).

The instantaneous positions and orientations of the two dipoles (multipoles) are usually unknown and may also rapidly change due to the thermal motion. Therefore, a time average of the Keesom energy of interaction, $\overline{W}_{ES}^K$, is sometimes more meaningful, and is given by (Nir, 1976):

$$\overline{W}_{ES}^K = \frac{\iint W_{ES}^K \exp(-W_{ES}^K/k_B T) d\Omega d\Omega'}{\iint \exp(-W_{ES}^K/k_B T) d\Omega d\Omega'} \quad (1.15)$$

In this formula, the integrals are evaluated over all orientations of the two dipoles, i.e., $d\Omega = \sin\theta d\theta d\varphi$, where θ is the angle made by the dipole with $\vec{R}$ and φ is the dihedral angle of the planes containing the vectors $\vec{R}$ and $\vec{p}$ and, respectively, $\vec{R}'$, and $\vec{p}'$. The Boltzmann factor $\exp(-W_{ES}^K/k_B T)$, introduced to allow for the fact that dipoles prefer the orientations with a lower potential energy ($k_B = 1.38 \times 10^{-23}$ J/K is the Boltzmann constant and T is the absolute temperature), can be approximated for $|W_{ES}^K| \ll k_B T$ by $1 - W_{ES}^K/k_B T$. After integrating (1.15), the average Keesom interaction energy is obtained (Nir, 1976; Israelachvili, 1992; Atkins and de Paula, 2002):

$$\overline{W}_{ES}^K(\vec{R}) = -\frac{1}{(4\pi\varepsilon_0\varepsilon_r)^2}\frac{2}{3k_B T}\frac{\vec{p}^2\vec{p}'^2}{R^6} \propto -\frac{1}{TR^6} \quad (1.16)$$

which states that the interaction depends inversely on the temperature and on the sixth power of the distance between the dipoles.

Observation: Keesom interactions are *attractive* interactions, as their average energy is always negative; they are also *short-range* interactions, since the energy decreases abruptly with the distance R.

To estimate the order of magnitude of Keesom interactions, let us consider two identical molecules possessing a dipole moment, $p = 1$ D, separated by a distance of 3 Å in a medium with $\varepsilon_r = 1$, at room temperature ($T = 298$ K). The Keesom energy of interaction between those two molecules given by equation (1.16) multiplied by the Avogadro's number and divided by 2 (for two molecules) gives the molar energy of ~ -0.7 kJ/mol. As seen, this energy is smaller than but on the same order of magnitude as the average thermal energy, $W_T = (3/2)RT = 3.7$ kJ/mol.

1.1.2.2 Debye (Induction) Interactions

As it was already pointed out above, Debye induction interactions are exerted between permanent dipoles (multipoles) and *induced* electrical dipoles (multipoles).

The electrical field produced by a dipole, locally characterized by the intensity $\vec{E}$, induces in a neighboring molecule an electric dipole, $\vec{p}_i$, given by the expression:

$$\vec{p}_i = \varepsilon_0 \varepsilon_r \alpha \vec{E} \quad (1.17)$$

where α is a positive quantity called *polarizability* of the molecule ($[\alpha]_{SI} = m^3$) and ε_r is the relative permittivity of the medium. Since the expression of the elementary energy of interaction between a dipole and an electrical field is:

$$dW = -\vec{p}_i \cdot d\vec{E} \qquad (1.18)$$

the expression of the global energy of induction interaction is:

$$W^D = -\varepsilon_0 \varepsilon_r \alpha \int_0^E \vec{E} \cdot d\vec{E} = -\frac{1}{2}\varepsilon_0 \varepsilon_r \alpha \int_0^E d\left(\vec{E} \cdot \vec{E}\right) = -\frac{1}{2}\varepsilon_0 \varepsilon_r \alpha E^2 \qquad (1.19)$$

To proceed further with our derivation of the energy of interaction, let us consider two macromolecules, each characterized by neutral ($q = q' = 0$) distributions of charges (see Fig. 1.3), but nonvanishing dipole moments, $\vec{p}$ and $\vec{p}'$, and polarizabilities, α and α' (i.e., the two molecules are polar). These charge distributions will generate electrostatic fields characterized by the potentials ϕ and ϕ' and the intensities $\vec{E}$ and $\vec{E}'$ given by the following expressions:

$$\vec{E}(R) = \frac{1}{4\pi\varepsilon_0\varepsilon_r}\left[-\frac{\vec{p}}{R^3} + \frac{3(\vec{p}\cdot\vec{R})\vec{R}}{R^5}\right], \text{ and} \qquad (1.20a)$$

$$\vec{E}'(R) = \frac{1}{4\pi\varepsilon_0\varepsilon_r}\left[-\frac{\vec{p}'}{R^3} + \frac{3(\vec{p}'\cdot\vec{R})\vec{R}}{R^5}\right] \qquad (1.20b)$$

Quiz 2. Prove relations (1.20) by using potentials of the type represented by (1.10) and using identities (1.12).

Under the above conditions, each dipole induces an instantaneous dipole into the other charge distribution with the instantaneous induction interaction energy given by equation (1.19). The induced dipoles are weaker than the permanent ones, and

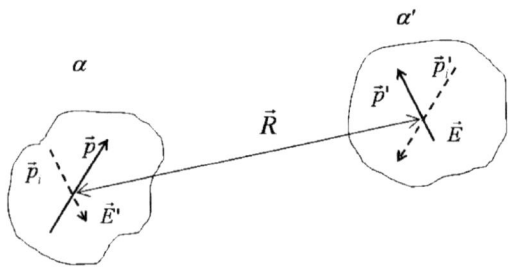

Fig. 1.3 Geometric and physical parameters of two charge distributions used for the description of Debye interactions. Note that in each pair formed by a permanent dipole (solid line) and an induced dipole (dashed line), the corresponding vectors are antiparallel ($\vec{p} \uparrow\downarrow \vec{p}'_i; \vec{p}' \uparrow\downarrow \vec{p}_i$). $\vec{E},\vec{E}'$ are electrical field intensities of the two charge distributions generated at their reciprocal positions.

1.1 Molecular Interactions

we ignore their effect on the other molecule. The total induction interaction energy is therefore given by the expression:

$$W^D = -\frac{\varepsilon_0 \varepsilon_r}{2}[\alpha' \vec{E}^2 + \alpha \vec{E}'^2] \tag{1.21}$$

By taking into account relations (1.20), equation (1.21) becomes:

$$W^D = -\frac{1}{32\pi^2 \varepsilon_0 \varepsilon_r R^6}\left[\alpha' \vec{p}^2 + 3\alpha' \frac{(\vec{p} \cdot \vec{R})^2}{R^2} + \alpha \vec{p}'^2 + 3\alpha \frac{(\vec{p}' \cdot \vec{R})^2}{R^2}\right] \tag{1.22}$$

To further analyze equation (1.22), it is convenient to replace the dot products between the instantaneous dipole moments and $\vec{R}$ in (1.22) by the relations: $\vec{p} \cdot \vec{R} = pR\cos\theta$ and $\vec{p}' \cdot \vec{R} = p'R\cos\theta'$, to obtain the instantaneous energy of interaction,

$$W^D = -\frac{1}{32\pi^2 \varepsilon_0 \varepsilon_r R^6}\left\{\alpha' p^2 + 3\alpha' [p\cos(\theta)]^2 + \alpha p'^2 + 3\alpha \left[p'\cos(\theta')\right]^2\right\} \tag{1.23}$$

which is dependent upon the instantaneous orientations of the dipole moments. Clearly the energy is negative, and hence the interaction is attractive, for any orientation of the permanent dipoles. In addition, the absolute value of energy is highest for parallel or antiparallel orientation of $\vec{p}$ ($\vec{p}'$) and $\vec{R}$, and lowest for perpendicular orientations.

The average energy of interaction obtained by inserting equation (1.23) into (1.15) and integrating over all orientations is [Nir, 1976]:

$$\overline{W}^D = -\frac{1}{16\pi^2 \varepsilon_0 \varepsilon_r R^6}[\alpha' p^2 + \alpha p'^2] \propto -\frac{1}{R^6} \tag{1.24}$$

Upon analyzing equation (1.24), one can easily see that, similar to Keesom interactions, the Debye interactions are *short-range* and attractive, but, unlike the Keesom interactions, the induction interactions are independent of temperature.

Observation: Expression (1.24) is the most general case of Debye interactions, and applies to distributions of charges (i.e., macromolecules) that possess permanent dipoles. In the particular cases where either $\vec{p} = 0$ or $\vec{p}' = 0$, particular formulas for the energy of interaction result, which correspond to the case of single dipole moments triggering the interaction between molecules.

1.1.2.3 London (Dispersion) Interactions

Some of the most interesting types of van der Waals interactions are the London dispersion interactions, sometimes also called *London-van der Waals interactions* (Glaser, 2001). They are attributed to charge fluctuations in a molecule. Dispersion interactions are the most common van der Waals forces, because they are present even in the absence of net charges or permanent dipoles. They play an important role in a wide range of phenomena: surface tension, adsorption, adhesion, wetting, flocculation (i.e., reversible aggregation) of colloidal particles, spatial structure of biological (e.g., proteins) and nonbiological (polymers) macromolecules, etc.

Even in nonpolar molecules, such as benzene, charge fluctuations may cause the center of positive charges to be shifted away from the center of the negative charges, the molecule becoming, for a very short time, a dipole. This instantaneous dipole will induce an instantaneous dipole into a neighboring molecule, and the two dipoles will exert attractive electrostatic forces upon one another. These forces are not directional, because the transient dipole induces a dipole in a neighboring molecule irrespective of its spatial orientation.

The mathematical expression of the dispersion interaction energy can be deduced in the framework of quantum mechanics, using the *second order perturbation theory*. This approach is beyond the scope of the present textbook and the reader is referred to the work of Nir (1976) for an adequate mathematical treatment. Herein, we limit ourselves to giving only the so-called "single-term" expression of the dispersion interaction energy between two molecules:

$$W^L(R) = -\frac{3\hbar}{2(4\pi\varepsilon_0\varepsilon_r)^2 R^6} \frac{\omega\omega'}{\omega+\omega'} \alpha\alpha' \propto -\frac{1}{R^6} \quad (1.25)$$

where $\hbar = h/2\pi$ is the reduced Planck constant, ω and ω' are the absorption angular frequencies ($\omega = 2\pi\nu$, ν *being the electronic absorption frequency*) of the molecules, while α and α' are their corresponding polarizabilities. Notice that we have retained in this equation the "primed" and "unprimed" convention we used before. A summary analysis of expression (1.25) indicates that London interactions are attractive, independent of temperature (being instantaneous), and, again, short-range (i.e., dependent upon R^{-6}).

Observation: Sometimes the expression of energy associated to London dispersive interactions (1.24) is given in terms of the ionization energy, $I_k = h\nu_k$ ($k = 1, 2$) as follows (Setlow and Pollard, 1962; Israelachvili, 1992):

$$W^L(R) = -\frac{3}{2(4\pi\varepsilon_0\varepsilon_r)^2 R^6} \frac{II'}{I+I'} \alpha\alpha'. \quad (1.26)$$

Relation (1.25) indicates that the greater the polarizabilities, the higher the dispersion interaction energy. This is the case of very large molecules (i.e., macromolecules) which tend to have large polarizabilities, due to their large number of electrons. However, since the dispersion interactions are significant only when the atoms of the interacting molecules are close enough to one another, a *structural complementarity* must exist between the interacting molecules. This structural (steric) complementarity puts many atoms of the interacting molecules in favorable spatial disposition (i.e., close together) thereby compensating for the weakness of the interaction between each atoms pair. Therefore, one can speak of the strength of the weak interactions, but only as exerted between many couples of atoms pertaining to the interacting molecules. This structural complementarity constitutes the basis of *specific* molecular *interactions* in biological systems. Two classical examples of

1.1 Molecular Interactions

this sort are represented by enzymes and their specific substrates, and antibodies with their antigens.

In conclusion, all types of van der Waals interactions are short-ranged, due to the $1/r^6$ dependence. The tight packing of proteins in their native state (i.e., as found in cells) is maintained by such short range interactions.

Observation: In general, the van der Waals interactions between molecules and especially between large particles do not follow the simple inverse-sixth power law, as one can find out from several publications (Nir, 1976; Israelachvili, 1992; Popescu, 1997; Glaser, 2001).

1.1.2.4 The Lenard-Jones Potential

The attractive van der Waals forces bring the molecules closer together, but further shortening of inter-atomic distances is prevented by quantum mechanical repulsions between the electrons of the non-covalently-bonded atoms, which cannot occupy the same space. These repulsions result from Pauli's exclusion principle, which states that any two electrons cannot occupy the same quantum state. These repulsions are very strong and have higher order dependence on distance ($1/r^{12}$) compared to van der Waals interactions, i.e., they are *very short-range interactions*.

By superimposing the attractive van der Waals interactions and repulsive quantum interactions we obtain the so called *Lenard-Jones potential energy*, $W_{LJ}(r)$:

$$W_{LJ}(r) = A/r^{12} - B/r^6 \qquad (1.27)$$

where A and B are positive constants. Figure 1.4 illustrates the distance dependence of the Lenard-Jones potential energy.

The Lenard-Jones potential energy provides to the interacting molecules an "energetic landscape" that presents a minimum, W_{min}, for a certain distance, r_{min}, between the two molecules. For $r \gg r_{min}$, the molecules attract each other, while for $r \ll r_{min}$, they repel each other.

Quiz 3. Using (1.27), show that the equilibrium distance between two interacting particles is $r_{min} = (2A/B)^{1/6}$.

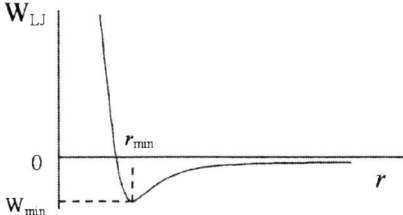

Fig. 1.4 Graphical representation of the Lenard-Jones potential energy. r_{min} is the distance corresponding to the equilibrium state, and W_{min} represents the energy minimum (i.e., the equilibrium state of the interacting particles).

1.2 Water and Polar Interactions

With its coverage of roughly three quarters of the surface of the Earth, water is the most widespread liquid on Earth and the most important liquid for life. According to currently accepted theories on the origin of life, the first primitive forms of life had appeared in an aqueous milieu. Life as we know it on Earth would not exist and manifest itself in the absence of water, which has been considered life's "*matter and matrix, mother and medium*" (Szent-Gyorgyi, 1972).

One may legitimately ask: what are the special properties that make water essential to life? It cannot be simply because of its abundance on the Earth that water constitutes the major component of any organism. The answer to our question seems to lie at the molecular level. For instance, all biochemical reactions take place in an aqueous environment (i.e., the cytosol of a cell, described in chapter 2). Also, water molecules act as suppliers of protons for charge transport, participate in the cellular metabolism, and contribute to the spatial organization (i.e., structure) of the living matter, as it will be seen later in the case of proteins, for instance. Note that this structural role is accomplished not by rendering the biological macromolecules rigid, but by preserving all of their freedom of motion needed for accomplishing their biological functions (Nagy et al., 2005).

The content of water in the living matter is a function of: (1) *species*, (2) type of *tissue*, and (3) *ontogenic development* of an organism, as detailed below.

1. In the case of bacterial spores the water percentage is ∼50%, while in the case of coelenterates (such as jellyfish), the content of water is about 97% (for this reason, it is sometimes said that the coelenterates are *living water*). The human beings and, generally, mammals are in the middle of this range, containing approximately 70% water in their bodies.
2. In the case of an adult human, the content of water varies from a few percents (for example, 4% in hair, and 9% in dentine – a major component in teeth) through 85% in nervous grey matter and even 93% in blood. It is known that the water content of a tissue is strongly correlated with its metabolic activity: the more intense is the metabolic activity of a tissue, the higher its content in water (e.g., the bones have mostly a structural role and contain 30% water, while the skeletal muscles have a water content of 76%).
3. As regards the ontogenic development, the content of water decreases with the age of an organism. For instance a human embryo has a content of 97% water, like the coelenterates (perhaps, not surprisingly, given the "primordial" medium in which it is formed), a new born, 76%, an adult, 70%, and an old female approximately 50%.

In pluricellular organisms, 70% of water is intracellular (as bound and bulk water), while 30% is extracellular water (as interstitial water and circulatory water).

1.2 Water and Polar Interactions

1.2.1 Physical Properties of Water

When compared to, e.g., hydrides of the elements of the sixth group of the periodic table (i.e., H_2S, H_2Se, H_2Te) and also to other substances (e.g., CH_4, alcohols, etc.), water manifests a series of properties which make it particularly suitable for life:

(i) Very high melting point ($t_m = 0°C$, $p = 1$ atm) and boiling point ($t_b = 100°C$, $p = 1$ atm). For comparison, H_2S melts at about $-86°C$ and boils at about $-61°C$, which means that H_2S (like other hydrides) is a gas at room temperature. Because of its melting and freezing temperatures, water is a liquid at temperatures that are usual on Earth.

(ii) The greatest surface tension coefficient ($\sigma \cong 73$ mN m^{-1}, $t = 20°C$) among all liquids, except for Hg, which has $\sigma \cong 425$ mN m^{-1}. (Note that Hg is a metal!) Other values include $\sigma \approx 64$ mN m^{-1} for glycerol and $\sigma \cong 23$ mN m^{-1} for methanol. Surface tension is defined as the energy necessary to increase a liquid's surface area by one unit; alternatively, σ may be defined as the stretching force necessary to break a film of unit length and depends on the strength of the forces of cohesion between a liquid's molecules, which are of a dipolar origin in the case of water. The high value of the surface tension coefficient of water has an effect on cellular membrane formation and stabilization, as well as on regulation of membrane permeability to various substances, as discussed later on in this book.

(iii) Relatively low dynamic viscosity coefficient ($\eta \cong 1.8$ cP at $0°C$ and $\eta \cong 1.00$ cP at $20°C$) (1 Poise $= 0.1$ N s m^{-2}). For comparison, the viscosity of methanol is 0.6 cP, while other substances, such as oils, have viscosities several orders of magnitude higher. The viscosity coefficient can be defined by considering the laminar flow of a fluid between two plates acted on by shear force F, according to the formula, $F = \eta A dv/dx$, where A is the surface area of the plates and dv/dx is the speed gradient in the direction perpendicular to the direction of flow.

> **Observation:** The SI unit for dynamic viscosity coefficient is cP=1 Pa s=1 N m^{-2}s. In CGS, the unit for dynamic viscosity coefficient is Poise (P). 1 P = 1 dyne cm^{-2}s. However, the centi Poise (cP) is more commonly used, because the dynamic viscosity coefficient for pure liquid water at $20°C$ happens to be 1 cP.

(iv) Very high specific heat capacity of 4.2 kJ kg^{-1} K^{-1} at $20°C$ (compare, e.g., to ~ 2 kJ kg^{-1} K^{-1} for alkanes and other hydrocarbons). For this reason, water presents a high "thermal inertia," that is, the absorption or release of heat, Q, by a water mass, m, leads only to a slight variation in its temperature, Δt, according to the formula:

$$\Delta t = \frac{Q}{mc} \quad (1.28)$$

Because the specific heat capacity of water is significantly higher than that of soil (which ranges from 0.8 to 2 kJ kg^{-1} K^{-1}, depending on the soil type) and rock (0.8 kJ kg^{-1} K^{-1}), coastal areas of great lakes, seas or oceans have a milder climate than the inland: If the cold or hot air currents pass over such huge masses of water, they are heated or cooled (depending on the temperature of the mass of water), thereby avoiding large variations of temperature from day to night and summer to winter. One can thus regard coastal water as *a heat reservoir*.

(v) Very high specific latent heat of melting/freezing ($\lambda_m = \lambda_f \approx 2.3\,\text{MJ/kg}$ at $p = 1\,\text{atm}$) and of vaporization/condensation ($\lambda_b = \lambda_c \approx 334\,\text{kJ/kg}$ at $p = 1\,\text{atm}$). Both of these values are about three times higher than those corresponding to ethanol, for example. The great value of the specific heat for vaporization explains why the perspiration evaporation is very effective in thermolysis (i.e., dissipation of heat).

(vi) A very peculiar and "anomalous" behavior of its dilatation coefficient with temperature: negative for the temperature range 0–3.98°C (i.e., the water contracts in this range!) and positive over the range 4–100°C. The liquid water has its greatest density at the temperature of 3.98°C. That means, for instance, that ice is less dense than water, which explains its floatation. Therefore, the water of rivers and lakes freezes starting from the surface, leaving liquid at the bottom, which protects aquatic plants and animals. At the same time, the ice layer acts like a thermal insulator diminishing further freezing and cooling of the water beneath it, again to the benefit of aquatic fauna.

Quiz 4. Prove that the molar concentration of pure liquid water at its highest density (i.e., at 3.98°C) is 55.55 M (1 Molar = 1 mol dm^{-3}).

(vii) A rather high molecular dipole moment, $|\vec{p}| = 1.85\,\text{D}$ (D = Debye), although comparable to other polar substances (such as methanol, $|\vec{p}| = 1.66\,\text{D}$). Due to this property, water has a great capacity to dissolve many substances, especially ionic and polar molecules. The partial negative charge of water oxygen interacts with cations (e.g., Na$^+$, K$^+$), while the partial positive charge of the two hydrogen atoms interacts with anions (e.g., Cl$^-$, HCO$_3^-$), creating a more or less thicker shell of water around the ions (the so-called *hydration shell*). The water molecules coming in direct contact with an ion form the *inner hydration shell*, which is more stable than the next shell, called *outer hydration shell*. At 20°C, for some ions of physiological interest (e.g., Na$^+$, K$^+$, Ca^{++}) the hydration layer exchanges water molecules with its environment at a high rate of (10^8–10^{10}) molecules s^{-1}, while for other ions like, for instance, Ni^{++}, Mg^{++} and Co^{++}, the exchange rate constants are significantly lower (10^4–10^6 molecules s^{-1}). This explains the high permeabilities of ion channels to the former compared to the latter type of ions (Hille, 2001).

In the case of polar substances, the charged parts of the water molecule interact with oppositely charged parts of the polar molecules. Polar molecules thus manifest a high affinity to water, and are therefore called *hydrophilic* (i.e., water-loving/liking) molecules. Uncharged and nonpolar molecules are not soluble in water, and are called *hydrophobic* (i.e., "water-hating") molecules, as we shall see in section 1.3.

1.2 Water and Polar Interactions

Quiz 5. Explain the high value of the surface tension coefficient of water in light of its large dipole moment. Does the high surface tension coefficient explain the high specific latent heat for vaporization?

(viii) A dielectric constant at normal room temperature ($\sim$81 at 20°C and 78.4 at 25°C) that is higher than that of most other constituents of the cell. For instance, the relative dielectric constant of the hydrocarbons of which the phospholipid tails are made (see below) is $\sim$2. This has significant implications for the formation of electrical layers nearby cell membranes, as we shall see in chapter 3.

1.2.1.1 Structure of the Water Molecule

The water molecule is composed of three atoms: two hydrogen atoms and one oxygen atom. However, the liquid water is not a "pure liquid" due to the existence of three hydrogen isotopes (i.e., protium, $_1^1H = H$, deuterium, $_1^2H = D$, and tritium, $_1^3H = T$) and three oxygen isotopes ($_8^{16}O = O$, $_8^{17}O$ and $_8^{18}O$). Therefore one could encounter, in principle, all possible combinations of hydrogen and oxygen isotopes in water. Because the isotopes $_1^1H$ and $_8^{16}O$ are naturally more abundant, the liquid water is dominated by the presence of water molecules (H_2O, common water) composed of these two isotopes. But, as mentioned, there also exist D_2O (heavy water), T_2O (radioactive tritiated water), HOD, HOT, DOT, etc.

In this chapter, we discuss properties and importance to life of the common water, H_2O, which is the most abundant of all types. As an example, the concentration ratio of heavy water and common water is $[D_2O]/[H_2O] \sim 1/6,000$ (i.e., approximately 1 kg of D_2O in 6 t of H_2O).

Quiz 6. Estimate the mass of heavy water, m_{hw}, in an adult human with a mass, $M = 80$ kg.

Quiz 7. Find the mass ratio between the "heaviest" and "lightest" species of water molecules.

The two valence bonds (O–H) form an angle of about 105° (Fig. 1.5) and have an average length of about 1 Å. Because the O atom is electronegative, while the H atoms are electropositive, the water molecule possesses a permanent electrical dipole moment, which gives liquid water a very polar character.

1.2.1.2 Electronic Structure of the Water Molecule

Most of the above physical and chemical properties of water molecules are explained on the basis of the *electronegative* character of oxygen and the *electropositivity* of hydrogen, which arise from the *electronic structure* of water molecules.

The water molecule possesses ten electrons (eight from the oxygen atom and two from the two hydrogen atoms). Two of these electrons are confined to the orbital 1s of the oxygen atom (i.e., they are "hidden" electrons), and do not contribute to

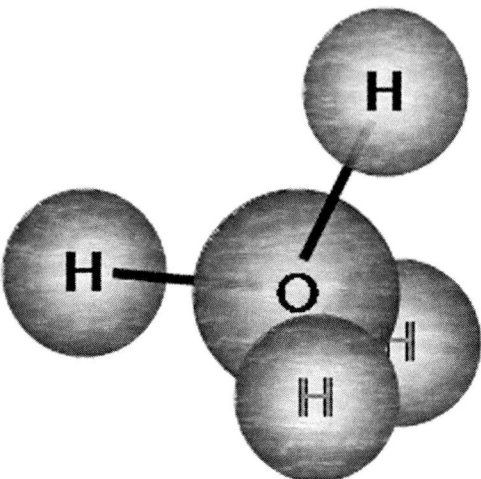

Fig. 1.5 The tetrahedral structure of a water molecule interacting with two other water molecules. The two unmarked hydrogen atoms are "bonded" through noncovalent (hydrogen) bonds to the O atoms of two adjacent molecules, as it will be explained below.

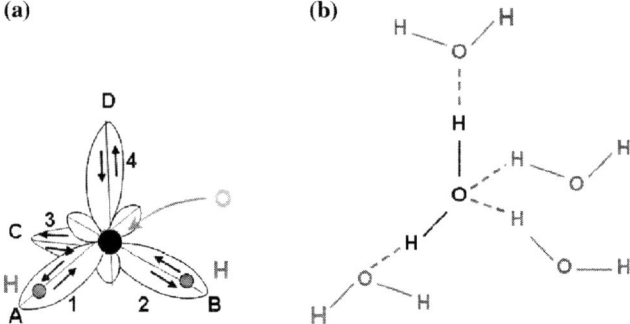

Fig. 1.6 (a) The electronic structure of the water molecule. Each of the four asymmetrically bilobate orbitals is occupied by two electrons with opposite spins; (b) hydrogen bonding in which a single water molecule (in the center of the figure) may be involved. The continuous lines are covalent bonds (with a length of about 1 Å) while the dashed lines are hydrogen bonds (with a length of about 1.8 Å).

the physical and chemical properties of the water molecule. The remaining eight electrons are shared by the three atoms, and are confined into four asymmetrically bilobate *orbitals*, whose symmetry axes are along the directions of the straight lines connecting the center, O, of a regular tetrahedron with its vertices (see lines OA, OB, OC, and OD in Fig. 1.6a).

> **Observation:** From a classical mechanics point of view, an orbital is a region of space in the vicinity of a nucleus, where one or two electrons spend about 90% of its/their lifetime(s). From a quantum mechanical point of view, an orbital is a region with a very high electronic density distribution, hence with a very high probability to find the electron(s) in it.

These four orbitals are generated by the *hybridization* (i.e., mixing of different types of orbitals) of the spherically symmetric orbital, $2s$, and the three bilobate orbitals: $2p_x$, $2p_y$ and $2p_z$ of the oxygen (having the rotation symmetry axes Ox, Oy, and Oz, respectively). The angle between any of the orbital symmetry axes is about $109°$. These hybrid orbitals provide room for the two electrons contributed by the two hydrogen atoms, so that each of the four orbitals accommodates two electrons (i.e., the maximum number, according to Pauli's *exclusion principle*), having their spins antiparallel.

1.2.1.3 Hydrogen Bonds and Association of Water Molecules

A direct consequence of the distribution of eight electrons over the four hybrid bilobate orbitals is the propensity of water molecules to associate with neighboring molecules. Indeed, the two electrons populating the orbitals 3 and 4 (Fig. 1.6a) of the water molecule are able to temporarily attract the protons (i.e., the hydrogen nuclei) from two neighboring water molecules, thereby establishing two coordinative bonds with them, known as *hydrogen bonds*. At the same time, the orbitals 1 and 2, containing the hydrogen nuclei, can be engaged in other two coordinative bonds with the oxygen atoms from other two neighboring molecules. Therefore, *a single water molecule* can potentialy be engaged, simultaneously, in *four hydrogen bonds* (Fig. 1.6b) with other water molecules. In the more general case, the water molecule can be engaged in hydrogen bonds with other molecules endowed with electronegative atoms (i.e., O, S, N) called *electron acceptors*; in all cases, H is called an *electron donor*.

The strength or the *binding energy* of a hydrogen bond can be defined as the energy necessary to destroy it. The binding energy of a hydrogen bond, W_{HB}, is an order of magnitude greater than that of the van der Waals, W_{VW}, interactions, but an order of magnitude smaller than binding energy of a covalent bond, W_{CB} (e.g., O–H):

$$W_{VW} \sim 1\text{kJ/mol} \ll W_{HB} = 20\text{kJ/mol} \ll W_{CB} = 460\text{kJ/mol} \qquad (1.29)$$

Observation: Hydrogen bonds act also intramolecularly (e.g., in proteins and nucleic acids) and contribute to maintaining the native 3D structure of macromolecules. Because the hydrogen atoms of macromolecules could be engaged also in hydrogen bonding with water molecules, the strength of a hydrogen bond realized inside a protein molecule is smaller, its binding energy being thus smaller than 20 kJ/mol.

As seen in Fig. 1.6, the four molecules of water occupy the vertices of an almost regulate tetrahedron having in its center a tetra-coordinated water molecule. However, this molecule association has a very short lifetime of the order of 50 fs (Cowan et al., 2005) due to continuous thermal motion. Therefore, the *antientropic* association process (i.e., for which the change in entropy between the final and the initial states is negative) driven by hydrogen bonds is permanently counterbalanced by the *entropic* process (i.e., for which the change in entropy is positive), tending to break the molecular associations. An indirect measure of the magnitude of the entropic effect is the dependence on temperature of the dynamic equilibrium between a state with many molecular associations and a state with very few associated molecules:

$$n(H_2O) \underset{breaking}{\overset{formation}{\rightleftarrows}} (H_2O)_n \qquad (1.30)$$

The maximum *degree of association*, n, has a greater value when $t \to 0°C$ and a lower value when $t \to 100°C$.

Observation: The water molecule association manifests *cooperative behavior*: association of two molecules by a hydrogen bond favors the attachment of other molecules, leading to a very short-lived water molecular cluster (in a sense, a microcrystal) of about $n = 100$ molecules, while the breaking of a single hydrogen bond from the cluster will engender the cascade breaking of other hydrogen bonds. The lifetime of this small molecular cluster is very short, however, due to the fact that the lifetime of a hydrogen bond itself is short too.

In solid water (i.e., ice) under normal pressure, the hydrogen bonding is saturated, each molecule being engaged in four hydrogen bonds that are stable (Fig. 1.7), due to very small thermal agitation energy ($3RT/2 = 3.7 \, kJ/mol$). X ray diffraction on ice crystals shows extensive hexagonal arrangements of water molecules which entail less dense packing than liquid water. In this way, one can explain why ice at $0°C$ is less dense than liquid water at the same temperature.

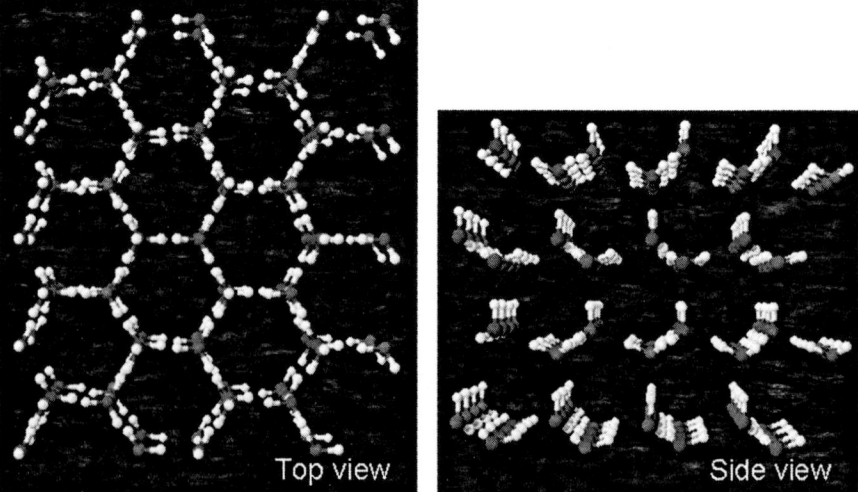

Fig. 1.7 Hexagonal ice (normal ice) structure (Figure generated with Jmol (http://www.jmol.org) using X-ray diffraction data from http://openpac.net/).

1.2 Water and Polar Interactions

If ice absorbs heat, the thermal agitation of its molecules increases, resulting in the breaking of some hydrogen bonds. The mobile molecules slip into the hexagonal "eyes" of ice, thus leading to a denser packing. The release of water monomers (single molecules) from the ice hexagonal network and their intrusion into ice interstices (i.e., ice hexagonal eyes) explain why the water density increases on the interval 0–3.98°C, which is considered to be an *anomalous behavior*. Beyond 3.98°C ($p = 1$ atm), the interstitial monomers leave their places due to the thermal agitation intensification, contributing to the water volume increase with temperature (which is considered to be a normal behavior).

In liquid water at 0°C ($p = 1$ atm), only 15% of hydrogen bonds are broken. For this reason, liquid water conserves its quasicrystalline structure, which becomes more and more fragile as the temperature increases.

In spite of the frequent molecular collisions, the liquid water still continues to be a mixture of monomers, dimers, tetramers, etc., of water molecules even at high temperatures; at lower temperatures highly associated clusters predominate, while at higher temperatures dimeric associations are more frequent. It is interesting to note that dimeric associations are very stable, due to the presence of two hydrogen bonds, being encountered even in gaseous phase of water.

Around the temperature of 40°C, about 50% of hydrogen bonds are broken, the water becoming roughly three times less viscous (compare 1.8×10^{-3} daP at 0°C to 0.7×10^{-3} daP at the human body temperature of 37°C). Due to this important increase of water fluidity (i.e., the reciprocal of viscosity) the temperature of 37°C is considered to be *the second melting point* of water. Perhaps it is not a pure coincidence that, during the evolutionary process on Earth, the stabilization of the body temperature of the homeothermic animals occurred in the range 35–41°C. In this range of temperatures, the proportion of monomers to dimers is very high, water becoming more chemically reactive. Under these conditions, plenty of water monomers become available for construction of hydrogen bond networks around biomacromolecules.

In gaseous phase, intense thermal fluctuations break almost all the water hydrogen bonds, except for those involved in some dimers that still persist in this phase.

The highly polar character of water molecules and their tendency to associate through hydrogen bonds can explain all the macroscopic physical properties of water enumerated above. In conclusion, the existence of the hydrogen bonds endows water with very peculiar properties that make it uniquely suited to sustain life on Earth.

1.2.1.4 Effects of Hydrogen Bonding on the Living State of Matter

As previously mentioned, in pure liquid water, the maximal degree of association, n, depends on the predominance of one of the two opposite tendencies: that of association promoted by hydrogen bonding and that of cluster breaking, due to thermal agitation. In the cytosol (i.e., the complex intracellular milieu consisting of molecular species of different sizes) there are competitive interactions of ions, small molecules and macromolecules with water dipoles, in addition to the interactions between water molecules. The consequences are not easy to anticipate and require specific investigations; in some cases the degree of water structuring is increased, but in others this degree is decreased.

Polar interactions lead to the following types of processes: (*a*1) Generation of *hydration shells* around free ions. In general cations are surrounded by thicker shells than anions. This is because cations tend to have smaller radii (due to missing an electron or more) than anions (which have one or more extra-electrons), and thereby present a stronger electrostatic potential that is capable of attracting more water molecules. (*a*2) Formation of the so called *bound water* around ionized and polar groups attached to proteins, nucleic acids, polysaccharides, macromolecular complexes. (*a*3) Formation of hydrogen bonds, which are invoked in explaining protein-protein interactions and protein-specific *ligand* interactions. (**b**) Formation of a water cage (a *crystal hydrate* or *clathrate*) maintained by a hydrogen bond network, around the so-called hydrophobic molecules or around the nonpolar groups of biomacromolecules (also hydrophobic). The relevance of these processes to the function of living matter is summarized below.

Observation: Some ions (e.g., Na^+, K^+, Cl^-) must lose their hydration shells when translocated across the cell membranes through specific channels (see chapter 4).

(*a*1) Water molecules possessing strong permanent electrical dipole moments are distributed around ions, as shown in Fig. 1.8. Thus the ions introduce some order in the water molecule distribution around them, thereby modifying the preexisting *bulk water* structure.

(*a*2) The dipolar water molecules can also cluster around ionized groups or around polar groups of macromolecules (e.g., proteins), realizing the so-called *bound water* layer. The degree of ordering is enhanced if there is a steric (conformational) fit between the lateral chemical groups of the macromolecules and the intermolecular distances between water molecules from the hydrogen bond network. This bound water is very important for a correct biological function of macromolecules, forming an integral part of native macromolecules. This kind of water is sometimes called *structured water*.

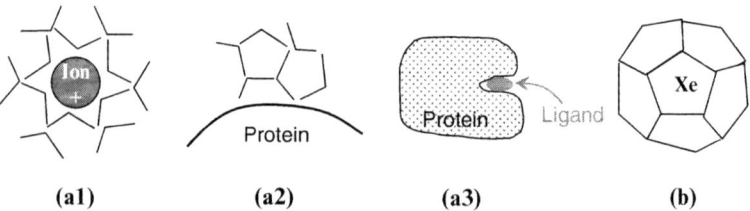

Fig. 1.8 (**a1**) Hydration shell around a positive ion; (**a2**) Bound water at the "surface" of a protein; (**a3**) Interaction of a protein with its specific ligand, during which the hydration shells of protein site and ligand are partially destroyed. (**b**) Formation of a water molecule cage (i.e., crystal hydrate/clathrate) around an insoluble molecule, in this case an inert gas (e.g., Xenon).

1.2 Water and Polar Interactions

(a3) Ligands are small molecules (e.g., polypeptides, nucleotides, sugars, lipids, diatoms, etc.), which interact specifically with a macromolecule, usually a protein or a nucleic acid (e.g., binding of O_2 or other diatoms by myoglobin or hemoglobin (Nagy et al., 2005)). Through the interaction between different partners in water, the order in the system increases, leading thus to a decrease in entropy following the interaction, namely:

$$\Delta S_{Assoc} \leq 0 \tag{1.31}$$

where the subscript *Assoc* stands for "Association". This seems to contradict the second law of thermodynamics, which requires an increase in entropy, S. However, this is not the case, as we shall see momentarily. The Gibbs free energy, $G = U + pV - TS \equiv H - TS$ (where all the symbols have their usual meaning in thermodynamics) also needs to decrease, in order for a reaction to take place spontaneously, that is:

$$\Delta G = \Delta H - T\Delta S \leq 0 \tag{1.32}$$

Simultaneous with ligand associations, another phenomenon takes place, which is able to generate an entropy excess, in order to counterbalance the negative variation due to molecular associations. This additional positive variation of entropy, $\Delta S_W \geq 0$, is attributed to disorganization of the molecular clusters, which were previously structured through hydrogen-bonded networks around the interacting molecules (especially near the contact regions between the interacting molecules, such as proteins and their hydrated ligands). If

$$|\Delta S_W| \geq |\Delta S_{Assoc}| \tag{1.33}$$

the total entropy variation, ΔS_T, obeys the *second law of thermodynamics*, namely:

$$\Delta S_t = \Delta S_{Assoc} + \Delta S_W \geq 0 \tag{1.34}$$

Observation: Relation (1.34) holds true both for *endothermic* ($\Delta H > 0$, where $\Delta H = \Delta U - p\Delta V$ is the enthalpy exchanged in the process) and *exothermic* ($\Delta H < 0$) reactions compatible with $\Delta G < 0$ (i.e., spontaneous reactions). In the case of endothermic reactions, the following condition is obeyed:

$$\Delta G = \Delta H - T\Delta S \leq 0 \tag{1.35}$$

which requires that the entropic term, $T\Delta S$, satisfies the relation:

$$T\Delta S \geq \Delta H \tag{1.36}$$

Therefore, endothermic interactions (i.e., for which $\Delta H > 0$) are *entropically driven*, or else they could not take place.

(b) In the case of molecules lacking ionized or polar groups (i.e., hydrophobic molecules) like oil, hydrocarbon chains of lipids, and inert gases, which are insoluble in water, a highly structured network of hydrogen-bonded water molecules is formed around them. These highly ordered water shells around the hydrophobic molecules are called *clathrates* or *crystal hydrates*.

As a consequence of this structure formation, the entropy of the water decreases, hence dispersion of hydrophobic molecules into liquid water is entropically disfavored (Atkins and de Paula, 2002). The somewhat rigid crystal hydrates formed around inert gases endows them with *anesthetic* properties, by blocking ionic channels of the axons (in the nervous system). The ion channels blocking prevents ionic diffusion and, consequently, the nervous impulse propagation; thus the anesthetic effect.

1.2.2 Overview of the Importance of Water for the Living Matter State

Due to its special properties, water presents critical importance practically in all biological processes. Next, we shall summarize some of the instances in which water plays a key role, both at molecular and supramolecular level.

(i) Water is a ubiquitous solvent for living matter, at the intracellular and interstitial level.
(ii) Through the hydrogen bonding and the hydrophobic effect, water helps formation and stabilization of the structure of macromolecules (e.g., proteins) and of biological membranes.
(iii) Water molecules directly participate in important intracellular reactions, such as hydrolysis (e.g., ATP hydrolysis), oxidation and condensation (e.g., peptide bond formation, as it is shown in chapter 2).
(iv) Water is a source of protons involved in intermolecular charge transfer and in the process of photosynthesis.
(v) Water is the transport medium for ions, small molecules, and macromolecules, as well as of cells (e.g., erythrocytes, leukocytes, lymphocytes) that circulate in higher organisms from one tissue to another. Such cells are usually involved in transport of substance by means of circulating fluids (e.g., blood, lymphatic liquid) or diffusion.
(vi) Water ensures an efficient mechanical protection by buffering the mechanical shocks, in the case of nervous central system (brain) and the fetus.
(vii) Water participates extensively into the body thermoregulation in the case of homeothermic animals. For instance, thermolysis (which reduces the body overheating) is mainly achieved through perspiration (followed by evaporation), respiration and intensification of peripheral blood circulation, which favor heat dissipation directly into the environment.
(viii) Water provides wetting of different mucosae, thus facilitating gas exchange between an organism and its environment. For example, oxygen is dissolved in wet alveolar mucosa of lungs, diffusing then through capillary walls and erythrocyte membranes, and reaching the hemoglobin molecules – the actual oxygen transporters in tissues.

1.2.3 The pH

All biological liquids, including the cell cytosol are aqueous electrolyte solutions, their characteristic ionic composition, which is essential for their biological role, being maintained constant in spite of numerous permanent perturbations. The role of many anions and cations will be discussed throughout this book. Here, we want to stress the involvement of hydrogen ions, hereafter called *protons*, in practically all biochemical and biophysical processes.

The protons (in fact, the hydrated protons or *hydronium ions*: H_3O^+) deserve a special attention due to their omnipresence (even in pure neutral water), their high mobility, and their electrostatic interactions with other charged chemical groups from proteins, nucleic acids and supramolecular structures, such as intracellular and cellular membranes. The knowledge of proton concentration in biological liquids is of major importance for physiologists, biochemists and biophysicists.

Because the molar concentrations of protons in solution may extend over many orders of magnitude, a logarithmic scale is generally used to express the proton concentration. The units of this logarithmic scale are known as **pH** (from, "**p**ower of **H**ydrogen"), and are defined by the simple formula (see, for instance, Stryer, 1988):

$$pH = -\log\left[H^+\right] \tag{1.37}$$

where $[H^+]$ represents the molar concentration of the proton, in $mol/dm^3 = M$.

One can easily deduce from (1.37) that a solution rich in free protons has a low pH (due to the negative sign) and vice versa. At room temperature, water molecules dissociate frequently, resulting in protons (H^+) and hydroxyl ions (OH^-), while the latter reassociate into water molecules. This happens while the *ionic product of water* (Atkins and de Paula, 2002),

$$K_w = \left[H^+\right]\left[OH^-\right] = 10^{-14} M^2 \tag{1.38}$$

remains constant.

Since $[H^+]$ and $[OH^-]$ should be equal (to preserve electrical neutrality), it follows from (1.38) that $[H^+] = 10^{-7}M$. Plugging this value into equation (1.37) gives pH = 7 for pure water, which is considered a *neutral* pH; an *acidic* solution has pH < 7, while *basic* solutions obey pH > 7. A completely dissociated (strong) acid with $[H^+] = 1M$ has pH = 0, while a completely dissociated (strong) base with $[OH^-] = 1M$, has a pH of 14. These constitute the practical pH limits for solutions.

The pH of biological systems is maintained constant via certain buffering systems, any change from the normal physiological proton concentration engendering diseases like *acidosis* and *alkalosis*. For instance, under normal physiological conditions, the pH of the blood is maintained at the slightly basic value of 7.4. If it drops below 6.8 or rises above 7.8, death may occur. Fortunately, there are natural buffers in the blood to protect against large variations in pH. These buffers are essentially constituted by *binary acid–base systems*, so that when an acid is added to a biological solution, the base captures the excess protons, while when a base is added, the acid neutralizes the OH⁻ by liberating protons; in this way, the proton concentrations remain practically constant even if biological processes may generate basic as well as acidic products.

1.3 Hydrophobic "Interactions" and Molecular Self-Association

1.3.1 The Hydrophobic Effect

As we have mentioned in section 1.2, non polar molecules (such as oil, hydrocarbon chains of lipids, and inert gases) do not interact attractively with the polar water molecules. Such molecules are called "*hydrophobic*" (i.e., "water-hating"), and their mixing with water or other polar solvents leads to very peculiar structures. Some of these structures have been described in section 1.2.1, but a large class of such hydrophobic "interactions," which are of paramount importance for biophysics, will be dealt with separately in this section.

Hydrophobic "interactions" are mainly the result of the preferential interactions between water molecules (through hydrogen bonds) and, to a lesser extent, of the preferential interactions between nonpolar molecules "dissolved" in water. Since each type of molecule (i.e., polar and non-polar) "prefers" to interact with a molecule of its own type, this leads effectively to different molecules avoiding one another. Note therefore that it is incorrect to regard the hydrophobic effect as true repulsion between water and nonpolar molecules, in spite of the fact that the term "hydrophobic" might suggest this. Unlike polar molecules (e.g., alcohols), nonpolar molecules are not able to form hydrogen bonds with water molecules. Because water molecules usually constitute the majority in polar-nonpolar mixtures of biological interest, the "hydrophobic" molecules will be insolated and surrounded by hydrogen bond networks. In this way, the nonpolar molecules are compelled to cluster together, reducing the total surface area of contact with water, and thus minimizing the energy of the entire system.

> **Observation:** If, on the contrary, the water molecules would be in minority in a mixture with "hydrophobic" oil molecules, they would be isolated and surrounded by oil molecules.

The hydrophobic interactions together with electrostatic and van der Waals interactions discussed in section 1.1 are involved in many molecular and supramolecular interactions (e.g., protein-protein interactions, protein polysaccharide interactions, protein-lipids interactions, etc.). As we shall see later, the hydrophobic effect plays a very important role in formation and stabilization of structures that are vital to biological cell functioning: the plasma and organellar membranes.

Among various hydrophobic substances, *hydrocarbons* are more relevant to the material covered in this book. To provide a measure of their hydrophobicity, in his book on the hydrophobic effect, Tanford (1973) proposed the use of the *free energy of transfer* from an aqueous solvent to a purely hydrocarbon solvent, which is the change in the *chemical potential*, μ, defined as the change in Gibbs free energy upon

1.3 Hydrophobic "Interactions" and Molecular Self-Association

addition of small numbers of molecules, i.e., $\mu = \left(\frac{dG}{dv}\right)_{T,p}$, where v is the number of moles. To calculate the free energy of transfer, let us start from the expression of the free energy of a hydrocarbon dissolved in water,

$$\mu^W = \mu_0^W + RT \ln X_W + RT \ln f_W \quad (1.39)$$

where R is the ideal gas constant, T is the absolute temperature, X_w is the concentration of the solute species in mol fraction units, μ_0^W is the standard chemical potential of the solute (hydrocarbon) in pure water (i.e., the chemical potential of the pure solute in the same state of aggregation as the solvent), and f_W is the activity coefficient at that concentration.

One can write an expression similar to (1.39) for the same hydrocarbon dissolved in a pure hydrocarbon:

$$\mu^{HC} = \mu_0^{HC} + RT \ln X_{HC} + RT \ln f_{HC} \quad (1.40)$$

If the hydrocarbon of interest is at equilibrium in the aqueous solution and then in the pure hydrocarbon solvent, the free energies expressed by equations (1.39) and (1.40) should be equal, which leads to:

$$\mu_0^{HC} - \mu_0^W = RT \ln X_W / X_{HC} + RT \ln f_W / f_{HC} \quad (1.41)$$

For low concentrations of the hydrocarbon of interest, the activity coefficients could be approximated by unity; therefore (1.41) becomes:

$$\mu_0^{HC} - \mu_0^W = RT \ln X_W / X_{HC} \quad (1.42)$$

Also, for a solution of hydrocarbon in a solvent of its own type (i.e., at *saturation*), both X_{HC} and f_{HC} in equation (1.40) would be equal to 1; further, if the concentration of hydrocarbon in water is low, $f_W = 1$. Therefore, equation (1.41) simplifies to:

$$\mu_0^{HC} - \mu_0^W = RT \ln X_W \quad (1.43)$$

Experiments show that at the transfer of hydrocarbons from water into themselves $\mu_0^{HC} - \mu_0^W$ is negative and depends linearly on the length (expressed in number of carbon atoms) of the hydrocarbon chain (for hydrocarbons of up to at least 22 carbon atoms) with a slope of about $-880\,\text{cal/mol}$ ($1\,\text{cal} = 4.19\,\text{J}$) per carbon atom (Tanford, 1973). This is an important result, which will be used momentarily in connection with *amphiphilic* molecules (see below).

1.3.2 Amphiphilic Molecules

There exists an important class of substances that are composed of a polar moiety (or polar *head*) and a nonpolar *tail* (usually, one or more linear hydrocarbon chains), which are called "*amphiphilic*" molecules or, simply, "*amphiphiles*."

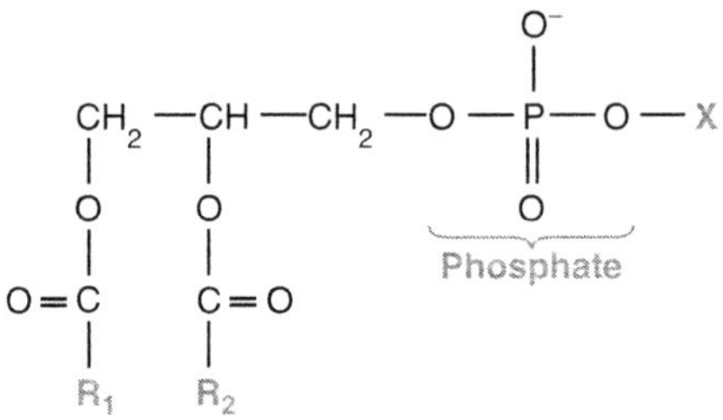

$$\begin{array}{c}\text{CH}_2-\text{CH}-\text{CH}_2-\text{O}-\underset{\underset{\text{O}}{\|}}{\overset{\overset{\text{O}^-}{|}}{\text{P}}}-\text{O}-\text{X}\\ |\quad\;\;|\\ \text{O}\quad\text{O}\\ |\quad\;\;|\\ \text{O}=\text{C}\quad\text{C}=\text{O}\\ |\quad\;\;|\\ \text{R}_1\quad\text{R}_2\end{array}$$

$\underbrace{}_{\text{Phosphate}}$

Fig. 1.9 The general formula for some of the most common phospholipids: phosphoglycerides. R_1 and R_2 are hydrocarbon chains (CH_3–CH_2–, etc.), while X can be positively or negatively charged, or zwitterionic (either positively or negatively charged, depending on pH). For X = H: phosphatidic acid; for X = $CH_2CH(NH_3^+)COO^-$: phophatidylserine; for X = $CH_2CH_2N(CH_3)_3^+$: phosphatidylcholine.

Surfactants – molecules with surface tension activity, which are popularly known as *detergents* – are among the most common amphiphiles. The free energy of transfer from water to pure hydrocarbon solvent varies linearly with the length of the hydrophobic tail (Tanford, 1973), roughly with a slope of -820 cal/mol per carbon atom (note the good agreement with the above result for pure hydrocarbon). This suggests that the polar head and the hydrophobic tail contribute to the free energy independently.

More common biologically are amphiphiles with two hydrocarbon tails, which are essentially composed of two fatty acids (hydrophobic tail) attached to a negatively charged phosphate group and a nitrogen-containing molecule or an alcohol (hydrophilic). Some of the most common phospholipids are *phosphatidylcholine* and *phosphatidylserine* (see Fig. 1.9).

1.3.2.1 Thermodynamics of Self-Association of Amphiphiles

An important feature of amphiphilic molecules is the fact that they can *self-associate* when dispersed in a polar or nonpolar solvent, so that the part that cannot be dissolved in a solvent can be "hidden" from it. Because of these properties, amphiphiles are essential components of a cell membrane and are also used in generating model synthetic membranes necessary in studies of membrane physical properties. The symbolic representation of a phospholipid is shown in Fig. 1.10.

The simplest type of amphiphilic aggregate is perhaps the *micelle* (see Fig. 1.10). Micelles are formed when a surfactant dissolved in water reaches a certain concentration, called *critical micelle concentration* (CMC), at which it becomes energetically more favorable to bring the hydrophobic tails together and position the polar heads towards the water solvent. Micelles formation can be treated formally as a chemical reaction of the type:

1.3 Hydrophobic "Interactions" and Molecular Self-Association

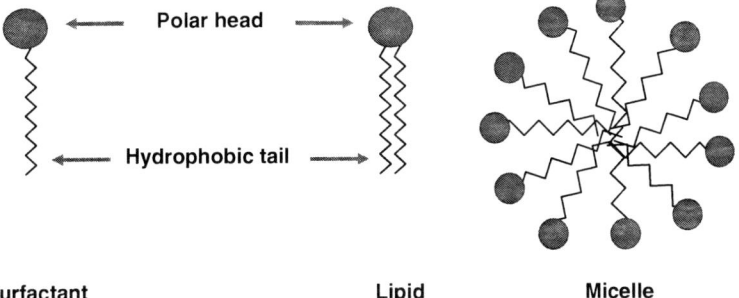

Fig. 1.10 Representation of two types of amphiphiles: surfactant (sometimes improperly called "detergent"), phospholipid and surfactant molecules self-assembled into a micelle.

$$\underbrace{A + A + A + \ldots}_{n(A)} \underset{k_n}{\overset{k_1}{\rightleftarrows}} (A)_n \qquad (1.44)$$

where k_1 and k_n are direct and inverse rate constants. At concentrations higher than the CMC, the rate of formation of micelles is given by:

$$v_1 = k_1 [A_1]^n f_1^n \qquad (1.45)$$

while the rate of dissociation is given by:

$$v_n = k_n \frac{[A_n]}{n} f_n \qquad (1.46)$$

where $[A_1]$ and $[A_n]$ are the concentrations of the two species, f_1 and f_n are their activity coefficients ($f \leq 1$), and n is the number of monomers per micelle. At equilibrium, according to the Law of Mass Action (see chapter 5), the rates of formation and dissociation are equal, thus:

$$k_1 [A_1]^n f_1^n = k_n \frac{[A_n]}{n} f_n \qquad (1.47)$$

or:

$$K \equiv \frac{k_1}{k_n} = \frac{[A_n] f_n}{n [A_1]^n f_1^n} = \exp\left[-n(\mu_0^n - \mu_0^1)/k_B T\right] \qquad (1.48)$$

where K is an equilibrium constant, while μ_0^n and μ_0^1 are the chemical potentials of single monomers in micelles and in aqueous solution, respectively (defined here with respect to the number of molecules).

If we make again the simplifications used to obtain equation (1.42) (concerning the activity coefficients), equation (1.48) leads to:

$$\frac{[A_n]}{n} = [A_1]^n \exp\left[-n(\mu_0^n - \mu_0^1)/k_B T\right] \qquad (1.49)$$

or, in logarithmic form:

$$\mu_0^n - \mu_0^1 = k_B T \ln \frac{n[A_1]^n}{[A_n]} \tag{1.50}$$

Noticing the different definitions of the chemical potentials (which correspond to either RT or $k_B T$ for the thermal energy), the latter equation is very similar to equation (1.42). This similarity also implies that superscripts "n" and "l" in equation (1.50) correspond to HC (the hydrocarbon phase) and W(water) in equation (1.42), and permits a direct comparison between the hydrophobicity of amphiphiles and their tendency to self-associate. In fact, this indirectly justifies the choice of the difference in chemical potentials for the transfer from water into hydrocarbon.

Either equation (1.49) or (1.50), together with the conservation of the total concentration of amphiphiles,

$$C = \sum_n^\infty [A_n] \tag{1.51}$$

completely define the system of water and amphiphilic molecules.

Since small micelles (such as $n = 2$) are thermodynamically unstable (Tanford, 1973; Israelachvili et al., 1976), an useful approximation of equation (1.49) is obtained when n is replaced with an average micellar size, $\bar{n}$:

$$\frac{[A_{\bar{n}}]}{\bar{n}} = [A_1]^{\bar{n}} \exp\left[-\bar{n}\left(\mu_0^{\bar{n}} - \mu_0^1\right)/k_B T\right] \tag{1.52}$$

where $\mu_0^{\bar{n}}$ is assumed size-independent (Tanford, 1973; Ruckenstein and Nagarajan, 1975).

Quiz 8. Use a computer program and equations (1.49) and (1.52) to show that concentrations of free and aggregated monomers depend on the total concentration of amphiphiles in the system, as shown in Fig. 1.11. Note: you will need to solve (numerically) a transcendent equation in $A_{\bar{n}}$ for each concentration C.

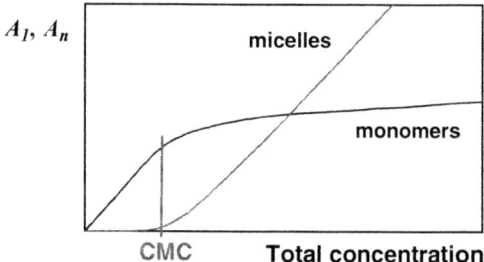

Fig. 1.11 (Associated to Quiz 8) The variation of the concentration of free and associated monomers as a function of the total concentration of amphiphiles in the system (arbitrary units) for $\bar{n} = 10$ and $\mu_0^{\bar{n}} - \mu_0^1 = 10 k_B T$ (arbitrary chosen). Note that under a certain critical concentration (called *critical micelle concentration*, CMC) the amphiphiles are present in the system as free monomers, while beyond CMC a fraction of the monomers begin to associate into micelles.

1.3 Hydrophobic "Interactions" and Molecular Self-Association

The average micelle size, $\bar{n}$, can be derived theoretically from:

$$\bar{n} = \sum_{n>1} n[A_n] \bigg/ \sum_{n>1} [A_n] = \sum_{n>1} n[A_n] \bigg/ \{C - [A_1]\} \tag{1.53}$$

while the standard deviation can be obtained from:

$$\sigma^2 = \overline{n^2} - (\bar{n})^2 = \frac{\sum_{n>1} n^2[A_n]}{C - [A_1]} - (\bar{n})^2 \tag{1.54}$$

It should be emphasized that a *minimum number* of amphiphiles is necessary to cooperate for a micelle to form; two or three monomers cannot form thermodynamically stable micelles as they cannot effectively eliminate the water-hydrocarbon interface. The process of micelle formation is therefore said to be *cooperative*. From a physical point of view this means that micelles do not present a purely statistical size distribution. Mathematically, one can accommodate this either by assuming that the summation index n in equations (1.53) and (1.54) is in fact greater than a certain value, m (which would be rather arbitrary) or, more naturally, to impose conditions on the chemical potential μ_0^n. We discuss this briefly in section 1.3.2.2.

1.3.2.2 Types of Amphiphilic Aggregates

Amphiphiles can self-associate into micelles as a result of two opposing tendencies: hydrophobic attraction between hydrocarbon tails, due to their tendency to decrease their interface with water, and hydrophilic repulsion between polar heads, which incorporates a steric contribution involving the hydration layer of the polar heads (Tanford, 1973; Israelachvili et al., 1976, 1977; Ruckenstein and Nagarajan, 1975). This situation is shown schematically in Fig. 1.12.

Mathematically, the repulsive and attractive contributions to the chemical potential can be represented, respectively, by γa and η/a, with a being the surface area per polar head (i.e., the total area of a micelle divided by the number of molecules

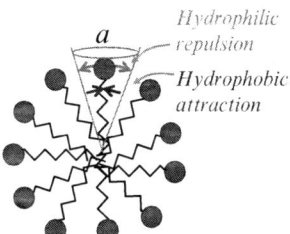

Fig. 1.12 Schematic representation of opposing forces that lead to formation and stabilization of a micelle. Note that the volume per amphiphilic molecules is described by an inverted cone with base area a, called in the text "the area per amphiphilic molecule".

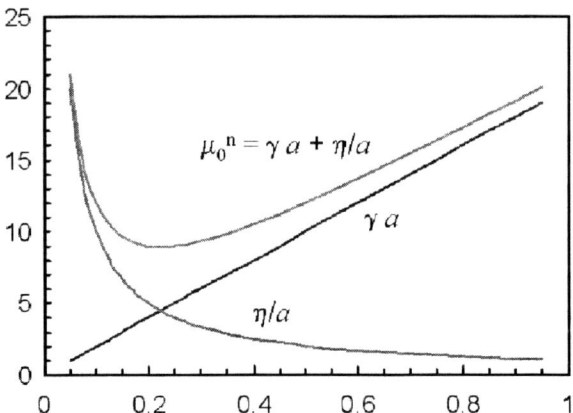

Fig. 1.13 Plot of interaction free energy as a function of the area per amphiphile head group.

per micelle), and γ and η two proportionality constants. The chemical potential of the amphiphiles associated in the micelles is:

$$\mu_0^n = \gamma a + \frac{\eta}{a} \qquad (1.55)$$

A plot of μ_0^n as a function of the area per polar head is shown in Fig. 1.13. It is seen that an increase in the area, a, leads to a minimization of the repulsive component of μ_0^n, while the attractive component minimizes the chemical potential as area tends to zero.

The constant γ (called surface tension coefficient) may be determined experimentally (and is of the order of $10\,\mathrm{mJ/m^2}$), while η may not. However, it is possible to determine η by noticing that μ_0^n reaches a minimum for a certain optimal value, a_0, of polar head area. By taking the first derivative with respect to the area as equal to zero, the optimum area and its corresponding minimum potential are obtained as:

$$\eta = a_0^2 \gamma, \text{ and} \qquad (1.56)$$

$$(\mu_0^n)_{\min} = 2\gamma a_0 \qquad (1.57)$$

which correspond to a micelle with an optimum number of monomers, m. Therefore, we can replace $(\mu_0^n)_{\min}$ by μ_0^m. Further, by substituting expressions (1.56) and (1.57) into (1.55), we obtain:

$$\mu_0^n = \gamma a + \frac{a_0^2 \gamma}{a} \qquad (1.58)$$

which can be conveniently rearranged as (Israelachvili, 1992):

$$\mu_0^n = 2\gamma a_0 + \frac{\gamma}{a}(a - a_0)^2 = \mu_0^m + \frac{\gamma}{a}(a - a_0)^2 \qquad (1.59)$$

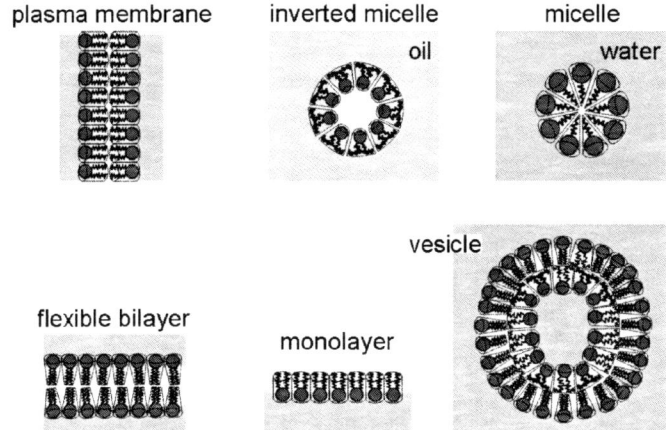

Fig. 1.14 Types of amphiphilic aggregates: monolayer, micelle, inverse micelles, planar bilayer, vesicle.

The last equation states that the chemical potential of the monomers aggregated into micelles reaches a minimum for an optimal area $a = a_0$ (as expected) and increases (i.e., becomes unfavorable thermodynamically) for larger and smaller values of a.

The above discussion suggests a way for predicting what type of aggregate can be formed by amphiphilic molecules with given geometry (such as the inverted cone in Fig. 1.12) and optimal surface area, a_0. As seen in Fig. 1.14, amphiphiles that can be enclosed within cones can form optimally small spheroidal micelles, amphiphiles that could be enclosed in truncated cones for which the polar head area is smaller than the hydrophobic tail area can form inverted micelles (i.e., with tails towards the solvent, which is a hydrocarbon), cylindrical shape conduces to planar bilayers, while slightly conical (truncated and inverted) shapes favor formation of flexible bilayers and vesicles (or spherical bilayers). Of all these possible structures, the planar bilayers and the vesicles are of greatest relevance to the living matter: the former represents a model system for the study of physical and chemical properties of bilayer membranes (such as substance transport, electrical properties, ions binding to the membrane, etc.), while the latter resembles the structure and shape of cell membranes.

References

Atkins, P. and de Paula J. (2002) *Atkins' Physical Chemistry*, 7th ed., Oxford University Press, New York

Cowan, M. L., Bruner, B. D., Huse, N., Dwyer, J. R., Chugh, B., Nibbering, E. T. J., Elsaesser T. and Miller, R. J. D. (2005) Ultrafast memory loss and energy redistribution in the hydrogen bond network of liquid H_2O, *Nature*, **434**: 199

Glaser, R. (2001) *Biophysics*, 5th ed., Springer-Verlag, Berlin/Heidelberg

Hille, B. (2001) *Ion Channels of Excitable Membranes*, 3rd ed., Sinauer, Sunderland, MA
Israelachvili, J. (1992) *Intermolecular and Surface Forces*, 2nd ed., Academic, London
Israelachvili, J. N., Mitchell, J. D. and Ninham, B. W. (1976) Theory of self-assembly of hydrocarbon amphiphiles into micelles and bilayers, *J. Chem. Soc. Faraday Trans II*, **72**: 1525
Israelachvili, J. N., Mitchell, J. D. and Ninham, B. W. (1977) Theory of self-assembly of lipid bilayers and vesicles, *Biochim. Biophys. Acta*, **470**: 185
Nagy, A. N., Raicu V. and Miller, R. J. D. (2005) Nonlinear optical studies of heme protein dynamics: Implications for proteins as hybrid states of matter, *Biochim. Biophys. Acta*, **1749**: 148
Nir, S. (1976) van der Waals Interactions between Surfaces of Biological Interest, *Prog. Surf. Sci.*, **8**: 11–58
Popescu, A. I. (1997) *Handbook of Molecular and Supramolecular Biophysics*, All Educational S.A., Bucharest (in Romanian)
Rein, R., Andre, J.-M. and Ladik, J. (1975) Interaction between polymer constituents and the structure of biopolymers. In: *Electronic Structure of Polymers and Molecular Crystals*, NATO Advanced Study Institute Series, Plenum, New York/London
Ruckenstein, E. and Nagarajan, R. (1975) Critical micelle concentration. A transition point for micellar size distribution, *J. Phys. Chem.*, **79** (24): 2622
Setlow, R. B. and Pollard, E. C. (1962) *Molecular biophysics*, Addison-Wesley, Palo Alto
Stratton, J. A. (1941) *Electromagnetic Theory*, McGraw-Hill, New York
Stryer, L. (1988) *Biochemistry*, 3rd ed., W. H. Freeman, New York
Szent-Gyorgyi, A. (1972) *The Living State with Observations on Cancer*, Academic, New York
Tanford, C. (1973) *The hydrophobic Effect: Formation of Micelles and Biological Membranes*, Wiley, New York

Chapter 2
The Composition and Architecture of the Cell

In chapter 1, we have seen how phospholipids can cooperatively assemble to form membrane structures, which resemble the membranes of biological cells. Now, let us take a look at the composition and organization of the cell, which will allow us to put the physics in its biological context. Some of the terms and concepts introduced in this chapter (e.g., proteins, cell membrane, etc.) will be dealt with in more details later on in this book, while others (such as those concerning the structure and function of cellular organelles, biochemistry of the cell, etc.) are explained in many excellent books that are available on molecular and cellular biology (e.g., Alberts et al., 2002; Lodish et al., 2004; Berg et al., 2002). The reader should therefore be aware that by no means do we intend in this section to provide a comprehensive review of subjects normally covered by cell biology and biochemistry textbooks. Instead, herein the biological information is reduced to its bare essentials, and it will be introduced and used only insofar as it can help the progression towards understanding of the biophysics concepts and principles presented in this book.

Broadly speaking, there are *two classes* of living systems: *viruses* and *uni-* or *multi-cellular organisms*. Of these, only cellular organisms present the two main distinguishing features of a living system, *self-reproduction* and *metabolism*, viruses not being endowed with their own metabolism. Both *deoxyribonucleic*-based (DNA) viruses and *ribonucleic*-based (RNA) viruses rely on the cellular metabolism of the host cell, in order to self-multiply.

Section 2.1 is concerned with a brief description of the organization of biological cells, which is considered to be the *fundamental morphological and functional unit* of living matter. Section 2.2 will provide a description of protein structure, folding and misfolding, while section 2.3 will discuss the DNA structure and replication (multiplication).

2.1 The Cell: An Overview

Biological cells fall into two main categories: *eukaryotic* and *prokaryotic* cells. *Eukaryotes* are complex, multi-compartmented cells with well distinguishable *sub-cellular organelles*, such as the *nucleus*, which contains the genetic material necessary for cell multiplication. *Prokaryotes* are some sort of *primitive cells* endowed only with a single compartment, separated from its surroundings by one or two membranes. Prokaryotes do not present a well defined nucleus, although they contain genetic material. Some members of this category are, for instance, bacteria, blue-green algae (which are also called cyanobacteria), etc. In the present chapter, we shall consider only eukaryotic cells, which comprise all the species of the animal and vegetal kingdoms, beginning with *protozoa* (such as *amoebas*) and *yeasts* (such as baker's yeast) and extending to *mammals* (such as humans).

> **Observation:** Human *erythrocyte* (i.e., red blood cell) is neither eukaryote nor prokaryote, since it presents no *nucleus* and no genetic material, being thus unable to divide itself; because of its rudimentary structure, the human erythrocyte is actually not a cell, although it looks like one.

2.1.1 Eukaryotic Cells

Some organisms are unicellular (e.g., *protozoa*), while others are multicellular with different degrees of complexity. Eukaryotic cells are encountered in tremendous varieties of shapes and sizes. In spite of this great diversity, one can identify some features that are common to all eukaryotes. Therefore, we first discuss the morphologically *"standard" type* of eukaryotes (see Fig. 2.1) and then point out possible variations from it.

Any cell is a three-dimensional structure bordered by a very thin closed shell called *plasma membrane* or, simply, *membrane*, which confines an aqueous medium, the *cytosol*. The cellular membrane consists primarily of a phospholipid bilayer (see section 1.3), which contains also proteins that play important roles, ranging from purely structural to important functional roles in, e.g., substance transport across membranes, or cell signaling. More in depth description of the membrane structure and function will be provided in chapter 3.

The interior of the cell (i.e., the *cytoplasm*) is *highly compartmentalized* by internal membranes forming the so-called *endoplasmic reticulum* (ER), which is also embedded in the cytosol. Some compartments of the ER are populated by different types of subcellular particles called *organelles*. The most prominent of them are: *nucleus*, provided with a *nucleolus, mitochondria, ribosomes, endoplasmic reticulum, Golgi complex, lysosomes, peroxysomes*, and *cytoskeleton*. We will provide next a brief description of the cellular organelles.

Nucleus is an essential cellular organelle, containing almost all the genetic material of the cell, either under the form of an unorganized *chromatin* (in the interphase – see below) or condensed into *chromosomes* during cellular division (mitosis or meiosis – see below). The number of chromosomes varies from one species to another. For example, there are eight chromosomes in the case of fruit fly (*Drosophila melanogaster*), and 46 in the case of humans.

Nucleus is delimited from the cellular milieu by a *double membrane* presenting many pores. Some portions of the outer nuclear membrane are directly connected

2.1 The Cell: An Overview

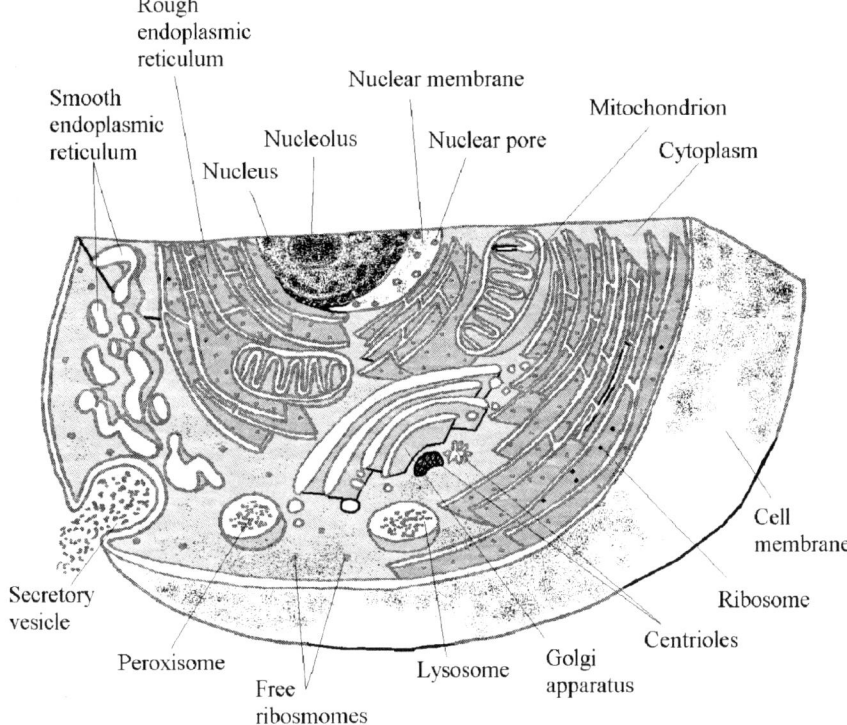

Fig. 2.1 3D schematic section through a eukaryotic animal cell.

to the endoplasmic reticulum. Synthesis of *messenger RNA* (mRNA) and *transfer RNA's* (tRNA's) – two major participants in the transmission of genetic information from DNA to the place of protein biosynthesis – occurs in the nucleus. Inside the nucleus, there is also a *nucleolus*, in which *ribosomal RNA* (rRNA) – another important contributor to the process of protein biosynthesis – is synthesized. These ribonucleic acids will be discussed later in this book.

Mitochondria are organelles involved in cellular energetics. They truly are the *energetic plants of the cells*, because they are the place where fatty acids and glucose are oxidized to generate energy, which is then stored by synthesizing adenosine three-phosphate (ATP) (Berg et al., 2002). ATP, also called a *macroergic molecule*, can be considered the *metabolic currency* of the cell, being used as a source of energy for all energy-consuming cellular processes. Mitochondria also contain small amounts of genetic material in the form of several copies of a circular DNA molecule. This is used in semi-autonomous reproduction of mitochondria inside the cell. Mitochondrial DNA is not contained in a nucleus and is not packed into chromatin. In fact, the overall structure of the mitochondria, including its DNA, resembles that of prokaryotic cells.

Ribosomes are small particles with an important role in protein synthesis. They are composed of rRNA and proteins, and consist of two subunits: one with a sedimentation rate of 60 S (Svedberg, see next observation) and another of 40 S. When these two particles bind to form the ribosome, a particle with sedimentation rate of 80 S is created. Ribosomes can be free in the cytosol or attached to the endoplasmic reticulum.

> **Observation:** Microscopic and submicroscopic particles, such as the ribosomes can be separated by centrifugation in high speed centrifuges or ultracentrifuges. The sedimentation constant, s, is defined as the ratio between the sedimentation speed, v, and the centripetal acceleration, $\omega^2 r$, where ω is the angular velocity and r is the distance from the center of rotation. Svedberg's theoretical model for sedimentation (Svedberg and Pedersen, 1940) provides the following formula for the sedimentation speed: $s = m(1 - \rho_s/\rho)/f$, where m is the particle mass, f is the shape-dependent drag coefficient, while ρ and ρ_s are the densities of the particle and solvent, respectively. Note that sedimentation can only occur if the densities are different. The unit of the sedimentation constant is Svedberg ($1\,\text{S} = 10^{-13}\,\text{s}^{-1}$) named after the Swedish scientist who has invented the ultracentrifuge (and received the Nobel Prize for Chemistry for his work in colloid chemistry). You may have already noticed that the sedimentation constants are not additive quantities. For instance, a 60 S ribosomal unit coupled to a 40 S unit will sediment at 80 S.

Endoplasmic reticulum (ER) is a membranous structure forming numerous sinuous channels and cisternae. When it is associated with ribosomes, ER forms the so called *rough endoplasmic reticulum*, which is involved in protein synthesis, processing and sorting.

Golgi complex is an organelle that catches the proteins synthesized in the endoplasmic reticulum in view of their progressive processing and sorting for secretion.

Peroxisomes are subcellular particles responsible for detoxification of the cell from free radicals (i.e., atoms or groups of atoms with unpaired electrons, which are highly reactive), and also involved in degradation of fatty acids.

In addition to the above organelles, animal cells present some specific organelles/structures, as follows.

Cytoskeleton is formed of a network of fibers spanning the whole cell, thus conferring a particular shape to each cell, and also participating in cellular movements.

Lysosomes are closed vesicles containing many *enzymes*, called *hydrolases*, which contribute to degradation of foreign particles or even particles belonging to the same cell. The pH inside the lysosome is usually two units lower (pH = 5) than the cytosolic pH.

Microvilli are thin membrane protuberances that increase the area of the interface between the cell and its microenvironment, for increased efficiency of molecular transport across the membrane. Such an increased membrane surface area is very important to some secretory cells such as hepatocytes (liver cells).

> **Observation:** Depending on their specific function, cells in a higher organism may differ in many regards from the "standard" cells outlined above. For instance, the muscle cells are rich in mitochondria, the macrophages are rich in lysosomes, the nervous cells contain neurofilaments, etc.

Eukaryotic plant cells, although presenting most of the morphological characteristics of the animal cells, are different at least in four aspects: (1) plant cells

2.1 The Cell: An Overview

contain special organelles, called *chloroplasts*, involved in *photosynthesis* (a process by which plants harvest and store energy from light); (2) present a mechanically resistant *wall* around the membrane; (3) have no *lysosomes*; and (4) present large *vacuoles*.

Chloroplasts. These are complex particles containing stacks of *thylakoid sacks*, the membranes of which embed protein-chlorophyll complexes that are directly involved in the process of *photosynthesis*.

Cell wall. Cells contain large concentrations of solutes compared to the external medium. As will be discussed in chapter 4, these concentration gradients lead to very high osmotic pressures being exerted on the cell membrane from inside the plant cell. The cell wall, composed of cellulose (a polysaccharide) that confers a rigid structure, protects plant cells against *mechanical stress* and their membranes against *osmotic lysis* (see the chapter 4). The cell wall also determines the shape of the cell. The bacterial cell wall has a different composition from that of plants but its role is basically the same.

Vacuoles. These organelles contain the cell's reserves of water, ions and nutrients. They have functions in degrading some macromolecules being somehow the correspondents of the animal cell lysosomes. Also, by controlling the content of poly-electrolytes in the vacuoles, the cell can regulate its own *internal osmotic pressure*. Other types of cells (besides plant cells), such as yeasts, also contain very large vacuoles, which are in general more prominent than the nucleus when observed under an optical microscope.

2.1.2 Cellular Self-Reproduction

The ability to self-reproduce is *the main feature of the living matter*. It ensures perpetuation of all animal and vegetal species, which otherwise will be extinguished. According to the biogenesis dogma, any living being has at least one natural *genitor*. In spite of all evident advances in molecular biology, genetics, biochemistry, biophysics, etc., it is not yet possible to assemble a cell starting from its molecular components. The only way of "creating" a cell is currently by starting from another cell.

There are two types of cell self-reproduction: *asexual reproduction* and *sexual reproduction*. The asexual reproduction is the simplest modality of reproduction, requiring only one genitor. It can be realized by *fission* or *mitosis* (in the case, for instance, of bacteria), or by *budding* (in the case, for example, of yeasts) as one can see in Fig. 2.2.

> **Observation:** *Cloning* is a special case of asexual reproduction, which is done in a laboratory. For cloning, a nucleus extracted from a *somatic cell* of a *donor* (male or female) is inoculated in a de-nucleated ovule (i.e., an ovule whose nucleus has been removed). This ovule is injected in an "adoptive mother" where it will start a process of successive divisions by fission, accompanied by cellular differentiation and morphogenesis generating an embryo and, later, an organism *similar to the donor*. This means that *any somatic cell* contains all the genetic information from the genitor.

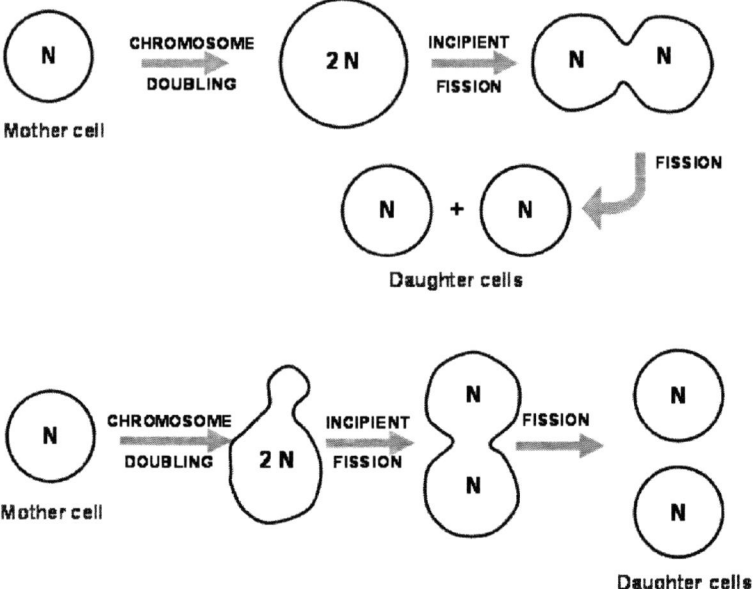

Fig. 2.2 Top: cell mitosis. Bottom: cell budding. N is the number of chromosomes.

The cellular life-cycle comprises many phases, culminating with *mitosis*, in which cell division occurs (Alberts et al., 2002; Lodish et al., 2004). Cell-cycle phases are illustrated in Fig. 2.3 and briefly described in its legend. The main subphases of mitosis are: *prophase, prometaphase, metaphase, anaphase and telophase* (Fig. 2.3). During these phases the actual division of the mother cell into two daughter cells is completed, each containing a full set of genetic material.

In the case of human cells with a high *proliferation rate*, the period, τ_c, of a cell cycle is $\sim$24 h (Fig. 2.3), mitosis occurring only briefly ($\tau_M = 0.5$ h). For yeast cells, τ_c is of the order of 2–4 h, with the mitosis lasting a few minutes. The longest phase is the *synthesis*, S, lasting about 10 h for human cells, while G_1 and G_2 last about 9 h, and 4.5 h, respectively (Lodish et al., 2004).

Observation: Some cells that are not normally proliferating (e.g., nerve cells) exit the G_1 phase and enter the G_0 phase, where they will spend most part of their life span. However, it was observed that even cells in this dormant state are sometimes able to re-enter the cellular cycle beginning to divide.

Sexual reproduction involves two *genitors* of opposite sex. Each genitor contributes a *haploid* cell with N chromosomes, called a *gamete*. The haploid gametes are formed in the two-step process of meiosis, during which a *diploid* cell containing 2N chromosomes divides to form four haploid cells with N chromosomes each. Meiosis contains all the sub-phases of mitosis described above, which are repeated in two steps of division occurring successively. For instance, in the case of humans, the female contributes an *ovule* having 22 *somatic chromosomes* and a *sex*

2.1 The Cell: An Overview

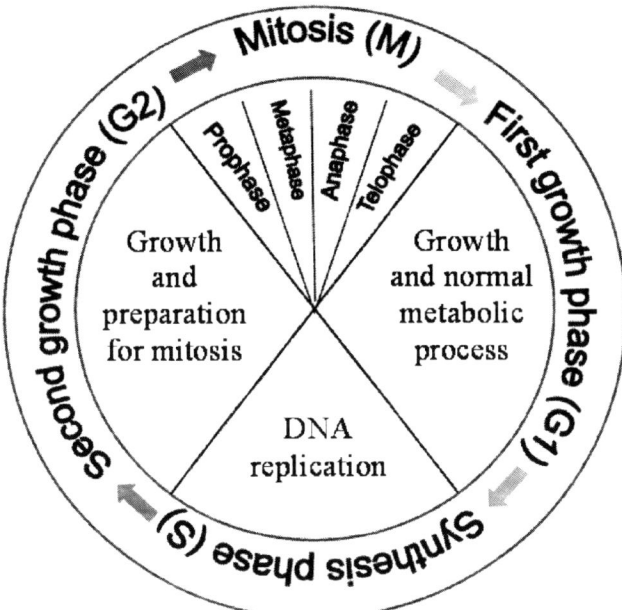

Fig. 2.3 Cell life-cycle: G_1 – cell cycle phase during which cells synthesize proteins and mRNA; S – the genetic material is multiplied (i.e., DNA replication and doubling of the number of chromosomes doubling); G_2 – dormant phase; M – mitosis. In its turn, mitosis is composed of four phases: Prophase – chromosome formation through condensation of chromatin and formation of the *spindle apparatus* to move the pairs of chromosomes around the cell; Metaphase – chromosomes separate and begin to be pulled towards the oppositely located centrioles; Anaphase – chromosomes continue moving towards the centrioles, which further move apart to elongate the cell; Telophase – reformation of cell structures including nucleus, and de-condensation of the chromosomes into chromatin. The duration of the main phases is given in the text for the case of human cells with high proliferation rate.

chromosome (denoted by X), while the male contributes a spermatozoid possessing 22 *somatic chromosomes* and a *sex* chromosome, either of type X or Y (Fig. 2.4).

In the process of fusion of two *haploid* gametes (i.e., an ovule and a spermatozoid, possessing $N = 23$ chromosomes each), a *zygote* is formed, that is a *diploid* cell (i.e., with $2N = 46$ chromosomes). Then, the zygote undergoes successive divisions (mitosis) followed by cell differentiation and morphogenesis, resulting in an embryo and, eventually, a fetus with *characteristics inherited* from both genitors (Fig. 2.4).

> **Observation:** During the very complex process of *morphogenesis*, in which division of a single cell leads not only to different kinds of tissue cells, but also to a well-defined spatial pattern (or supracellular architecture), some cells must die so as to properly generate functional organisms. This process of cellular death is programmed and is called *apoptosis*. For instance, in the case of a fetus in a mother's womb, it is important that cellular material disappears for the fingers to be formed (similar to wood carving!).

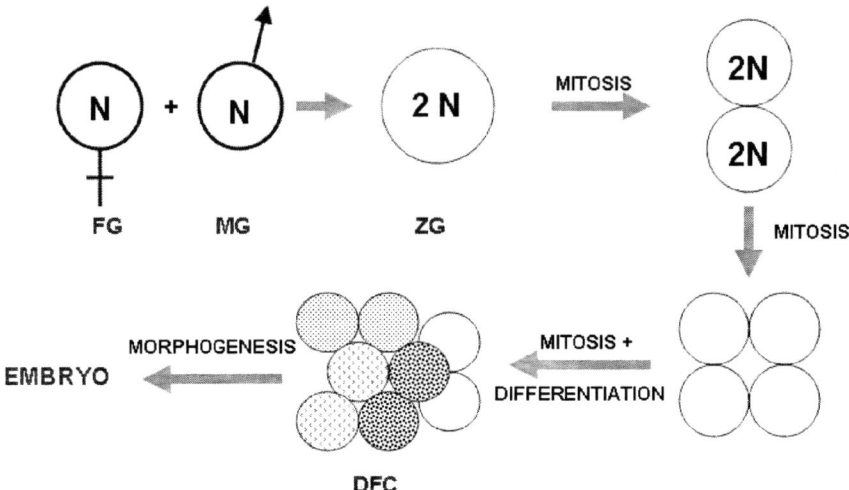

Fig. 2.4 Sexual self-reproduction followed by mitosis (asexual division), differentiation and morphogenesis. FG – female gamete (i.e., ovule, which is a haploid cell); MG – male gamete (i.e., spermatozoon, a haploid cell); ZG – zygote (a diploid cell); DFC – differentiated cell species.

2.1.3 Cellular Metabolism

As mentioned above, the second basic feature of the living matter, after the self-reproduction, is the *metabolism*. The concept of metabolism includes all the biochemical transformations taking place in a given unicellular or multicellular organism. The metabolism consists of two antagonistic (i.e., opposite) processes: (1) *anabolism*, which includes all *biosynthesis processes*; and (2) *catabolism*, which includes all the *degradation processes*. From an energetic point of view, anabolic biochemical reactions are *endergonic* processes, that is, energy-consuming, while the catabolic reactions are *exergonic* processes, that is, energy-generating processes.

According to the second law of thermodynamics, the direction of evolution of a process is given by the variation of the *Gibbs free energy*, ΔG. For *endergonic* reactions, $\Delta G_{en} > 0$ (which means that the process is not spontaneous), while in the case of *exergonic* reaction, $\Delta G_{ex} < 0$, which means that the process is spontaneous and runs in the forward direction.

In the language of nonequilibrium thermodynamics, the anabolism and catabolism are *coupled processes*. Catabolic processes drive the anabolic ones, so that, globally, the metabolism is an exergonic process, i.e.:

$$\Delta G_{en} + \Delta G_{ex} < 0 \qquad (2.1)$$

On the other hand, one may characterize the metabolism by the variation of *entropy* in the process. Accordingly, anabolic processes, leading to assembly of myriads of small disordered molecules into fewer and more ordered macromolecules (such as proteins, nucleic acids, polysaccharides) create order by consuming entropy, $\Delta S_{en} < 0$.

As Erwin Schrödinger, one of the founders of quantum mechanics, has put it, living organisms feed on negative entropy (Schrödinger, 1992).

Catabolic processes generate molecular disorder, hence entropy, $\Delta S_{ex} > 0$. Such exergonic processes are necessary in the economy of the cell because they supply the energy necessary for driving biosynthesis processes, which are anabolic. However, according to the second law of thermodynamics, the global entropy variation must be positive,

$$\Delta S_{en} + \Delta S_{ex} > 0 \tag{2.2}$$

Observation: The variation of total entropy of any organism is negative, because biological organisms are open systems that can reduce their entropy by "exporting" an excess of entropy into their environment. Thus, the total entropy will thereby increase, so as to satisfy inequality (2.2). This continuous *"fight"* of living organisms against entropy (i.e., against disorder) is won by biological organisms, at least temporarily, by incessantly consuming free energy from outside (light energy or food). Thus living biological systems are *dissipative*, in terms of nonequilibrium thermodynamics.

2.2 Proteins

Proteins are ubiquitous biological macromolecules that are essential to all cellular processes. Indeed, all metabolic biochemical reactions in the cell, including *degradation* (catabolism) of some molecules and *biosynthesis* (anabolism) of others are assisted by proteins. Proteins are also involved in the replication of the deoxyribonucleic acids, as well as in their own biosynthesis. For all these reasons, proteins could be considered as the *executive power* of the cell and, in fact, of life itself.

Proteins present in any unicellular or multicellular organism can be classified in two main categories: (a) *functional proteins*, and (b) *structural proteins* with a morphological role. Functional proteins include, for instance, *enzymes* (involved in metabolic reactions), *immunoglobulins* [with a role in the so-called humoral immunity (Martini, 2004; Walter et al., 2004)], *membrane receptors* (involved in cellular recognition), ion *pumps, carriers, and channels* (performing active and passive ion transport across membranes), etc. As for the structural proteins category, one may cite: *collagen* (a basic component of conjunctive tissue, cartilage, bone and skin), *elastin* (a component of the skin), *keratin* (from nails), etc. We will discuss physical mechanisms underlying specific functions of these types of proteins (and others) in separate sections of this book. In the following sections, we will discuss structural properties of proteins in general. For further details, the interested reader is referred to of excellent books published on the subject (e.g., Berg et al., 2002; Lodish et al., 2004).

2.2.1 Protein Structure

As we shall see later, protein expression in cells follows precise "recipes" encoded in certain sequences of DNA, called *genes*. All genes of an organism constitute practically a "cookbook" for that organism – which is called an organism's *genome*. According to recent results, the *Human Genome* comprises fewer than 25,000 protein-encoding genes (International Human Genome Sequencing Consortium, 2004), while the number of proteins in one organism is much larger (of the order of hundred of thousands). The question as to how could this be possible currently is an active area of research.

Considering that there are millions animal and vegetal species on the Earth, each with its own set of proteins, it appears that the total number of different proteins in Nature is huge. However, in spite of their great variety, all proteins are made up of a very small number of *elementary building blocks*: 20 types of natural *amino acids* (AAs). Therefore, all proteins are in fact *biopolymers*, i.e., sequences (*heteropolymers*) of natural amino acids.

2.2.1.1 Amino Acids as Universal Building Blocks of All Proteins

An amino acid (AA) is a simple molecule having in its chemical structure an amino (**–NH$_2$**) and an acidic (**–COOH**) group, grafted on to a so-called alpha carbon (**C$_\alpha$**) as shown in Fig. 2.5. A hydrogen atom and a *side chain* or *residue*, **R**^{i}($i = 1$–20), are linked to same carbon atom.

An amino acid can take three different ionic forms, depending on the pH of the medium. At a certain pH, called *isoelectric* pH or *isoelectric point* (pI) and which is specific to each amino acid, both amino and acidic groups are ionized, the AA becoming a *zwitterion* (Fig. 2.6). In a more *acidic medium* (pH $<$ pI) the amino group captures a proton, thus becoming *positively charged* (**–NH$_3{}^+$**), and in a *basic medium* (pH $>$ pI) the acidic group looses a proton, becoming *negatively charged* (**–COO$^-$**).

The identity of an AA is given by its *side chain*, **R**^{i}, which also confers a hydrophilic or hydrophobic character to the amino acid. Figure 2.7 presents the side chains of some AAs, with their name, standard three-letter and one-letter code notations and also the notations used for the side-chain atoms. The structures and notations for all known 20 AAs are provided in an Appendix.

Some of the simplest AAs are *Glycine*, for which the side chain is **H**, and *Alanine*, whose side chain is the methyl group (**–CH$_3$**). From Appendix 1 one can see that 15 AAs have linear (sometimes aliphatic) side chains, while 4 AAs have cyclic side chains: *Histidine* (His: H), *Phenylalanine* (Phe: F), *Tyrosine* (Tyr: Y) and *Tryptophan* (Trp: W). The single *atypical* AA is *Prolyne* (Pro: P), which has the *amino*

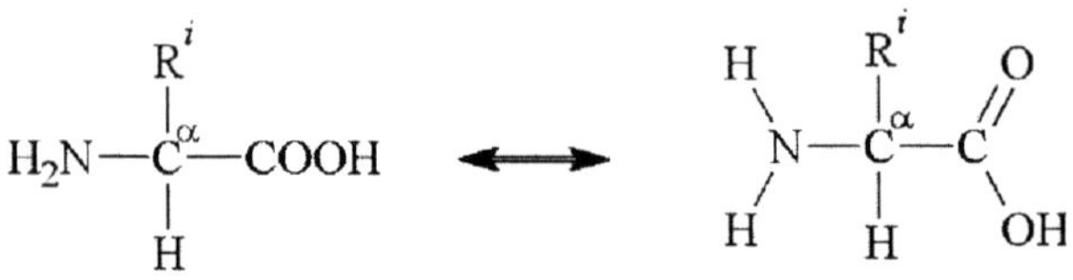

Fig. 2.5 General structure of an amino acid. **R**^{i} is the amino acid side chain ($i = 1$–20).

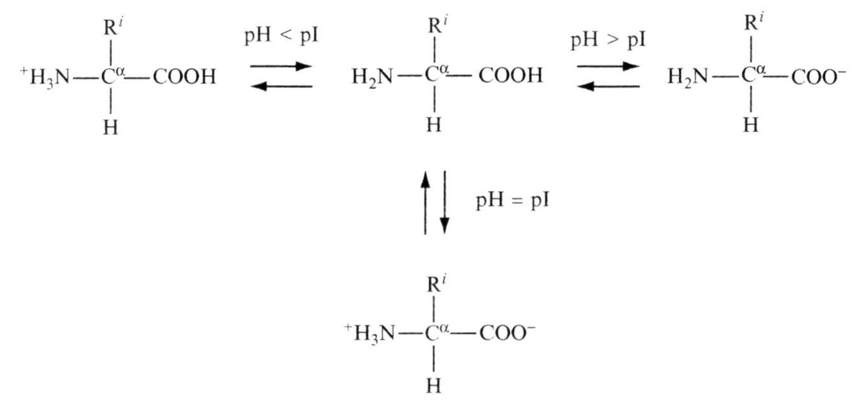

2.2 Proteins 49

Fig. 2.6 Ionic forms of an amino acid at various pH values, relative to the pI.

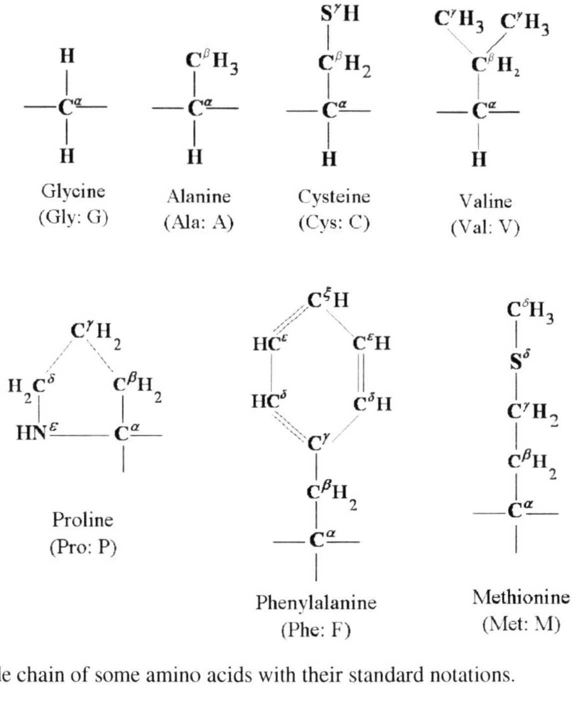

Fig. 2.7 The side chain of some amino acids with their standard notations.

group included in a cycle. For this reason its presence in a protein strand will induce important steric constraints, related to spatial disposition of atoms in a molecule (see below).

The atoms (H, C, O, N, S) in a protein are spatially distributed in a *relatively invariant* structure specific to each protein type. In this structure one can distinguish elements of lower ranks, such as *primary* and *secondary* structures, as well as higher level of organization represented by *tertiary* and *quaternary* structures; all these will be briefly described below.

2.2.1.2 Primary Structure of Proteins

This is the lowest level of protein structure and consists of a shorter or longer chain of amino acids bound together by strong covalent (or peptide) bonds.

If two AAs (residues represented by subscripts *i* and *j* in Fig. 2.8) chemically interact (an *endergonic* process, driven by a source of energy) a water molecule is eliminated and a larger molecule is formed, which is stabilized by a covalent bond, known as *peptide bond* (Fig. 2.8).

Observation: The atoms (O, C, N, H) involved in a peptide bond are, to a first approximation, *coplanar*, forming a somewhat rigid configuration.

If another AA, with its side chain, $\mathbf{R}^k$, covalently binds to the dipeptide in Fig. 2.8, a *tripeptide* will result (Fig. 2.9). The variable angle between the plane ($\mathbf{HNC}^{\alpha}$) and the plane ($\mathbf{NC}^{\alpha}\mathbf{R}^{\mathbf{j}}$) is denoted by Φ_j, and the angle between the plane ($\mathbf{R}^{\mathbf{j}}\mathbf{C}^{\alpha}\mathbf{C}$) and the plane ($\mathbf{C}^{\alpha}\mathbf{CO}$) is usually denoted by Ψ_j. These angles may vary between $-180°$ and $+180°$, although in real proteins their values are limited by the secondary structure in which the residue is involved.

If the process of polymerization continues, one obtains *tetrapeptides, pentapeptides*, etc., and generally, *polypeptides*. If the number, *n*, of AAs in a polypeptide chain (Fig. 2.10) is very small, the sequence is called an *oligopeptide*, if it is larger, the sequence is called a *polypeptide* and if this number is (arbitrarily chosen) *more than 50*, one speaks of *proteins*.

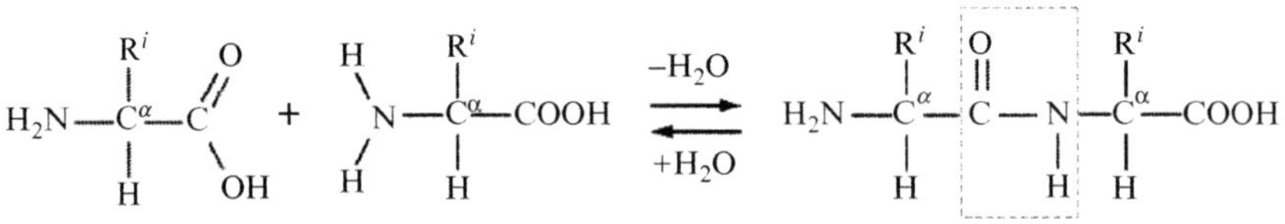

Fig. 2.8 Synthesis of a "dipeptide" from two AAs. The atoms participating in the peptide bond are included in a rectangle.

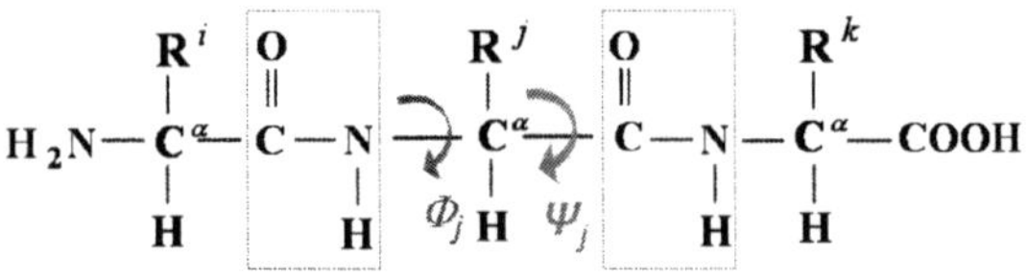

Fig. 2.9 The primary structure of a tripeptide. The rotation angles, Φ_j and, Ψ_j around the bonds $\mathbf{N}$–$\mathbf{C}^{\alpha}$ and $\mathbf{C}^{\alpha}$–$\mathbf{C}$, respectively are called the Ramachandran angles.

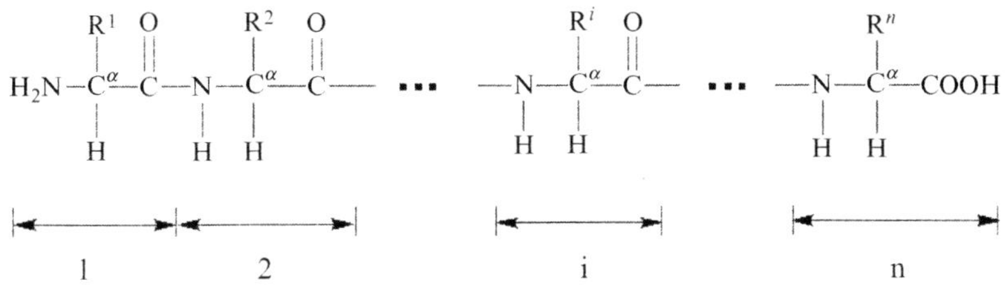

2.2 Proteins

$$H_2N-\underset{H}{\overset{\overset{\displaystyle R^1}{|}}{C^\alpha}}-\overset{\overset{\displaystyle O}{\|}}{C}-\underset{H}{N}-\underset{H}{\overset{\overset{\displaystyle R^2}{|}}{C^\alpha}}-\overset{\overset{\displaystyle O}{\|}}{C}-\cdots -\underset{H}{N}-\underset{H}{\overset{\overset{\displaystyle R^i}{|}}{C^\alpha}}-\overset{\overset{\displaystyle O}{\|}}{C}-\cdots -\underset{H}{N}-\underset{H}{\overset{\overset{\displaystyle R^n}{|}}{C^\alpha}}-COOH$$

|←—— 1 ——→|←—— 2 ——→| |←—— i ——→| |←—— n ——→|

Fig. 2.10 A polypeptide (protein) chain formed by AAs polymerization.

H_2N —(V)(L)(W)---(H)(F)(Y)---(G)(M)(R)— COOH

Fig. 2.11 A simplified pictorial representation of a protein strand.

A polypeptide chain presents a *backbone* formed by the atoms $NC_\alpha C$, starting from the amino group and ending at the acidic (carboxyl) group, to which there are attached n side residues. The number, n, of AAs in natural proteins may vary from 50 to about 4,000 (Lodish et al., 2004). In a more simplified way, one could represent the primary structure of a protein as a chain of beads, indicating the name of each AA (Fig. 2.11). The protein strand is oriented, conventionally, from the amino (start) group to carboxyl (end) group.

Due to the fact that the peptide bonds are very strong, thermal or other physical conditions do not affect the primary structure of proteins. For instance, by boiling an egg, the primary structure of its albumin (an egg protein) is not affected at all, but higher order structures (see below) are.

Quiz 1. If one changes the sequence of AAs in a given protein, so that R^n becomes R^1, R^{n-1} becomes R^2 and so on, until R^1 becomes R^n, one obtains a protein with the same AA composition. Is this new protein different from the original protein?

Observation: For any protein, the primary structure (i.e., the AA sequence) is uniquely determined by the *codon* sequence in its associated *gene*. If a single AA is replaced by another one, a different protein is synthesized. Even by exchanging reciprocally the places of two AAs in the same protein strand, one obtains another protein. In genetic engineering one changes, on purpose, one AA with another one, obtaining thus, a *mutant* of the wild type protein, in order to check if the respective AA is important or not in a cellular process.

We conclude the discussion on the primary protein structure by stating that the AA sequence (i.e., primary structure) of many proteins has been deciphered already, either with the aid of sequencing machines or, more recently, by deducing it directly from the codon sequence of the gene that encodes the protein information. New primary protein structures are being deposited every day in data banks which can be accessed over the Internet, such as the Protein Data Bank (http://www.rcsb.org/pdb/Welcome.do), SwissProt (http://ca.expasy.org/sprot), etc.

2.2.1.3 Secondary Structure of Proteins (Local Folding)

Even during their synthesis, protein chains do not remain linear; instead, due to a multitude of physical interactions (e.g., ionic interactions, *intra-chain* hydrogen bond formation, van der Waals interactions, etc.) a protein thread will adopt different *secondary structures*, the process being known as *local folding*. Very often, the intra-chain hydrogen bridges are made between the –NH group of the *ith* AA and the –C = O group of the $i + 4th$ AA.

There exist mainly four types of secondary structures: (1) alpha-helix (α-*helix*); (2) beta-strand (β-*strand*), occasionally generating β-*sheets*; (3) beta-turn (β-*turn*); and (4) amorphous (*random coil*) structure. We shall briefly describe each type of secondary structure below (Eisenberg, 2003).

***Alpha-helix* structure**. In this case, the atoms of the protein backbone (i.e., **N**, **C$^\alpha$**, and **C**) are situated on an ideal, usually right-handed helix, with a turn (pitch) of 5.4 Å, a diameter of about 12 Å and with 3.6 amino acids per turn (equivalent to 18 amino acids per 5 turns). The number of atoms covalently bound per pitch is 13. For these reasons, the most common α-*helix* structure is denoted by 3.6_{13}, in order to distinguish it from other α-*helix* types. The secondary structure of an α-*helix* is represented in Fig. 2.12a. The AA side chains are located at the exterior of the imaginary cylinder in which the helix is inscribed.

Quiz 2. By projecting the C_α perpendicularly to the imaginary axis of the helix, find out the distance, *d*, between two successive projections.

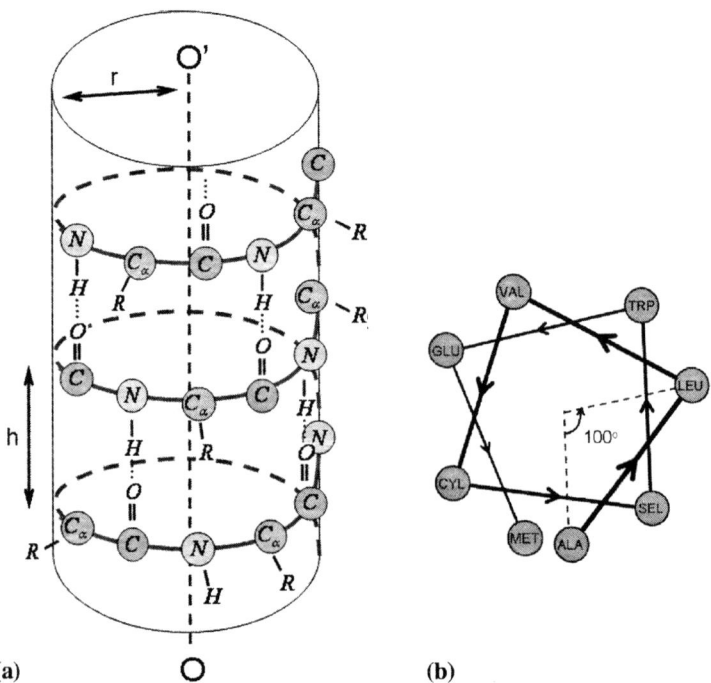

Fig. 2.12 (a) The α-helix structure (of the 3.6_{13} type) of a protein. (b) Wheel representation of lateral residues (see text for details).

One can use also the so-called wheel representation of the α-*helix* side chains (Fig. 2.12b). In this figure, the wheel is situated in a plane perpendicular to the helix axis, each AA being represented by a circle. Two successive AAs are connected by a straight line, forming an isosceles triangle (with an angle $\delta = 100°$) with the two radii connecting the wheel center and the two AAs. This type of representation could help to see, for instance, if there is segregation of *hydrophilic* and *hydrophobic* lateral chains on one side or another of the protein α-*helix*.

Beta-strand/Beta-sheet structures. The β-*strand* is not as compact as the α-*helix*, the distance between two C^α being of the order of 3.5 Å. The β-*strands* of the protein thread are usually aggregated in two or more strands, forming the so-called β-*pleated sheets*, or simply, β-*sheets*, due especially to the forming of many inter-strand hydrogen bridges (Fig. 2.13). The name of β-*pleated sheet* comes from the disposition of the protein backbone atoms in an imaginary pleated sheet. In contrast to the case of α-*helix* structures, the hydrogen bonds in β-*sheets* form between the –NH and –C = O groups pertaining to AAs from different fragments of the protein thread.

There are two subclasses of β-*sheets*: parallel β-*sheets*, denoted by β_p, and antiparallel β-*sheets*, denoted by β_a (Fig. 2.13). From an energetic point of view, the β_a-*sheets* are more stable than the β_p-*sheets*, so that their frequency of occurrence in the protein structure is higher.

Beta-turn and random coil structure. β-*turn* structures consist of a quite small number of AAs that look like kinks in the protein thread, being also known as *hairpin turns* (Fig. 2.13c). There are seven types of β-*turn* structures, all of which are usually encountered as connecting elements between α-*helices* and/or β-*strands*. Although there exist several types of turns, β-turns are among the most common ones. In some proteins, one may also find loops with no regular structure, called random coils. These structures confer flexibility to the protein chain, unlike α-*helices* and β-*strands* that provide rigidity.

> **Observation:** Some proteins have only one type of secondary structure, besides β–*turns*, while other proteins have all types of secondary structures. Therefore, based on their secondary structures, proteins fall into the following categories: (1) *all* α, possessing only α-*helices*, connected by β-*turns* (e.g., myoglobin, hemoglobin); (2) *all* β, possessing only β-*strands* (e.g., β-plastocyanin); (3) α/β, when the α-*helices* and the β-*strands* are alternating in the protein chain (e.g., taka-amylase); (4) α and β, when the α-*helices* and β-*strands* are segregated in their protein chain (e.g., α-lytic protease); (5) *random coil* proteins.

In general, the total percentage of α-*helices* and β-*strands* in a protein chain is about 60%, the rest being in the form of β-*turns* and *random coils* (Lodish et al., 2004).

2.2.1.4 Tertiary Protein Structure (Long-Range Global Folding)

The elements of secondary structure described above may fold into a *3D compact structure*, due to reciprocal interactions between chains through, e.g., formation of

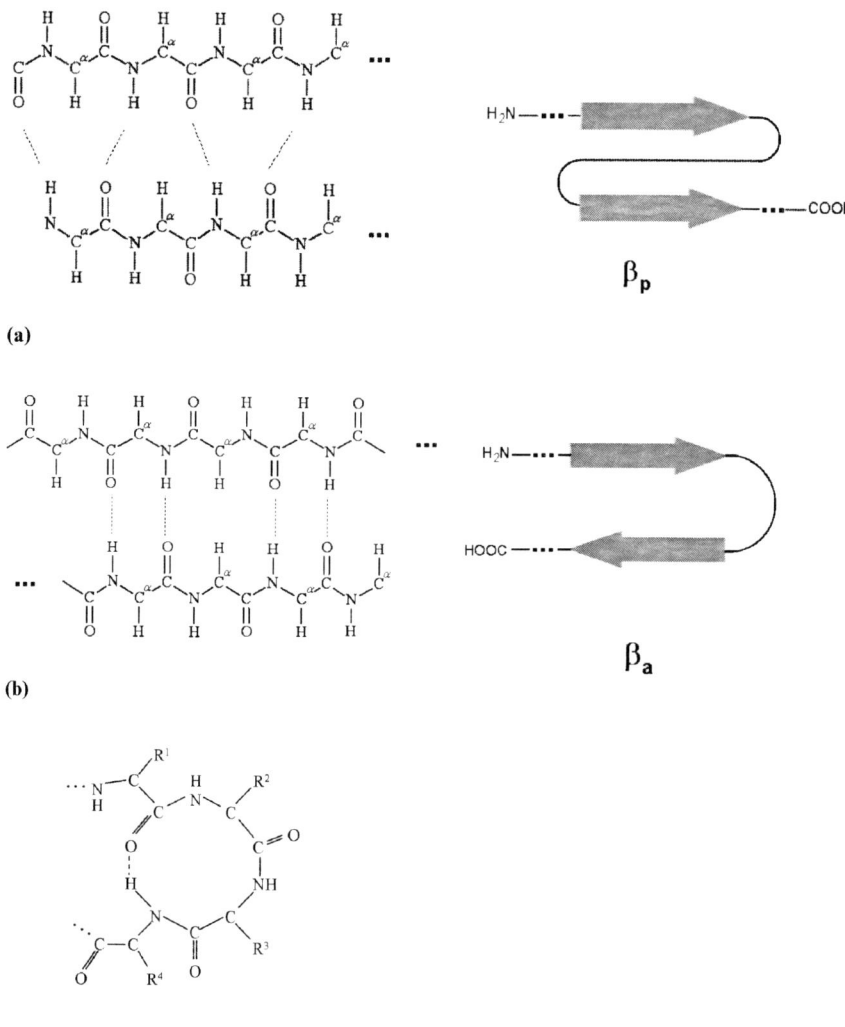

Fig. 2.13 β-secondary protein structure: (**a**) parallel β-sheet; (**b**) antiparallel β-sheet; (**c**) β-turn. The diagrams on the right hand side are ribbon representations of those on the left.

disulfide bridges (covalent bonds between two sulfur atoms on two side chains of Cysteine), as well as to interactions with water molecules (mostly, *hydrophobic interactions*). In this structure, it is possible for some elements of the secondary structure, formed by AAs situated close to the amino end of the protein, to reach into the close vicinity of the elements of secondary structure, involving the AAs on the carboxyl end of the protein chain, justifying the qualification of this type of folding as a *long-range* folding.

In general, proteins' three dimensional structures are compact, many proteins adopting a more or less globular form, with the hydrophobic AAs buried inside and

2.2 Proteins

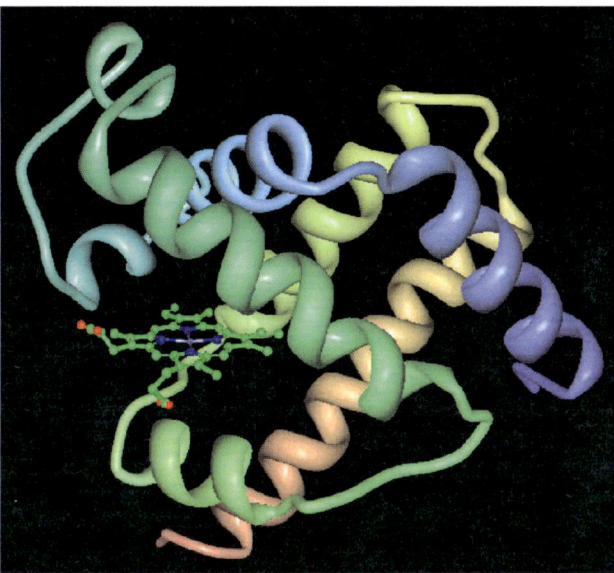

Fig. 2.14 The tertiary structure of myoglobin. Note how α-helix domains are linked by "amorphous" loops or β-turns. A prosthetic part (i.e., the heme) is drawn in green, blue and red. This figure has been prepared using "The Molecular Biology Toolkit" tool provided at http://www.rcsb.org/pdb/Welcome.do (Moreland et al., 2005).

the hydrophilic AAs exposed on the external "surface" of the globule to water molecules. This level of spatial structure of a protein is known as the *tertiary structure* (Fig. 2.14).

If a protein is formed only from AAs, it is called an *apoprotein*. But many globular proteins, in order to become functional, are provided with a hydrophobic pocket in which a *prosthetic* (non-proteic) molecule is embedded (e.g., myoglobin which contains a particular porphyrin ring called *heme*). Such proteins are *called holoproteins* (Fig. 2.14). The prosthetic parts are, as a general rule, the *active centers* of these proteins. The binding site of an enzyme, for instance, is considered to be its active center. In large proteins, one or more regions can exist with a high *affinity* for different other molecules, called *ligands*. These regions are known as specific *binding sites* of the protein.

The tertiary structure of a protein implies a unique 3D topology. In this state the protein is fully functional, and therefore it is called the *native state* of the protein. In contrast with the primary structure, this state is very sensitive to physical and chemical denaturants (e.g., heat, and chemical treatments).

There are many ways of representing the 3D protein structure. Here we present three of them (Fig. 2.15): the *ribbon model*, which puts in evidence the secondary structure elements (i.e., α-helices β-sheets, β-turns), space-filling van *der Waals spheres*, and the protein backbone represented by sticks (showing all atomic bonds, except for those involving hydrogen atoms).

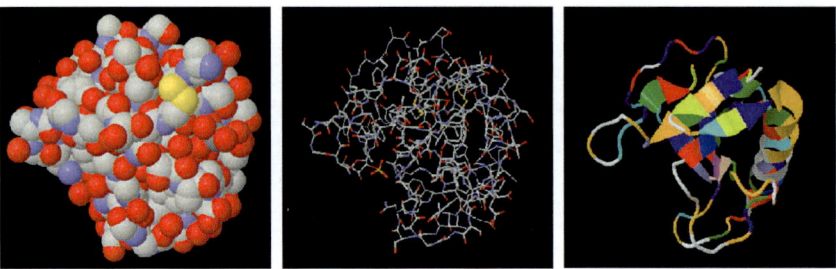

Fig. 2.15 Different modalities of representing the three-dimensional structure of Ribonuclease T1 V89S mutant. In order, from left to right: atoms, sticks, ribbon. Figures produced using Jmol (http://www.jmol.org) and structural data obtained from http://www.rcsb.org/pdb/Welcome.do.

Observation: In the process of folding (which could be either spontaneous or assisted by other proteins called *chaperonins*) deviations from the native state (for which the protein reaches a local minimum in its energetic landscape) are possible. This misfolded state can induce the protein to malfunction, thereby being able to provoke various diseases in the host organism. Such proteins folded in a tertiary non-native structure, called *prions* (Dobson, 2002) are responsible for instance for the *mad cow-disease* in cattle and for the *Kreutzfeld-Jacob disease* in humans.

As we already stated above, new primary structures of proteins are recorded daily in the Protein Data Bank, but the methods to experimentally determine their 3D structures (e.g., X-ray diffraction, NMR) and/or theoretically predict them (e.g., sequence homology or neural network-based methods) are unable to keep the pace with the huge flux of protein primary structures.

2.2.1.5 Quaternary Protein Structure (Multimeric Organization)

There is a large number of proteins that attain only the tertiary level of organization, as in the case of myoglobin, ribonuclease, etc. Yet, there are many other proteins that are functional only if they are associated with other molecules of their own or different kind (*monomers*), to form a *multimer*. Such a *multimeric structure* is known as a *quaternary structure* of a protein.

A multimer could be formed by two, three, four, etc., monomers of different size. For example, *horse alcohol dehydrogenase* is a dimer, *hemagglutinin* is a trimer, *hemoglobin* is a tetramer, *aspartate transcarbamylase* is a hexamer, etc (Lodish et al., 2004).

The association of "monomers" in multimers is usually the result of weak but numerous interactions (e.g., hydrogen bonds, hydrophobic interactions) and also of stronger ionic interactions. Analogous to the tertiary structure, the quaternary structure is also very sensitive to the action of different denaturing agents. Generally speaking, there are three classes of 3D protein structures: *globular* [e.g., hemoglobin (Fig. 2.16), elastin, etc.], *fibrous* (e.g., collagen) and *membranar* (e.g., rhodopsin, ion pumps, ion channels, etc.).

2.2 Proteins

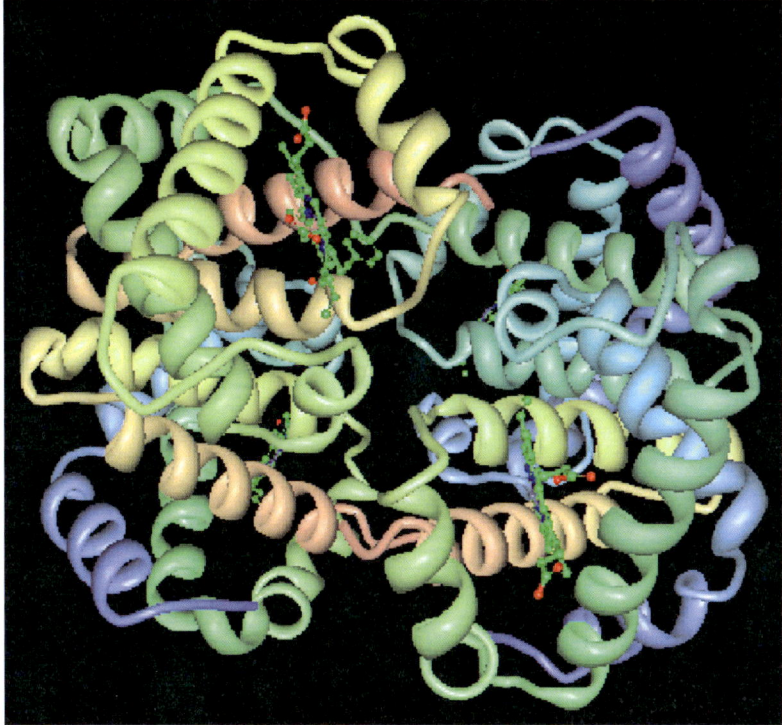

Fig. 2.16 Quaternary structure of hemoglobin – a tetramer of units consisting of two α chains (in red) and two β chains (yellow-green), similar to myoglobin. The four prostetic groups (hemes) are shown in green. Figure prepared using the "The Molecular Biology Toolkit" tool provided at http://www.rcsb.org/pdb/Welcome.do (Moreland et al., 2005).

Protein folding may sometime go beyond the quaternary. Some proteins, in order to become fully functional, form supramolecular assemblies by associating, for example, with lipids in membrane, with nucleic acids in chromosomes and ribosomes, with polysaccharides, etc.

2.2.2 Protein Folding

As it could be inferred from their chemical formulas, amino acids present both polar (hydrophilic) and hydrophobic parts, which may lead to self-association, similarly to amphiphilic molecules (see section 1.3). Hydrophobic interactions are not the only interactions that govern protein folding. Long-range Coulombian interactions between the charged amino and carboxyl groups, as well as short-range van der Waals interactions and quantum mechanical repulsion (section 1.1) also play their roles. In addition, hydration of various parts of the protein practically control the above mentioned interactions. All these interactions may be incorporated into a potential energy function of a protein. The differing energies associated with positioning amino acid residues relative to one another or to the solvent lead to some structures being more stable than others. If random mutations in the sequence of amino acids occur, kinetic traps may appear, which could lead to misfolded but structurally stable forms of a protein (Clementi and Plotkin, 2006).

Given the right (physiological) conditions, most proteins fold into their native, biologically functional state spontaneously and without any external help, since all of the needed information is encoded in the amino acid sequence (i.e., primary structure). This remarkable fact has been well documented experimentally (Anfinsen, 1973). By contrast, folding of large proteins is usually facilitated by "molecular chaperones," which prevent misfolding.

The problem of protein folding presents great scientific interest and poses equally big challenges. It can be formulated simply as: how does the amino acid sequence specify the native folded state of a protein? A related question is how is it possible for proteins to rapidly fold (i.e., within microseconds) into their native states, given the extraordinary large number of misfolded and, yet, potentially stable states that a protein could assume [$\sim 10^{95}$ for a chain with a mere 100 residues (Rose et al., 2006)]? The answers to such questions are as yet not fully known, although significant advancements have been made recently (Rose et al., 2006; Baker, 2000).

In modeling protein folding, small proteins are assumed to present only two global states – *folded* and *unfolded* –, since all the intermediate states (partly folded) are unstable. Under this simplification, the transition between folded and unfolded states can be represented as an equilibrium reaction,

$$U \underset{-k_f}{\overset{k_f}{\rightleftarrows}} F$$

with a rate of folding, k_f, given by (Mirny and Shaknovich, 2001):

$$k_f = C \exp\left(-\frac{\Delta G_{F-U}}{k_B T}\right) \qquad (2.3)$$

where C is a constant and ΔG_{F-U} is the Gibbs free-energy barrier for the folding–unfolding $(F-U)$ process starting from an unfolded protein. Due to the exponential dependence, small changes in Gibbs free energy lead to very abrupt changes in the conformation state of the protein. In fact, the transition state is not a single protein conformation but rather an ensemble of conformations, which is called *transition state ensemble* (TSE).

In computational studies of protein folding kinetics, a TSE is derived from a one-dimensional free energy profile of a certain *order parameter*, A (Mirny and Shaknovich, 2001; Saven et al., 1994):

$$\Delta G(A) = -k_B T \log f(A) \qquad (2.4)$$

2.2 Proteins

where $f(A)$ is the frequency of observing the order parameter in the range $(A, A + \Delta A)$, and the transition state is identified with the maximum of $\Delta G(A)$ at some value A_{trans}. The order parameter may be defined as proportional to the number of amino acids that are in their native conformation (if the native conformation is known), or the number of native contacts between amino acids.

Over the past several decades, computational biophysicists have tried to simulate protein structures starting from first physics principles and known amino acid sequences, but all efforts have met with limited success (Pande, 2003). However, significant insight has been gained into the distinction between *random amino acid sequences* and *naturally selected sequences* (i.e., proteins that are found in biological cells). It has been found, for instance, that the interactions between protein segments in natural proteins are mutually supportive, or cooperative (Onuchic and Wolynes, 2004). The global energetic landscape associated with different conformations of naturally occurring amino acid sequences (i.e., proteins) is funneled (Fig. 2.17), unlike that of proteins with arbitrary amino acid sequence. Therefore, in their "search" of a native conformation, proteins do not have to sample all possible states in the conformational space, but are rather "guided" towards reaching a global minimum.

Another important realization was that the backbone hydrogen bonds dominate the folding process (Rose et al., 2006). This leads to a reduced number of secondary structures, which are known to exist: helices, β-strands, β-turns and loops (described above). The average number of such structures in polypeptides with 100 residues was estimated by Rose et al. to about 10, which leaves only 2^{10} or $\sim 1,000$ possible elementary stable domains to be explored by the protein during the folding process, and not 10^{95}, as estimated before. A particular folded state that confers

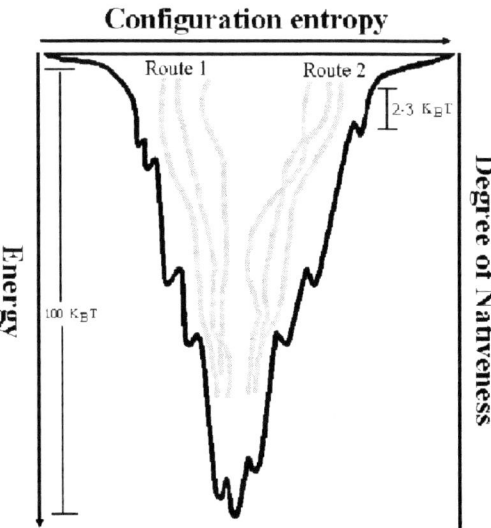

Fig. 2.17 Energetic landscape of protein conformations. Figure redrawn from Onuchic and Wolynes (2004), with permission form Elsevier.

the protein biological functionality is selected from among the 1,000 possibilities seems to be governed by the presence of the side chains, though the details of how that happens remain to be explored.

2.3 Deoxyribonucleic Acid (DNA): The Cell's Legislative Power

2.3.1 DNA Structure

While proteins could be considered the *executive power* of the cell, due to their intervention in all biochemical processes, *deoxyribonucleic acid* (DNA) could be considered the *legislative power* of the cell. This is because, on one hand, DNA is *the depositary of the genetic information* to be transmitted to the next generation and, on the other hand, it encrypts the guiding rules for the intricate *gene-expression machinery* performing to protein biosynthesis. Therefore, one can say that the whole set of DNA molecules of an organism constitutes a real data bank containing all the information necessary both for (auto)reproduction and for specifying the rules necessary for proper functioning of the organism.

Therefore, DNA, which is sometimes called metaphorically *"the helix of life,"* proves to be the most important and (why not?) ingenious macromolecule on Earth. Despite their diversity and complexity, DNA molecules are *heterobiopolymers* composed of only four types of *elementary blocks*, or monomers, called *nucleotides*. We will introduce them next.

2.3.1.1 DNA Nucleotides

A *nucleotide* is a complex molecule obtained from covalent binding of three smaller molecules: a *nitrogenous (or nitrogen-containing) base* that we denote by **B**, a sugar, **S**, called *deoxyribose*, and *phosphoric acid*, H_3PO_4, denoted by **P**. Because any nucleotide contains both the sugar and the phosphoric acid components, *the fingerprint* of a nucleotide is conferred by its *nitrogenous bases* alone. Therefore we shall firstly focus our attention on their structure.

There are two types of nitrogenous bases: *pyrimidines* (with a single atomic cycle) and *purines* (with two atomic cycles). One can graft various atomic groups onto these cycles, thus generating a variety of molecular nitrogen-containing bases. There are five major bases in biological cells: three pyrimidinic bases (*Cytosine*: **C**, *Thymine*: **T**, and *Uracil*: **U**) and two purinic bases (*Adenine*: **A** and *Guanine*: **G**), but only four of them (**C, T, A, G**) enter into DNA composition. The chemical structures of these four bases are given in Fig. 2.18.

The description of the building blocks of a nucleotide is completed by the chemical structures of *deoxyribose* (with the generally accepted atom notation) and of the *phosphoric acid* with its *phosphate radical* (Fig. 2.19).

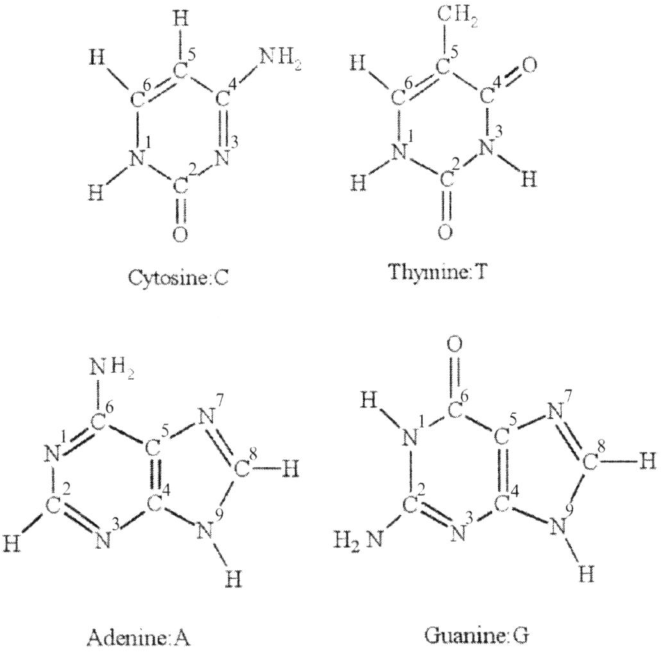

2.3 Deoxyribonucleic Acid (DNA): The Cell's Legislative Power

Fig. 2.18 Chemical structure of the four major bases encountered in DNA.

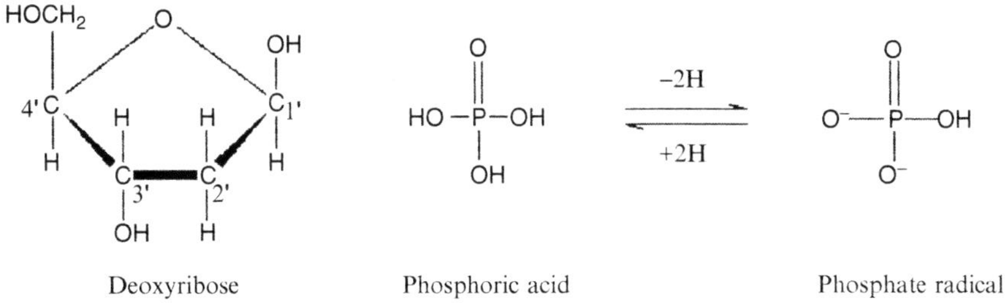

Fig. 2.19 Chemical structures of deoxyribose, phosphoric acid and its phosphate radical. Positions $3'$ and $5'$ of deoxyribose carbon atoms are very important for DNA synthesis. Note: if the H atom at the $2'$ carbon is replaced by OH, a ribose is formed, which enters the structure of ribonucleic acid, RNA.

When the three molecules, **B**, **S** and **P** interact chemically with one another to form covalent bonds, an *X monophosphate nucleotide* (**XMP**) is formed, where **X** can be any of the four nitrogen-containing bases: **C, T, A** and **G**. Synthesis of a particular nucleotide, *Adenosine Monophosphate* (**AMP**) is depicted in Fig. 2.20.

Addition of another phosphate radical to **AMP** (through the so-called *phosphodiester bond*), results in *Adeonosine Diphosphate* (**ADP**), and yet another phosphate

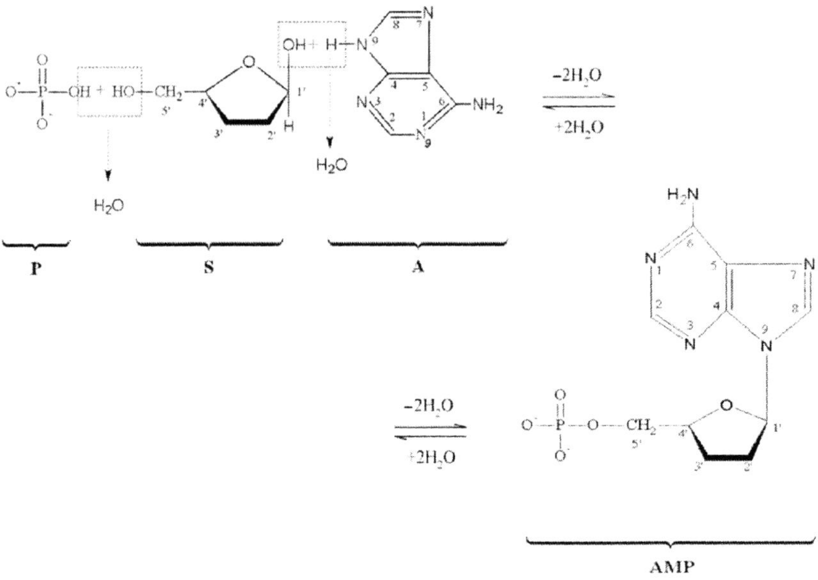

Fig. 2.20 Synthesis of *Adenosine Monophosphate* (AMP) from adenine (A), deoxyribose (S) and phosphoric acid (P).

addition results in *Adenosine Triphosphate* (**ATP**), as represented by the following reactions:

$$\mathbf{AMP} + \mathbf{P} \leftrightarrow \mathbf{ADP},$$
$$\mathbf{ADP} + \mathbf{P} \leftrightarrow \mathbf{ATP}.$$

As we mentioned in a previous section, **ATP** is a *macroergic* molecule, since its splitting (by hydrolysis of the first radical phosphate) into **ADP** and **P** produces free energy ($\Delta G < 0$) that becomes available for endergonic cellular processes. Because of this, **ATP** can be considered as an *energetic currency* of the cell.

Some cellular endergonic processes are: **DNA** and protein biosynthesis, active transport across membranes, muscle contraction, etc., some of which will be dealt with in separate chapters. To illustrate the degree to which **ATP** is involved in cellular processes, we remark that, although a human body contains only $\sim$100 g **ATP** at any given time, a quantity equal to half the mass of a human body ($\sim$40 kg **ATP**) is synthesized and hydrolyzed in 24 h, or even more (to exceed the body mass) during intense physical exercises (Berg et al., 2002).

2.3.1.2 Primary DNA Structure

Two nucleotides bonded through a *phosphodiester bond* form a *dinucleotide* (Fig. 2.21). The bonding is *always in the "direction"* $5' \rightarrow 3'$ of the *deoxyribose* carbons.

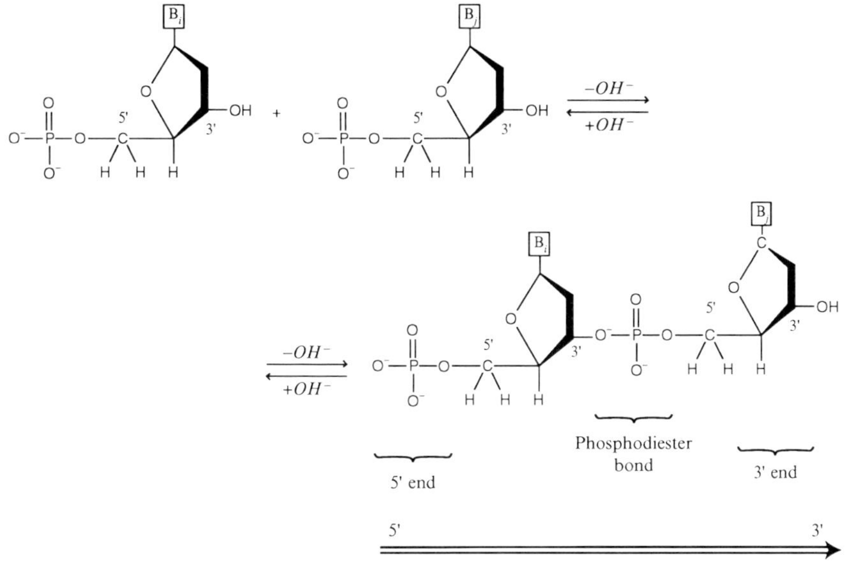

2.3 Deoxyribonucleic Acid (DNA): The Cell's Legislative Power

Fig. 2.21 Synthesis of a dinucleotide. Notice the growth in the $5' - 3'$ direction. B_i and B_j are nitrogenous bases.

By continuing to link nucleotides, one obtains, successively: *trinucleotides, tetranucleotides*, etc. (*polynucleotides*).

Polymerization of a large number of nucleotides leads to long strands of **DNA**, which could be composed of as many as millions of nucleotides. The backbone of such single **DNA** strands is formed by the sugar, **S**, and the phosphoric acid radical, **P**. Further, two **DNA** strands couple together through hydrogen bonds to form a double strand (Fig. 2.22a). It is an extraordinary property of the living matter that the *primary structure of DNA* consists of *two complementary strands*. This endows **DNA** with essential properties, including its *ability to replicate*, as we shall see below.

> **Observation:** Although the sequence of nucleotides and, implicitly, of the nitrogen bases in the **DNA** strand may seem random, it is in fact very *deterministic*, this order being essential for proper storage and transmission of information encrypted in the DNA base sequence.

The two **DNA** strands *are complementary* in the sense that: (*a*) they are coupled in *opposite directions* (see below under 2.3.1.3) and (*b*) *the purinic bases* on a chain are always coupled by hydrogen bonds to *the pyrimidinic* ones on the other chain, and vice versa. An additional restriction in constructing the two **DNA** strands is that **T** is always coupled by two hydrogen bonds to its complementary basis, **A**, while **C** is coupled through three hydrogen bonds to **G**. It is also established that the complementary bases are co-planar (Malacinschi, 2003), this alignment being favored by the hydrogen bonds between them (Fig. 2.22b).

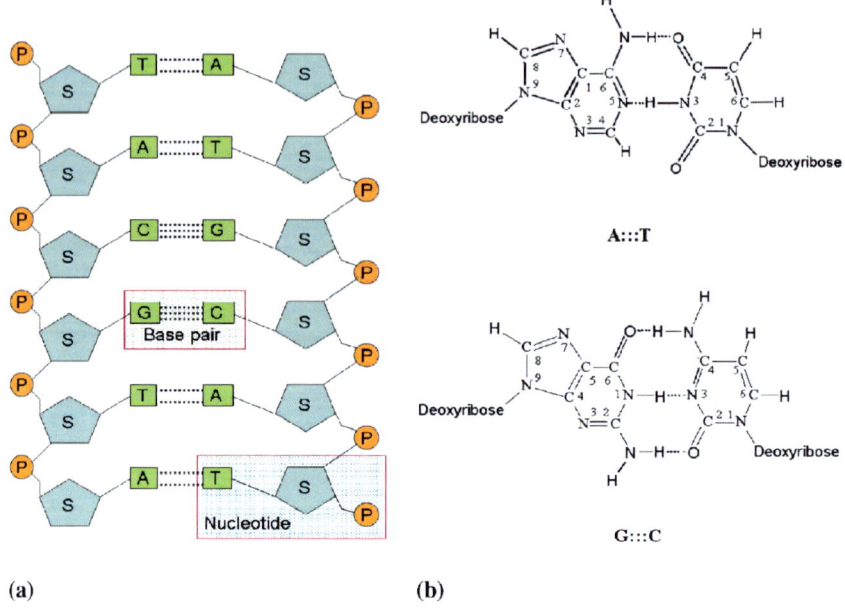

Fig. 2.22 (**a**) Primary structure of a DNA double strand. Standard notations for the nucleotide components were used, while chemical structures were ignored, for simplicity. (**b**) Hydrogen bonds between complementary nitrogen-containing bases (base pairs).

2.3.1.3 Secondary DNA Structure: The Double Helix

As mentioned above, the DNA double strand is very long. For instance, a human chromosome is formed by a single molecule containing up to 237×10^6 nucleotides (i.e., roughly 8 cm in length!). Therefore from an energetic point of view, DNA cannot remain as a linear ladder, but will adopt a more energetically favorable three-dimensional structure: the famous DNA *double helix* (Fig. 2.23). The most important features of the DNA double-helix are summarized below:

1. The backbones of the two nucleotide strands, formed by the sugar-phosphate radical sequence, are *antiparallel*, the *phosphodiesteric binding* $5' \rightarrow 3'$, running in opposite directions on the two strands.
2. The two winding strands form a *right-handed double helix* with a pitch of 34 Å, including *10 nucleotides per turn*. The two strands are not in real opposition, but are winding together around the helix axis (offering room for a third chain).
3. The two helix backbones are inscribed in an imaginary cylinder with a diameter of 20 Å.
4. The bases are located in the interior of the double helix and their planes are *almost perpendicular onto the helix axis*. In a projection of these base planes onto a plane containing the helix axis, two successive bases are a distance of 3.4 Å apart.

2.3 Deoxyribonucleic Acid (DNA): The Cell's Legislative Power

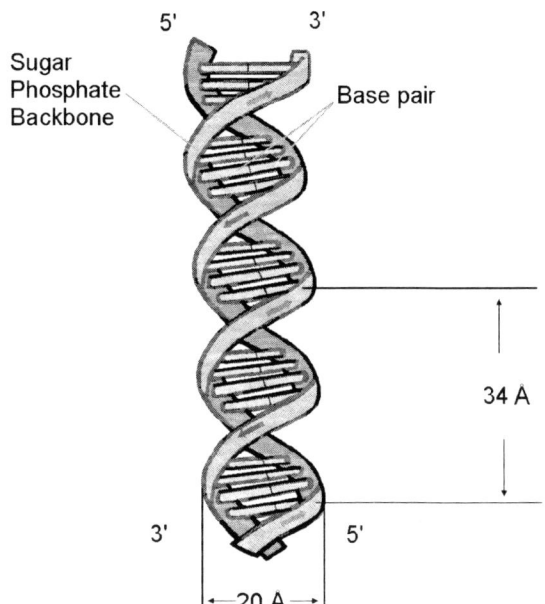

Fig. 2.23 The simplified model of the DNA double helix.

5. The double helix is maintained not only by the multitude of *hydrogen bonds* between their complementary bases, but, more importantly, by *van der Waals interactions* between their parallel pairs of bases, as well as by *hydrophobic interactions*.
6. The planes of the deoxyribose molecules, *situated at the exterior* of the double helix, are almost parallel with the helix axis.
7. The double helix presents *two helical spaces* between the intertwined strands, called *grooves:* a *minor groove* and a deeper *major* grooves. In this way, two types of **DNA** binding surfaces are accessible from the double helix exterior. Different **DNA**-binding proteins will sense the edge atoms of the bases, either on the minor groove or on the major groove surfaces.
8. The backbones of the nucleotides are electrically charged, due to the presence of the phosphate radicals. To prevent the separation of the two strands, which would be caused by the strong electrostatic repulsion, electrolytes (i.e., free ions) are required in the medium to shield the charges thus neutralizing the repulsive effect of the like charges.

Quiz 3. The co-planar pairs of complementary bases are separated along the double helix axis by 3.4 Å, but, at the same time, their projections on a plane perpendicular on the helix axis rotate with an angle, δ. Find this angle.

The double helix model has been established by Watson and Crick (1953) by analyzing and integrating all the biochemical data existing at that time and those obtained from X-ray diffraction diagrams on crystallized **DNA**. For their important model, Watson and Crick have been awarded the Nobel Prize in 1959. Watson and

Crick's results have relied on the prior theoretical work on alpha-helix protein structure (by Pauling and Corey), and, very importantly, on experimental diffraction data from Maurice Wilkins (shared Nobel Prize with Watson and Crick) and Rosalind Franklin. The latter, unfortunately, died prematurely (see, e.g., Maddox, 2003, for a more detailed history of this remarkable discovery).

2.3.1.4 Higher Order DNA Structure

Although the double helix attains a lower energy than the corresponding extended linear double strand, it cannot remain as a rigid stick, being forced by different types of interactions to adopt lower energetic states. As a consequence, the double helix may fold into *tertiary structures* as *circles, loops* or *superhelices* (*i.e., helices of higher order*). Thus, in the case of *prokaryotic* cells, **DNA** double helix will adopt a *circular shape* (i.e., with the two ends joined together). In some circumstances, a **DNA** ring can adopt a structure of *twisted superhelix*, due to *torsional stress* inside the double helix.

Even in the case of eukaryotic cells, **DNA** makes many long loops with their ends attached to the chromosome scaffold. On the other hand, the **DNA** double helix is complexed by histone and nonhistone proteins, attaining different levels of "condensation" into chromosomes.

More details on these very interesting structures can be found in biochemistry textbooks (see, e.g., Berg et al., 2002). We shall conclude this section by remarking that during a sub-phase of the cellular *mitosis* (i.e., the *metaphase*), the whole human **DNA** double helix, which is about 2 m long, is highly compacted in 46 microscopic *chromosomes*.

2.3.2 DNA Replication (Gene Autoreproduction)

The smallest units of **DNA** that preserve biological functionality, called *genes*, are responsible for encoding all the biological features of an organism. They are transmitted to a cell's (or a superior organism's) progenies, from generation to generation through a process that is facilitated by **DNA** replication (multiplication). This complex process of replication is *programmed* and takes place just before a cell enters the *mitotic division*. Because, through mitosis, two daughter cells are formed which must have the same genetic inheritance as their mother, *the quantity of **DNA** in the mother cell must be doubled* prior to mitosis. Otherwise, after a short series of successive divisions without **DNA** replication, the **DNA** quantity per daughter cell will dramatically decrease to complete extinction.

Because **DNA** replication is not a spontaneous process, it must be assisted by "molecular actuators" like, for instance, the **DNA**-*polymerase* (an enzyme), – and fuelled by an *energy source*: the cellular reserve of **ATP**. It is interesting to

Fig. 2.24 Semiconservative replication of **DNA**: a snapshot of the replication fork. The newly formed chains are drawn in dark color.

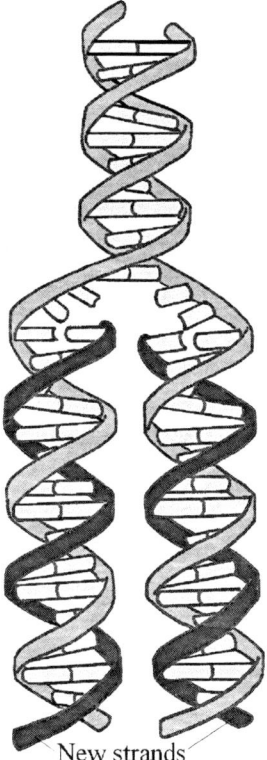

New strands

remark that the **DNA**-*polymerase* itself is synthesized according to the information contained in **DNA** genes. Enough amounts of building blocks – *the four types of nucleotides* (**AMP, TMP, GTP, CTP**) entering **DNA** composition – are also necessary in the cytosol.

The key element of gene transmission to the offspring without alteration resides precisely in the *structural complementarity* of the two **DNA** strands. In the process of **DNA** multiplication, some specific enzymes – *helicases* and **DNA** *polymerase* – attack the **DNA** double strand at one of its unique points, called *origin* (usually very rich in A-T sequences) and begin to unwind the double helix, breaking the hydrogen bonds between the complementary bases (Fig. 2.24). The place where the helicase and **DNA** polymerase act at a given moment is suggestively called "*replication fork*".

Following the action of the *helicase* and of the DNA-*polymerase*, each strand exposes its template of nitrogenous bases to the cytosol, where all kinds of nucleotides are already present. Through random collisions of these free cytosol nucleotides with the base template offered by each single strand, *the complementary nucleotides attach to each strand*, and realize covalent *phosphodiester bonds* between them, with the "assistance" of the **DNA**-polymerase. The **DNA**-polymerase acts like a real *molecular motor* fuelled by **ATP** to assemble the two new **DNA** strands on the two uncoiled complementary templates. (For more details, see Malacinschi, 2003; Berg et al., 2002; Lodish et al., 2004; Alberts et al., 2002).

DNA-polymerase assists in assembling complementary bases *only in the direction* $5' \rightarrow 3'$ of one strand (*leading strand*), the other strand being in opposite direction (*lagging strand*). For the lagging strand, base attachment is also in the direction $5' \rightarrow 3'$, but on small fragments (called *Okazaki fragments*) that will subsequently form covalent phosphodiester bonds, aided by a **DNA** *ligase* enzyme. The whole process of synthesis results in *two identical double strands*, that is, a double ammount of **DNA**.

This type of **DNA** replication is called *semiconservative*, because each new double strand has inherited *half of the initial amount* of **DNA**, that is, one **DNA** strand from the parental **DNA** molecule.

> **Observation:** Apart from this **DNA** *programmed synthesis*, an unscheduled (*unprogrammed*) **DNA** synthesis process also takes place quite frequently. This type of synthesis occurs when one of the **DNA** strands is damaged by mutagenic agents (e.g., ionizing radiations, non-ionizing **UV** radiations, chemical reagents, etc.), which triggers the action of a battery of enzymes that cut the damaged strand and replace it with a new, identical chain. This **DNA** repair process is *a molecular mechanism of defense* that preserves the genetic material for the next generations of individuals.

2.4 Determination of Molecular Structure

The example of DNA has illustrated the crucial relation of the three-dimensional (3D) atomic-scale structure of a biomolecule to its function. The same applies to enzymes and other proteins. Consequently, one of the primary goals of biochemists and molecular biologists is to know in detail the native spatial folding of bio(macro)molecules, in order to better understand their biological functions.

A prime source of information about 3D biomolecular structures is X-ray crystallography (Drenth, 1994), which is briefly described in chapter 7. In this technique a beam of hard X-rays (with wavelength of the order of interatomic distances) is directed into a crystal, and the structure of the repeat unit (in the case of a protein, a small number of molecular units) determined by the angular variation of the scattered X-rays. Mathematical analysis of the results produces high resolution static maps of the distribution of the atoms within the crystallized protein. In one of its more recent variants, the X-ray diffraction method is used to actually determine different states of the same molecule involved in a biological function (Schmidt et al., 2004).

Of the approximately 750,000 proteins whose amino acid sequences have been found, the 3D structure of less than 6% have been elucidated and recorded in the Protein Data Bank. Membrane proteins account for 1% of all structures determined. Yet, 70% of drugs on the market today are aimed at modifying the functions of membrane proteins.

The primary reason for the bottleneck is the difficulty of protein crystallization. An X-ray free electron laser (XFEL) such as one currently under construction in

2.4 Determination of Molecular Structure

the US, Europe, and Japan (Normille, 2006) which promises pulsed coherent X-rays of peak brightness 10 to 11 orders of magnitude greater than that available at the brightest of current X-ray sources (third-generation synchrotrons) may allow a bypassing of this bottleneck by offering the intriguing possibility of being able to measure scattering by individual molecules. The molecules may be presented to the X-rays as a beam created by electrospraying or Rayleigh droplet formation (Fenn, 2002; Spence et al., 2005) of sufficiently low flux to permit no more than a single hydrated molecule to be illuminated by an individual X-ray pulse. Current technology would allow ~ 100 such single-molecule diffraction patterns to be read out per second. Provided the duration of the X-ray pulses is kept down to less than ~ 50 fs, computer simulations (Neutze et al., 2000) have suggested that useful diffraction patterns may be recorded before radiation-induced disintegration of the molecule. The random orientations of different copies of the molecule in the X-ray beam at any one time may even be an advantage for 3D structure determination, provided a suitable computer algorithm may be devised for deducing the structure from the large number of noisy diffraction patterns (Shneerson et al., 2007).

Following the seminal work by Kurt Wütrich's group in 1980s, it has also become possible to determine *de novo* protein structures from measurements of nuclear magnetic resonance (NMR) of proteins in solution (Kline et al., 1988; Williamson et al., 1985). NMR spectroscopy interrogates the spins of atomic nuclei (e.g., ^{1}H and ^{13}C) placed in constant magnetic field by applying radiofrequency fields (see chapter 8). The NMR method for determination of protein structure consists of the following four steps (Wütrich, 2001):

(i) Use of multidimensional NMR techniques for data collection
(ii) Determination of the interproton distance by using the nuclear Overhauser effect (NOE) – a dipole-dipole interaction phenomenon between different nuclei, which depends on the inverse sixth power of the distance between interacting nuclei
(iii) Assign resonances to sequences of amino acids, starting from the fact that each amino acid residue represents an interacting spin system, in which the backbone hydrogen atoms and the side chain protons are coupled via covalent bonds
(iv) Use computational tools for extracting structural information from experimental NMR constrains, such as NOE data, ^{1}H chemical shifts (see chapter 8), etc.

The structures of small proteins (<10 kDa or up to ~ 100 amino acids) can be usually obtained from NOE signal alone, while for larger proteins (~ 10 kDa or more) additional constraints are required, which are based on labeling of amino acids with ^{13}C or ^{15}N, which are isotopes of ^{12}C and ^{14}N that present nuclear spin.

Unlike diffraction studies, NMR methods give more than a single structure, and usually about 20 different sets of structures that fit the NMR data. This may not necessarily mean reduced accuracy, and may come from the fact that peripheral segments of folded proteins are allowed to assume several possible positions, which are more tightly constrained in the rigid structure of crystals used for X-ray diffraction. We will see in chapter 8 that the ability to determine these conformational variants may be used to advantage in protein-protein interactions studies.

A significant fraction of the protein structures posted on PDB have been determined by using solution NMR. A significant increase in the number of structures thus determined seems currently difficult to achieve, due to the fact that application of this method is limited to proteins with masses less than about 500 kDa. The primary reason for this limitation is loss in resolution with increase in molar mass of the protein, which is due to resonance line broadening. Another notable limitation cited in the literature is the long time required for data acquisition, which is sometimes of the order of days (Serdyuk et al., 2007).

References

Alberts, B., Johnson, A., Lewis, J., Raff, M., Roberts, K and Walter, P. (2002) *Molecular Biology of the Cell*, 4th ed., Garland Science/Taylor & Francis, New York
Anfinsen, C. B. (1973) Principles that govern the folding of protein chains, *Science*, **181**: 223
Baker, D. (2000) A surprising simplicity to protein folding, *Nature*, **405**: 39
Berg, J. M., Tymoczo, J. L. and Stryer, L. (2002) *Biochemistry*, 5th ed., W. H. Freeman, New York
Clementi, C. and Plotkin, S. S. (2006) The effects of nonnative interactions on protein folding rates: Theory and simulation, *Prot. Sci.*, **13**: 1750
Dobson, C. M. (2002) Getting out of shape, *Nature*, **418**: 730
Drenth, J. (1994) *Principles of X-Ray Crystallography*, Springer, New York
Eisenberg, D. (2003) The discovery of the α-helix and β-sheet, the principal structural features of proteins, *Proc. Natl. Acad. Sci. USA*, **100**: 11207
Fenn, J. B. (2002) Electrospray ionization mass spectrometry: How it all began, *J. Biomol. Tech.*, **13**: 101
International Human Genome Sequencing Consortium (2004) Finishing the euchromatic sequence of the human genome, *Nature*, **431**: 931
Kline, A. D., Braun, W. and Wütrich, K. (1988) Determination of the complete three-dimensional structure of the α-amylase inhibitor tendamistat in aqueous solution by nuclear magnetic resonance and distance geometry, *J. Mol. Biol.*, **204**: 675
Lodish, H., Berk, A., Matsudaira, P., Kaiser, C. K., Krieger, M., Scott, M. P., Zipursky, S. L. and Darnell, J. (2004) *Molecular Cell Biology*, 5th ed., W. H. Freeman, New York
Malacinschi, G. M. (2003) *Essentials of Molecular Biology*, 4th ed., Jones and Bartlett, Boston, MA/London
Maddox, B. (2003) *Rosalind Franklin: The Dark Lady of DNA*, Harper Collins, New York
Martini, F. H. (2004) *Fundamentals of Anatomy & Physiology*, 7th ed., Benjamin Cummings, San Francisco
Mirny, L. and Shaknovich, E. (2001) Protein folding theory: from lattice to all-atom models, *Annu. Rev. Biophys. Biomol. Struct.*, **30**: 361
Moreland, J. L., Gramada, A., Buzko, O. V., Zhang, Q. and Bourne, P. E. (2005) The Molecular Biology Toolkit (mbt): A modular platform for developing molecular visualization applications, *BMC Bioinformatics*, **6**: 21
Neutze, R., Wouts, R., van der Spoel, D., Weckert, E. and Hajdu, J. (2000) *Nature* **406**: 752
Normille, D. (2006) Japanese latecomer joins race to build a hard X-ray laser, *Science* **314**: 751
Onuchic, J. N. and Wolynes, P. G. (2004) Theory of protein folding, *Curr. Opin. Struct. Biol.*, **14**: 70
Pande, V. S. (2003) Meeting halfway on the bridge between protein folding theory and experiment, *Proc. Natl. Acad. Sci. USA*, **100**: 3555
Rose, G. D., Fleming, P. J. Banavar, J. R. and Maritan, A. (2006) A backbone-based theory of protein folding, *Proc. Natl. Acad. Sci. USA*, **103**: 16623

References

Saven, J. G., Wang, J. and Wolynes, P. G. (1994) Kinetics of protein folding: The dynamics of globally connected rough energy landscapes with biases, *J. Chem. Phys.*, **101**: 11037

Shneerson, V. L., Ourmazd, A. and Saldin, D. K. (2007) Crystallography without crystals. I. The common-line method for assembling a three-dimensional diffraction volume from single-particle scattering, *Acta. Cryst. A*, **64**: 303

Schmidt, M., Pahl, R., Srajer, V., Anderson, S., Ren, Z., Ihee, H., Rajagopal, S. and Moffat, K. (2004) Protein kinetics: structures of intermediates and reaction mechanism from time-resolved x-ray data, *Proc. Natl. Acad. Sci. USA*, **101**: 4799

Schrödinger, E. (1992) *What Is Life?: The Physical Aspect of the Living Cell With Mind and Matter and Autobiographical Sketches*, Cambridge University Press, Cambridge

Serdyuk, I. N., Zaccai, N. R. and Zaccai, J. (2007) *Methods in Molecular Biophysics. Structure, Dynamics, Function*, Cambridge University Press, Cambridge/New York/Melbourne

Spence, J. C. H., Schmidt, K., Wu, J. S., Hembree, G., Weierstall, U., Doak, B. and Fromme, P. (2005) Diffraction and imaging from a beam of laser-aligned proteins: resolution limits, *Acta Cryst. A*, **61**: 237

Svedberg, T., Pedersen, K. O. (1940) *The Ultracentrifuge*, Clarendon, Oxford

Walter, F., Boron, E. L. and Boulpaep, M. D. (2004) *Medical Physiology: A Cellular and Molecular Approach*, Saunders/Elsevier, Philadelphia

Watson, J. D. and Crick, F. H. C. (1953) Molecular structure of nucleic acids. A structure for the deoxyribose nucleic acid. *Nature*, **171**: 737

Williamson, M. P., Havel, T. F. and Wütrich, K. (1985) Solution conformation of proteinase inhibitor IIA from bull seminal plasma by ^{1}H nuclear magnetic resonance and distance geometry, *J. Mol. Biol.*, **182**: 295

Wütrich, K. (2001) The way to NMR structures of proteins, *Nature*, **8**: 923

Chapter 3
Cell Membrane: Structure and Physical Properties

The *cell membrane* (or *plasma membrane*) is a thin *closed sheet* that fulfils a double role: (a) *morphological* – delimitates the cell from its external microenvironment and confines all of its subcellular organelles; (b) *functional* – regulates the exchange of substance between internal and external media, maintains actively the ionic asymmetry between its sides, and intermediates internalization or externalization of *physical* and *chemical signals* important for cell functions.

The plasma membrane undergoes continual changes both in its molecular composition and its structure (i.e., spatial distribution of its components), although during the entire lifespan of the cell its global architecture remains the same. It plays an important role in the economy of the cell, exerting a *selective control* on the entire traffic of ions, water, and molecules.

The membrane is involved also in intake (*endocytosis*) and secretion (*exocytosis*) of large particles. For example, *macrophages*, involved in the immune defence system, are able to engulf and destroy microbes and other foreign particles, this complex cellular process being called *phagocytosis* (see chapter 4). Being placed at the exterior of a cell, the membrane is also the first target of physical, chemical, and biological agents such as thermal and mechanical stress, toxins, hormones, viruses, microbes, etc. The membrane of specialized cells, such as neurons, is involved in propagation of *nervous signals* (see chapter 6) towards other neurons in the brain or muscle and glandular cells. Finally, the plasma membrane participates actively in the process of *cellular recognition* during the complex process of *morphogenesis*, when some types of differentiated (i.e., specialized) cells are segregated to form different types of tissues.

3.1 Membrane Structure

The cell membrane has a very complex anisotropic composition and spatial structure. This allows it to perform a wide variety of general as well as specialized tasks. Although some membrane characteristics may vary from cell to cell, certain general properties are the same for every cell, as we shall see below.

3.1.1 Chemical Composition of the Plasma Membrane

The main building blocks of all membranes are: *lipids, proteins, glycoproteins, lipoproteins, water* and *ions*.

Lipid molecules are the most abundant components in the membrane. *Lipid* is a generic term which includes a broad class of molecules, the most representative being *phospholipids* and *cholesterol*. The chemical and physical properties of phospholipids have been discussed in some detail in chapter 1, in particular their amphiphilic character, which leads to their *self-association* into micelles and bilayer membranes. The molecule of cholesterol is a special type of amphiphile with a single hydrocarbon tail which induces some rigidity into a lipid bilayer and consequently, into a membrane.

Proteins are among the most important components of the cell membrane. There exists a large variety of membrane proteins (e.g., channels, carriers, ion pumps, etc.). Membrane proteins, although present in a smaller numbers than lipids, represent approximately 50% of the whole membrane mass (due to their large molecular mass). While membrane proteins could play an important role in the membrane 3D structure, they are involved especially in the membrane specific functions, as we shall see later. In fact, most of the membrane specific functions are associated with certain membrane proteins.

Glycolipids are formed, as their names suggest, by combinations between simple *sugars* and *lipids* (e.g., glucosylcerebroside, in which *glucose* is covalently attached to the lipid *sphyngosine*). Cell membranes contain 2–10% lipids complexed with different kinds of sugars.

Glycoproteins are formed by chemical interactions between membrane proteins and sugars, the latter being exposed exclusively towards the extracellular side of the membrane (such as in the case of glycophorin) forming the so-called membrane *glycocalix*.

Lipoproteins are represented by proteins chemically attached to membrane lipids, and can be found on both sides of a membrane.

Water is the solvent for all molecules in the living matter, as we have already seen in previous chapters, and it comes to no surprise that cell membranes too incorporate water molecules, either bound to the polar groups exposed on the polar side of the membrane molecules (also called "structured water") or unbound (i.e., bulk water) within pores and some ionic channels traversing the membranes from one side to another.

Ions are associated with membranes, either through simple adsorption to the two membrane surfaces or by simply transiting the membrane through ionic channels (membrane proteins) or ion pumps (membrane proteins with enzymatic character; see chapter 4). Some of the ions most often encountered in association with the membrane structure and functions are: H^+, Na^+, K^+, Cl^-, Ca^{++}, HCO_3^-.

3.1.2 Spatial Architecture of the Plasma Membrane

Over the past 70 years numerous *structural models* of the membrane have been proposed mostly based on interpretation of the complex physical properties that natural membranes exhibit. The great majority of models included as a common characteristic the existence of a *lipid bilayer*, which confers an intrinsic thermodynamic stability to the membrane. We shall skip over the intricate history of membrane model evolution over time, and shall present only the currently accepted structural model, which can explain many of the membrane properties and functions and is able to accommodate new experimental findings. This model, called the *fluid mosaic model of membrane*, was elaborated by Singer and Nicolson (1972).

According to the *fluid mosaic model*, the *basic structural frame* of the cell membrane is provided by a *lipid bilayer* in which all kinds of proteins and other complex molecules mentioned above are embedded (Fig. 3.1). According to this model, the proteins embedded in the phospholipid bilayer confer to the membrane a mosaic-like aspect, while the *fluid* character is provided by the ability of all membrane components to diffuse laterally in the bilayer "plane" (the membrane can be considered practically a 2D structure).

The main structural and physical features of the fluid mosaic model of the membrane are listed below and should be considered in conjunction with Fig. 3.1.

1. The energetic stability of the membrane structure is mainly ensured by noncovalent *hydrophilic* and *hydrophobic* interactions exerted between membrane molecules, and between them and the aqueous medium, as has been discussed in detail in chapter 1. Other types of interactions are also involved (e.g., electrostatic, hydrogen bonds, and van der Waals interactions).

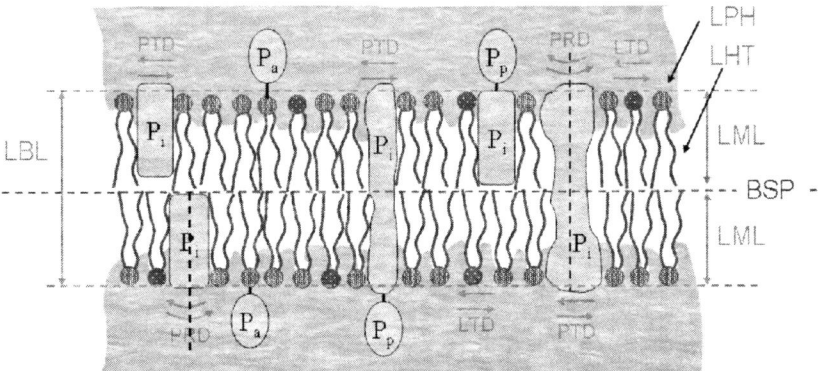

Fig. 3.1 Fluid mosaic model of the cell membrane. Significance of the symbols: LBL – lipid bilayer; LML – lipid monolayer; BSP – bilayer symmetry plane; LPH – lipid polar heads; LHT – lipid hydrophobic tails; P_i, P_p, P_a – integral, peripheral and, respectively, anchored proteins; LTD – lipid translational diffusion; PTD – protein translational diffusion; PRD – protein rotational diffusion.

2. The quasilinear lipid molecules are almost perpendicular to the membrane surfaces, so that their polar heads, involved in *hydrophilic interactions*, are facing the external and internal aqueous phases, while their tails (i.e., hydrocarbon chains) are buried inside the membrane, forming a quasiliquid *hydrophobic core* of the membrane. Generally roughly speaking, a lipid *monolayer* (also called *leaflet*) could be considered as a mirror image of the other one. (This is perfectly true only in the case of a planar bilayer composed of a single lipid species). The thickness of the plasma membrane bilayer varies from case to case between 4 and 6 nm (Berg et al., 2002). The thickness of the plasma membrane has been first estimated by Fricke (1927) from the relatively high electrical capacitance $(0.81\,\mu F/cm^2)$ of the plasma membrane as obtained from dielectric measurements of suspensions of cells in the audio/radio-frequency range. By assuming the permittivity of the hydrophobic layer relative to free space $(\varepsilon_0 = 8.854 \times 10^{-12}\,F/m)$ to be about 3, Fricke obtained a value of 3.3 nm for the thickness of the membrane, which is in good agreement with the currently accepted values, as mentioned above. More on the results of this kind of studies will be presented in sections 3.3 and 3.4.
3. The membrane components *can diffuse laterally* in the membrane plane, their diffusion coefficients (see chapter 4) depending on particles sizes and on their interactions with other particles. The translational diffusion in the membrane has been evidenced even before the fluid mosaic model was proposed, in the case of the so called hybrid "supercells" obtained by fusion of two different cell species (Frye and Edidin, 1970). The lateral (i.e., translational) diffusion, can also be easily evidenced using the technique of *fluorescence recovery after photobleaching* (FRAP) of fluorescently-labeled membrane proteins (Goodwin et al., 2005). The membrane components *can also randomly rotate* around their axes perpendicular to the membrane plane.
4. Lipid molecules can undergo *flip-flop movements*, in which they can jump from one lipid monolayer to another. These movements are energetically unfavorable and, for this transition to occur, lipids are assisted by an enzyme (*flippase*) with consumption of an ATP molecule (Lodish et al., 2004). This movement may play a role in controlling the composition in lipids of the two membrane layers.
5. Proteins associated to the membrane are of three types (Lodish et al., 2004): *integral* proteins, P_i, which are inserted into the membrane, *lipid-anchored* proteins, P_a, and *peripheral* proteins, P_p, which are weakly bound to the membrane (Fig. 3.1).

    Integral proteins can be either transmembranar, crossing the membrane from one face to another (e.g., the channels, carriers and ionic pumps) or may be embedded more or less into only one monolayer. There exist integral proteins that cross the bilayer only once (e.g., glycophorin A) (Lodish et al., 2004; Berg et al., 2002), or several times. For instance the K^+ channel (see chapter 7) crosses the bilayer two times, while the mammalian glucose transporter crosses the bilayer twelve times (Fig. 3.2). In all these cases, the intramembrane strands are organized as α-helices, but there are also proteins (e.g., porins) that are organized only

3.1 Membrane Structure

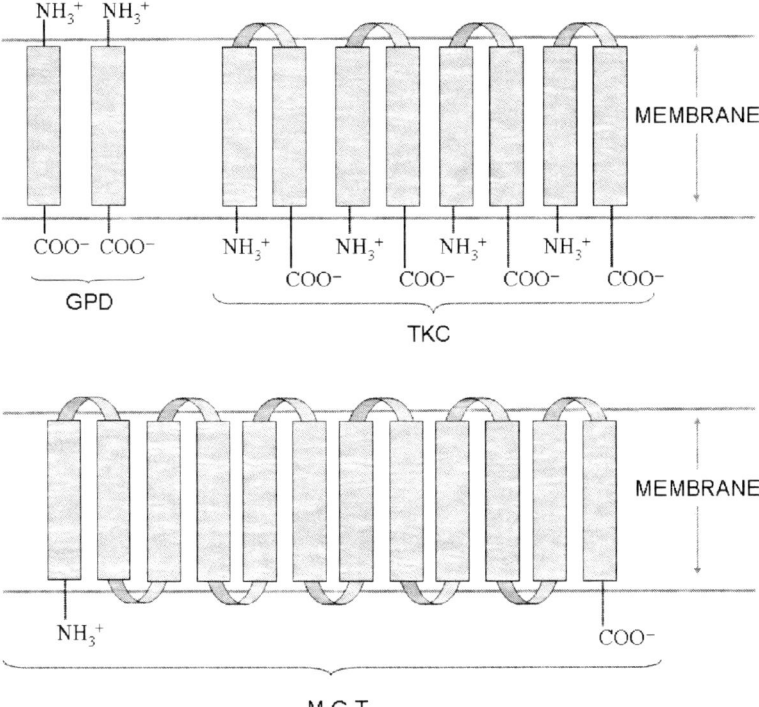

Fig. 3.2 Various kinds of spatial relations between integral proteins and the bilayer frame. Significance of the symbols: GPD – Glycophorin dimer; TKC – tetrameric potassium channel; MGT – Mammalian glucose transporter.

as β-strands and are able to traverse many times the lipid bilayer, forming, in the case of bacteria (e.g., *Escherichia coli*), membrane channels for disaccharides, phosphates and other small molecules. The segments of the integral proteins embedded in the membrane are composed especially of hydrophobic amino acids, while their parts exposed out of the lipid bilayers are predominantly formed from hydrophilic amino acids.

Peripheral proteins can be easily removed from the membrane. They are very important in the economy of the cell, for ensuring the transmission of many specific signals, either from the exterior to the interior of the cell or conversely, *via* the integral proteins.

6. There is always a non-uniform distribution of proteins and protein-complexes among the two phospholipids monolayers, which leads to an *asymmetry* of the membrane. For instance, *glycoproteins* are only associated with *the outer monolayer*, forming the so called cellular *glycocalix*, while in the particular case of erythrocytes, a peripheral protein called *spectrin* is associated only to the *internal face* of the membrane. Moreover, integral proteins are always exposing on

the two membrane surfaces different portions of their strands, thereby contributing to the structural asymmetry of the membrane. Unlike lipids, proteins do not undergo flip-flop motion, the asymmetric protein distribution being permanently maintained in the membrane.
7. Due to the thermal motions of lipids, some pores could appear transiently at random positions in the lipid matrix of the double layer, permitting a direct communication between the interior and exterior of the cell (Popescu et al., 2003; Movileanu et al., 2006). Due to the hydrophilic and hydrophobic interactions, these pores are rapidly resealed.

In conclusion, the cell membrane is a complex and dynamic structure which accomplishes essential functions in the cell, as we shall see later, in next chapters.

3.2 Surface Charges

3.2.1 Origin of the Surface Charges

When microscopic or macroscopic objects are immersed in an aqueous electrolyte solution, their surfaces become *electrical charged*, except for the particular case of the so-called *isoelectric pH*, when their net surface charge is zero. The electrical charge is due to the *adsorption* of anions and cations onto the body surfaces. As we have seen in chapter 1, both types of ions are hydrated, but the cations have a thicker hydration shell. As a result of this, the cation charges are more screened than those of anions and their interaction with the immersed surfaces is weaker. This leads to a preferential adsorption of anions as compared with cations. Alternatively, one can say that the anions present greater *polarizabilities* and, consequently, are better adsorbed. This mechanism of electrical charging of the surface is called *extrinsic charging mechanism*, being induced by the immersion medium.

> **Observation:** If the solution pH is decreased, the surface charges are strongly modified, on one hand, due to H^+ electrostatic interaction with already adsorbed anions and, on the other hand, due to direct H^+ adsorption onto the surface. Thus, beyond the isoelectric pH the surface charge changes its sign. It is important to note that H^+ is more easily adsorbed onto neutral sites of a membrane surface, being associated only with one water molecule (i.e., forming the so called *hydronium* ion: H_3O^+).

In the case of biological membranes, besides the extrinsic charging mechanism, an *intrinsic charging mechanism* is also acting, due to the electrical *dissociation* of the chemical groups on the membrane surface. Thus, at neutral pH, which characterizes the great majority of biological liquids, most of the dissociable chemical groups generate mainly negative charges. For instance, the phospholipid head groups can dissociate to generate $-H_2PO_4^-$, $-HPO_4^{-2}$, $-PO_4^{-3}$, while the sialic acid associated with integral proteins generates $-COO^-$ groups. It is also possible for some amino groups to become positively charged ($-NH_3^+$). However, in physiological solutions, the contribution of the positive groups is overwhelmed by that of the negative groups, so that the *intrinsic mechanism* too leads to a net negative surface charge. This is supported by the experimental evidence that cells migrate towards the *anode* when subjected to an external electrical field.

> **Observation:** Although the net surface charge of biological surfaces is negative, there may be patches on the cell surface that are positively charged. Therefore one can speak of a mosaic of electrical charges on the cell membranes both concerning their nature and their charge signs. These surface charges form a *dynamic electrostatic landscape* with an irregular pattern.

3.2.2 Electrical Double Layer

Electrical charges on the surface of biological particles exert an antientropic effect (i.e., $\Delta S_a < 0$) on the populations of ions located in the vicinity of the surface. This leads to a tendency to organize the nearby charges spatially, the *counter-ions* (i.e., ions with opposite charge to surface) being electrostatically attracted, and the *co-ions* (i.e., ions with same sign as the surface) being repelled. As a consequence, the surface charges lead to generation of an *electrical double layer*, which roughly consists of a layer of charges pertaining to the surface, and a layer of counter-ions at a small distance.

By contrast to the antientropic effect of the surface charges, thermal agitation of the ions has an entropic effect ($\Delta S_e > 0$), disrupting the organization of the surface charges, affecting thus the structure of the double layer (compared to what it would be, for instance, at very low temperatures).

Due to these two opposite tendencies, a "tradeoff structure" of the ion populations near the surface is attained, in which the organizing effect of the surface charges prevails near the surface (i.e., $\Delta S_a + \Delta S_e < 0$), while leading to an overall increase in entropy in the bulk solution ($\Delta S_b > 0$). With this, the second law of thermodynamics is obeyed, because $\Delta S_a + \Delta S_e + \Delta S_b > 0$.

The tradeoff electrical double layer has two components (Fig. 3.3): a *compact* or *Helmholtz double layer* (CDL) (Glaser, 2001), whose counter-ion charges do not completely neutralize the surface charges, and a *diffuse double layer* (DDL), whose counter-ion charges neutralize the charges left uncompensated by the CDL. Within DDL, the counter-ions concentration decreases with distance from the surface, while the concentration of co-ions increases. The specific dependence of ion concentrations on distance will be derived below.

3.2.3 Gouy-Chapman-Stern Theory of the Electrical Double Layer

Gouy (1910) and Chapman (1913) independently developed a diffuse double layer model in which ions are free to move (Brett and Oliveira Brett, 1993). In 1924, Stern combined the Helmholtz model for CDL, with the Gouy-Chapman model for DDL

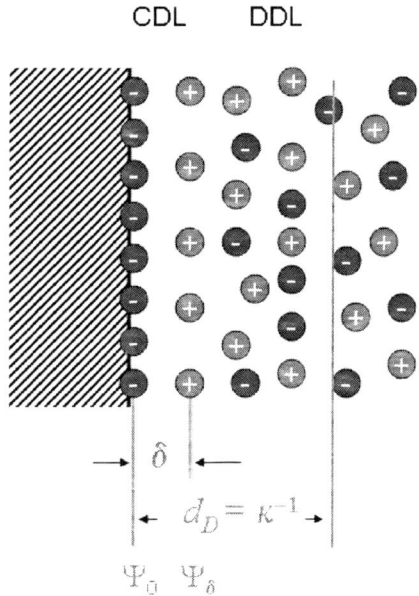

Fig. 3.3 Schematic representation of an electrical double layer near a planar surface of a solid particle (hatched). CDL – Helmholtz compact double layer; DDL – diffuse double layer; δ – compact layer thickness; d_D – Debye length, Ψ_0 – surface potential; Ψ_δ – Stern potential (see text for the physical meaning of the parameters).

(Atkins and de Paula, 2002). Later, in 1947, Grahame developed a more complex model, considering also the specifically adsorbed ions onto the solid surface (Brett and Oliveira Brett, 1993).

3.2.3.1 Problem Formulation and Simplifications

In the case of smooth surfaces, the geometry of the double layer is rather simple: if the surface has a planar or curved shape the double layer will have a planar or curved shape too. By contrast, most biological particles present irregular surfaces, which lead to electrical double layers with very complicated geometrical shapes that lead to very elaborated mathematical models. For this reason, and for many other reasons that will become clearer as we proceed with the mathematical treatment of the double layer, we will make the following simplifying assumptions:

(a) For spheroidal particles (such as cells), the radii of the hydrated ions are much smaller then the local radius of curvature of the surface (i.e., a few Ångstroms vs. $>5\,\mu m$); therefore, the ions are considered as *point charges*.
(b) The double layer is located near the cell surface.
(c) The surface charges are assumed to be distributed uniformly over the cell surfaces. The lipid bilayer fluidity has the tendency to lead to uniform distribution of the electrical charges on the surface, while the cytoskeleton has the opposite tendency.
(d) Partly due to approximation (a), the mathematical treatment can be applied to a very small area of the particle surface, so that even for spherical particles a planar model will provide a good approximation. From an electrical point of view,

3.2 Surface Charges

this amounts to assuming that the *electrical potential* to the "left" (Fig. 3.3) of the double layer and *inside the spherical particle is constant*. This assumption greatly simplifies the mathematical treatment of the physical model.

In addition to the geometrical simplifications suggested above, we shall also make some assumptions concerning the electrolyte composition (less drastic than the ones above). The aqueous medium is electrically neutral, and is composed of p kinds of ions (both, anions and cations) of electrovalence, z_k ($k = 1, 2, \ldots, p$), having very low concentrations (so that the solution could be considered dilute). An infinitesimal volume, dV, centered on a point, P, located inside the double layer has a partial net charge, $dq_k = z_k e \, dN_k$, where dN_k is the number of ions of species k contained in dV. We can now express the partial electrical charge density, ρ_k, of the element of volume dV as:

$$\rho_k(P) = \frac{dq_k}{dV} = n_k(P) z_k e, \tag{3.1}$$

where $n_k = dN_k/dV$ represents the numeric concentration of the ionic species k (i.e., the number of ions of the unit volume). The total charge density, $\rho(P)$, at a point, P, is given by the sum of each ion contribution:

$$\rho(P) = \sum_{k=1}^{p} n_k(P) z_k e. \tag{3.2a}$$

Since, according to the approximations above, the double layer may be considered planar, the charge density at any point contained in a plane parallel to the particle surface and situated at the distance, x, from the surface will be always the same. This means that the mathematical description of the double layer can be reduced to a single spatial dimension, x. Thus the total charge density, at a distance, x, is given by the sum of over all ion contributions:

$$\rho(x) = \sum_{k=1}^{p} n_k(x) z_k e. \tag{3.2b}$$

3.2.3.2 Poisson-Boltzmann Equation and the Electrical Potential

Having defined the volume charge density, we can now relate the electrical potential, Ψ, at any point in the double layer, to the charge density, according to the *Poisson equation*:

$$\Delta \Psi(x) = -\rho(x)/\varepsilon, \tag{3.3}$$

where Δ ($= \nabla^2$) is the second order differential *Laplacean operator*, and ε is the permittivity of the medium.

> **Observation:** The permittivity varies with the distance from the double layer. However, the exact form of its variation is usually unknown, so one has to make certain approximations, as we shall see below.

By combining equations (3.2b) and (3.3) and writing explicitly the one-dimensional character of our problem, we obtain:

$$\frac{d^2\Psi(x)}{dx^2} = -\frac{e}{\varepsilon}\sum_{k=1}^{p} n_k(x)z_k. \qquad (3.4)$$

In order to solve the Poisson equation, we need to know how n_k varies with distance. In the Gouy-Chapman theory, it is assumed that the energetic states of each ion, k, are described by a Boltzmann distribution, given the thermal equilibrium that is normally attained by the ion populations. Therefore, one can write:

$$n_k(x) = n_k(\infty)\exp\left[-\frac{z_k e \Psi(x)}{k_B T}\right] \qquad (3.5)$$

where $n_k(\infty)$ represents the concentration of k ionic species in the bulk solution (i.e., at $x \to \infty$) and k_B is the Boltzmann constant.

Combining equations (3.4) and (3.5), one obtains the so-called *Poisson-Boltzmann equation* of the planar double layer:

$$\frac{d^2\Psi(x)}{dx^2} = -\frac{e}{\varepsilon}\sum_{k=1}^{p} z_k n_k(\infty)\exp\left[-\frac{z_k e \Psi(x)}{k_B T}\right], \qquad (3.6)$$

which is a nonlinear differential equation that could be only numerically solved.

However, by making further approximations, equation (3.6) can be analytically solved. Indeed, if we assume that the electrical potential at the particle surface, $\Psi(0) = \Psi_0$, is less than about 25 mV, the exponent in equation (3.6) is smaller than 0.01. Therefore, we can expand (3.6) in Taylor series around $x = 0$ and retain only the first order term, to obtain:

$$\frac{d^2\Psi(x)}{dx^2} \cong -\frac{e}{\varepsilon}\sum_{k=1}^{p} z_k n_k(\infty)\left[1 - \frac{z_k e \Psi(x)}{k_B T}\right]. \qquad (3.7)$$

In this expression, the first part of the sum is $\sum_{k=1}^{p} e z_k n_k(\infty) = \rho(\infty) = 0$, because of the neutrality of the solution. With this, equation (3.7) further simplifies to:

$$\frac{d^2\Psi(x)}{dx^2} \cong \kappa^2\, \Psi(x), \qquad (3.8)$$

where we have introduced the customary notation:

$$\kappa = \left[\frac{e^2}{\varepsilon k_B T}\sum_{k=1}^{p} z_k^2 n_k(\infty)\right]^{\frac{1}{2}}. \qquad (3.9)$$

Here, κ represents the *Debye-Hückel parameter*, which depends on the physical properties of the electrolyte, and is independent of the nature of the particle surface.

3.2 Surface Charges

For the moment, we shall limit ourselves to stating that the Debye-Hückel parameter depends on the square root of the *ionic strength, I,* of the electrolyte solution, given by:

$$I = \frac{1}{2}\sum_{k=1}^{p} z_k^2 c_k(\infty) = \frac{1}{2N_A}\sum_{k=1}^{p} z_k^2 n_k(\infty), \qquad (3.10)$$

where $c_k(\infty)$ is the molar concentration of the k species, that is, $c_k(\infty) = n_k(\infty)/N_A$ (N_A being Avogadro's number).

Observation: The ionic strength, I, depends on the second power of the electrovalences, z_k. Therefore two solutions with same concentration can present quite different ionic strengths and thus can have quite different influences on the surface charge screening. Indeed, if a solution contains, e.g., Al^{3+} and another contains Na^+, the contribution of the trivalent ions to the sum (3.10) will be nine times greater than that of the monovalent sodium ions.

By using equation (3.10), equation (3.9) becomes:

$$\kappa = \left[\frac{2e^2 N_A}{\varepsilon k_B T}I\right]^{\frac{1}{2}} = \left[\frac{2F^2}{\varepsilon R T}I\right]^{\frac{1}{2}}, \qquad (3.11)$$

where the following relations have been used: $eN_A = F$ (Faraday's number) and $k_B = R/N_A$.

For x ranging over the interval $(0, \delta)$, there are no electrical charges, and the Poisson-Boltzmann equation (3.8) reduces itself to the *Laplace equation*, which can be solved by successive integrations to give:

$$\Psi(x) = ax + b, \qquad (3.12)$$

where a and b are two constants that can be determined from the boundary conditions at $x \to 0$ (which gives $b = \Psi_0$) and $x \to \delta$ (giving $a = (\Psi_\delta - \Psi_0)/\delta$). Thus,

$$\Psi(0 < x < \delta) = \frac{\Psi_\delta - \Psi_0}{\delta}x + \Psi_0, \qquad (3.13)$$

and therefore in the *Stern space*, $0 < x < \delta$, the absolute value of electrical potential *decreases linearly* with the distance from the charged surface.

The general solution of equation (3.8) for $x \in [\delta, \infty)$ is:

$$\Psi(x) = A\exp(-\kappa x) + B\exp(+\kappa x). \qquad (3.14)$$

By imposing the following boundary conditions $\Psi(x \to \delta) = \Psi_\delta$ (*Stern potential*) and $\Psi(x \to \infty) = 0$ (that is, far enough from the surface, the potential vanishes, due to electroneutrality of the medium), it results $B = 0$, so that the particular solution of the equation (3.14), having a physical meaning, is of the form:

$$\Psi(x \geq \delta) = \Psi_\delta \exp[-\kappa(x - \delta)]. \qquad (3.15)$$

Therefore, the absolute value of the electrical potential *decreases exponentially* with the distance for $x \geq \delta$. This means that the electrical potential generated by the surface charges is screened by the counter ionic atmosphere. Moreover, this screening depends on the ionic strength, which is included in κ (Debye-Hückel parameter) through equation (3.11), the solutions with a greater ionic strength being more effective in charge screening.

At this point, we can discuss the physical meaning of the Debye-Hückel parameter. By convention, because the DDL has no precise boundaries, the "thickness" of the DDL is defined as the distance, d_D, at which the electrical potential decreases by e (e being the base of the natural logarithm, or the *Euler number*) as compared to Ψ_δ, namely

$$\Psi(d_D) = \Psi_\delta e^{-\kappa(d_D-\delta)} = \Psi_\delta/e. \tag{3.16}$$

Thus, $1/\kappa = d_D - \delta \approx d_D (\delta \ll d_D)$, that is, the *reciprocal* of *Debye-Hückel parameter* is equal with the *thickness* of the *diffuse double layer* (Figs. 3.3 and 3.5) and called *Debye length*.

The Debye length is a measure of how far the effect of surface charges may be felt into the bulk of the solution. It depends on the ionic strength of the solution. Thus, in Ringer solution (a solution of salts in water that prolongs the survival time of excised tissue), the Debye length is 7.8 Å (Hille, 2001), while in physiological saline (i.e., 0.145 M NaCl) it is 8 Å.

Therefore the effect of surface charges on the ions in physiological solutions extends into the double layer over distances much shorter than the size of the macromolecules (Hille, 2001).

Observation: In the case of the natural plasma membrane, the Stern layer (Fig. 3.3) is also populated with ions entering into or exiting from the membrane (due to membrane permeability to ions), and relation (3.16) does not hold any more.

Quiz 1. Plot the Debye length as a function of concentration for solutions of NaCl having the following concentrations: (a) 1 μM; (b) 10 μM; (c) 100 μM; (d) 1 mM; (e) 10 mM; (f) 100 mM.

3.2.3.3 Measuring the Electrochemical Potential Through Electromigration

Particles migrating through an aqueous milieu carry along an ionic shell of thickness a. The sum of the particle radius and the shell thickness is referred to as the *electrokinetic radius*, r_{ek}. The surface of separation, at $x = a$, between the moving and non moving part of the charge cloud is called the "slipping plane" or shear plane (Fig. 3.4).

Depending on their surface charges, biological particles (cells, organelles, viruses, etc.) suspended in physiological solutions can migrate in an electrical field, E, with different *mobilities*, u, defined as:

$$u = v/E \tag{3.17}$$

where v is the relative velocity of the particle driven by the electrical field. This phenomenon of migration of electrically charged particles in an electrical field is called

3.2 Surface Charges

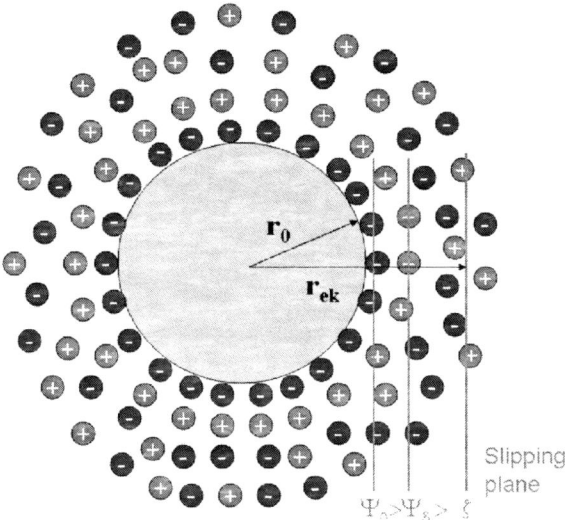

Fig. 3.4 Geometrical and electrokinetic radii of an electrically charged particle. The potential measured in the slipping plane is called the *zeta* potential (ζ), and can be determined experimentally from measurements of electrokinetic mobility, i.e., from electrophoresis (see text below); r_0 is the particle radius; r_{ek} is the effective radius of the particle including part of the counterion cloud.

electrophoresis and has practical applications in biophysics and colloidal chemistry, as we will discuss briefly below.

Figure 3.5 shows an example of variation of electrical potential with distance from the charged surface, as predicted by a combination of equations (3.13) and (3.15). Generally, one can find the following relation among the three defined potentials:

$$|\zeta| < |\Psi_\delta| < |\Psi_0|. \tag{3.18}$$

The electrical potential, at the distance $x = \delta$ (i.e., at the slipping plane) is called *electrokinetic potential* or *zeta potential*, $\Psi(\delta) = \zeta$, while the other potentials have been defined in the discussion above. The zeta potential is the only potential related to the electrical double layer that can be measured directly (by electrophoresis), and as such, it provides an indication of the order of magnitude of the other potentials.

In the case of spherical particles of geometrical radius, r_0, the *most general theory* giving mathematical relation between ζ and u (electrophoretic mobility), is due to *Henry*.

According to the *Henry theory* (Hunter, 1987), the relation is:

$$u_{HR} = \frac{2}{3}\frac{\varepsilon \zeta}{\eta} f(\kappa r_0) \tag{3.19}$$

where ε represents the dielectric constant of the solution, η is its dynamical viscosity coefficient, and $f(\kappa r_0)$ is the empirical Henry function, taking values over the range [1, 3/2], for $\zeta < 25$ mV.

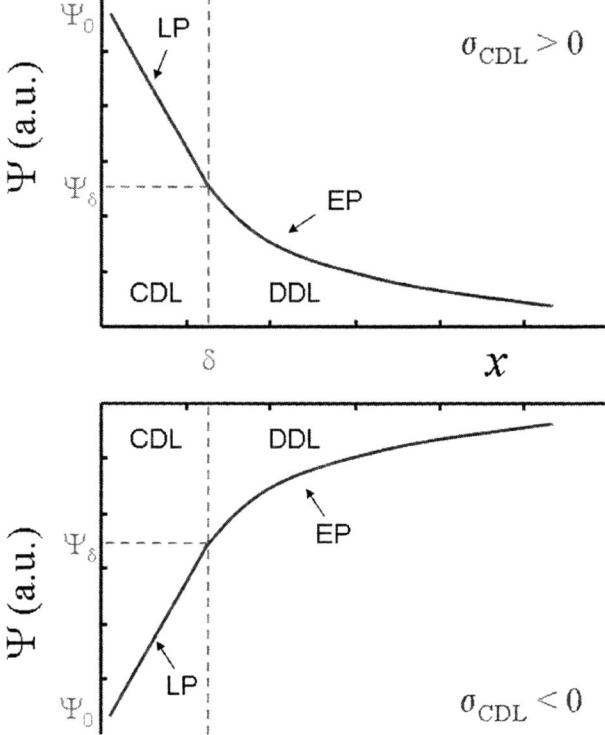

Fig. 3.5 Dependence of electrical potential in the double layer on the distance x from the surface for positive (top panel) and negative (bottom panel) charge density of the compact double layer. Significance of the symbols: CDL – compact double layer; DDL – diffuse double layer; SP – Stern plane; SS – slipping plane; Ψ_0 – surface electrical potential; Ψ_δ – Stern electrical potential; LP, EP – linear and respectively, exponential portion of the potential's spatial dependence.

For the largest biological particles suspended in relatively dilute electrolyte (i.e., cells), $\kappa r_0 > 100$. In this situation, $f(\kappa r_0) = 3/2$, and (3.19) reduces to the simpler *Helmhlotz-Smoluchowski* formula:

$$u_{HS} = \frac{\varepsilon \zeta}{\eta}. \tag{3.20}$$

In the opposite case when *smaller biological particles* are taken into account (e.g., viruses), $\kappa r_0 < 0.1$, and $f(\kappa r_0) = 1$, and (3.19) reduces to the *Hückel* formula:

$$u_{HK} = \frac{2}{3}\frac{\varepsilon \zeta}{\eta}. \tag{3.21}$$

For the intermediate cases, $0.1 < \kappa r_0 < 100$, one must employ the more general *Henry* formula.

Observation: The electrical double layer around biological particles is also involved in the electrostatic interaction between cells suspended in a physiological solution.

3.2 Surface Charges

Practically, the mobility of single cells subjected to external electrical fields can be determined under an optical microscope; the technique is called *microelectrophoresis* and can be used for determination of the zeta potential.

Electromigration of ultra-small biological particles is used widely in biological and biochemical laboratories for separation and identification of fluorescently stained macromolecules or fragments of macromolecules, such as for DNA fragment identification in gel-electrophoresis.

3.2.3.4 Surface Charge Density and the Electrical Capacitance of the Double Layer

In this section, we will derive a simple relationship between a specific electrical capacitance of the electrical double layer and the Debye length, thereby providing a means for determining the later. We begin by writing the surface charge density of the diffuse layer as the line integral of the total volume charge of the counter ion layer, namely:

$$\sigma_{DDL} = \sum_k \int_\delta^\infty z_k e n_k(x) \, dx = \int_\delta^\infty \rho(x) dx. \quad (3.22)$$

The integral can be also obtained by integrating the Poisson equation (3.3). Thus:

$$\sigma_{DDL} = -\varepsilon_{DDL} \int_\delta^\infty \frac{d^2\Psi}{dx^2} dx = -\varepsilon_{DDL} \left.\frac{d\Psi}{dx}\right|_\delta^\infty = -\varepsilon_{DDL}\left(0 - \left.\frac{d\Psi}{dx}\right|_\delta\right). \quad (3.23)$$

From (3.15), $\dfrac{d\Psi}{dx} = -\kappa \Psi_\delta \exp[-\kappa(x-\delta)]$, and therefore:

$$\sigma_{DDL} = -\varepsilon_{DDL} \kappa \Psi_\delta. \quad (3.24)$$

Similarly, for the singularity at $x = 0$ (i.e., the fixed surface charge), we obtain:

$$\sigma_{CDL} = -\varepsilon_{CDL} \frac{\Psi_\delta - \Psi_0}{\delta}, \quad (3.25)$$

where we have used the approximation (d) in section 3.2.3.1, to set the potential at $x = 0$ equal to a constant, Ψ_0. This approximation is also valid for a planar lipid bilayer with symmetrical electrical double layers.

Since $\sigma_{DDL} = -\sigma_{CDL}$ (due to electroneutrality of the system), we get:

$$\Psi_\delta = \Psi_0 \frac{\varepsilon_{CDL}}{\varepsilon_{CDL} + \varepsilon_{DDL}\kappa\delta} = \frac{\Psi_0}{1 + \frac{\varepsilon_{DDL}}{\varepsilon_{CDL}}\kappa\delta}. \quad (3.26)$$

For small concentrations ($c \to 0$, i.e., weak electrolytes) or, equivalently, $\kappa\delta \ll 1$, equation (3.26) gives $\Psi_\delta \approx \Psi_0$. The charges on the Stern layer are compensated only at large distances from the surface. On the other hand, if the concentration is large (i.e., strong electrolytes), we have $\kappa\delta \gg 1$, and equation (3.26) gives $|\Psi_\delta| \ll |\Psi_0|$. In this case, the Stern layer almost completely neutralizes the adsorbed charges.

By using equations (3.24) and (3.26), we can now determine the specific electrical capacitance (i.e., capacitance divided by the surface area) of the electrical double layer, simply as the ratio between the surface charge density and the potential difference detected between an electrode placed at $x = 0$ and another one for $x \to \infty$, namely:

$$C_{DL} = \frac{\sigma_{DDL}}{\underbrace{\Psi(\infty)}_{0} - \Psi_0} = \frac{-\varepsilon_{DDL}\kappa\Psi_\delta}{-\Psi_0} = \frac{\varepsilon_{DDL}\kappa}{1 + \frac{\varepsilon_{DDL}}{\varepsilon_{CDL}}\kappa\delta}. \quad (3.27)$$

As we mentioned in the preceding sections of this chapter, in general, the permittivity varies with the distance from the particle surface and into the bulk of the electrolyte in a manner unspecified. We will take this variation into account by distinguishing between the three main layers, namely $\varepsilon_{CDL} < \varepsilon_{DDL} < \varepsilon_{water}$.

For small concentrations, $c \to 0$ and $C_{DL} \to \varepsilon_{DDL}\kappa$. If we consider, for example, a 10 mM KCl solution, and $\varepsilon_{DDL} \approx 30\varepsilon_0$, we obtain a capacitance of the double layer $C_{DL} \approx 100 \mu F/cm^2$. Conversely, this capacitance can be determined from electrical measurements on lipid bilayers, and the Debye length can thus be determined experimentally. We will discuss methods for measuring electrical capacitances of membranes in the next sub-section.

3.3 Static Electrical Properties of Planar Membranes

As it was mentioned in section 3.1, the thickness of the plasma membrane has been determined indirectly from measurements of permittivity and conductivity of cell suspensions subjected to alternating fields in the audio/radiofrequency range (Fricke, 1927). This method, generically known as *dielectric* or *impedance spectroscopy*, has since been applied to the study of electrical properties of artificial as well as natural membrane bilayers. By selecting the range of frequencies of the applied field, one can obtain information about different layers of the cell membrane as well as of other membranes internal to the cell. In this section, we will introduce the reader to the principles of the *dielectric spectroscopy method* and will discuss its application to the determination of the dielectric properties of the main layers (electrical double layer, the polar head region, and the hydrophobic core) of artificial lipid bilayers, as well as of the plasma membrane of the cell. The main goal will be to illustrate that, in spite of the difficulties one faces in trying to directly observe the plasma membrane, there are very good reasons to believe that the membrane model introduced in section 3.1, which considers that the membrane consists of a lipid-bilayer matrix, is correct.

3.3.1 Electrical Parameters as Complex Quantities

Throughout section 3.3, we will employ the concepts of *complex permittivity* and *complex conductivity*, which will be defined momentarily.

3.3 Static Electrical Properties of Planar Membranes

To begin with, let us define an "alternating" electric field, $E(t)$, as

$$E(t) = E_0 e^{j\omega t} \tag{3.28}$$

where E_0 is the constant part (amplitude) of the field, $\omega (= 2\pi f$ with f being the frequency) is the angular frequency of the field and $j = \sqrt{-1}$.

The general expression for an electrical current density $[J = I/(Surface area)]$ is given by the Ohm law for variable fields,

$$J(t) = \sigma E(t) + \frac{\partial D(t)}{\partial t} \tag{3.29}$$

where D is electrical displacement $(= \varepsilon E$ with ε the *permittivity*).

The above equation suggests that the *conductivity*, σ, is related to instantaneous motion of charges (either translation of rotation), while the *permittivity*, ε, is related to a delay in the particle response to the applied field.

For a constant permittivity, equations (3.28) and (3.29) give:

$$J = (\sigma + j\omega \varepsilon) E \stackrel{def}{=} \sigma^* E \tag{3.30}$$

where we have introduced the notation, σ^*, which represents the *complex conductivity*. It is also possible to introduce a quantity called *complex permittivity*, defined by the following relations:

$$\varepsilon^* \stackrel{def}{=} \frac{\sigma^*}{j\omega} = \varepsilon - j\frac{\sigma}{\omega} \stackrel{def}{=} \varepsilon - j\varepsilon'. \tag{3.31}$$

Both σ^* and ε^* are very useful in the theory of dielectrics, because they offer a synthetic way of dealing with the true permittivity and conductivity at the same time.

Materials characterized both by permittivity and conductivity (i.e., present both free and bound charges) are generically referred to as dielectrics.

Let us assume that such a dielectric material is placed between the plates of a parallel-plate capacitor and subjected to a voltage U. A quantity, called admittance, Y, will be measured, which relates to the complex conductivity through:

$$Y \stackrel{def}{=} \frac{I}{U} = \frac{JS}{Ed} = \sigma^* \frac{S}{d}, \tag{3.32}$$

where S is the surface area of each of the two identical plates, and d is the separation between them. Y is a complex number and can be rewritten, by taking equation (3.30) into account, as:

$$Y = \sigma \frac{S}{d} + j\omega \varepsilon \frac{S}{d} = G + j\omega C, \tag{3.33}$$

where G represents the *conductance* and C the *capacitance* of the dielectric material. Equation (3.33) corresponds to a parallel combination of a conductance, G, with a capacitance, C. One can also define the inverse of the admittance or the *impedance*:

$$Z \stackrel{def}{=} \frac{1}{Y} = \frac{1}{G+j\omega C}. \tag{3.34}$$

Observation: If a system consists of a series combination of two admittances Y_1 and Y_2, the equivalent admittance is:

$$\frac{1}{Y} = \frac{U_1+U_2}{I} = \frac{1}{Y_1} + \frac{1}{Y_2}, \text{ or } Z = Z_1 + Z_2 \tag{3.35}$$

where U_1 and U_2 are two potential differences applied to each circuit of admittance Y_1 and Y_2, respectively. If $C_1 = 0$ and $G_2 = 0$, equation (3.34) gives:

$$Z = \frac{1}{G_1} + \frac{1}{j\omega C_2} = R_1 + \frac{1}{j\omega C_2} \tag{3.36}$$

which upon a re-notation gives the definition of the impedance for a *series combination* of a resistor and a capacitor,

$$Z = R + \frac{1}{j\omega C}. \tag{3.37}$$

3.3.2 Dielectric Relaxation of a Dielectric Multi-Layer

3.3.2.1 Interfacial Maxwell-Wagner Polarization

Let us consider a system formed of two stacked dielectric layers "sandwiched" between the plates of a parallel-plate capacitor. Each of the two layers (Fig. 3.6) can be represented by a parallel combination of capacitance and conductance (i.e., complex permittivity multiplied by a geometrical factor – see equation (3.33), for the case of a parallel-plate capacitor).

The measured admittance of the circuit is:

$$Y = \left[1/(G_1+j\omega C_1) + 1/(G_2+j\omega C_2)\right]^{-1}. \tag{3.38}$$

Rearranging the right-hand side of this equation, one obtains (Hanai, 1960):

$$C^* = C_h + \frac{C_l - C_h}{1+j\omega\tau} - j\frac{G_l}{\omega}, \tag{3.39}$$

where the following convenient notations have been introduced:

$$C_h = \frac{C_1 C_2}{C_1+C_2},$$

$$C_l = \frac{C_1 G_2^2 + C_2 G_1^2}{(G_1+G_2)^2},$$

$$\tau = \frac{C_1+C_2}{G_1+G_2},$$

3.3 Static Electrical Properties of Planar Membranes

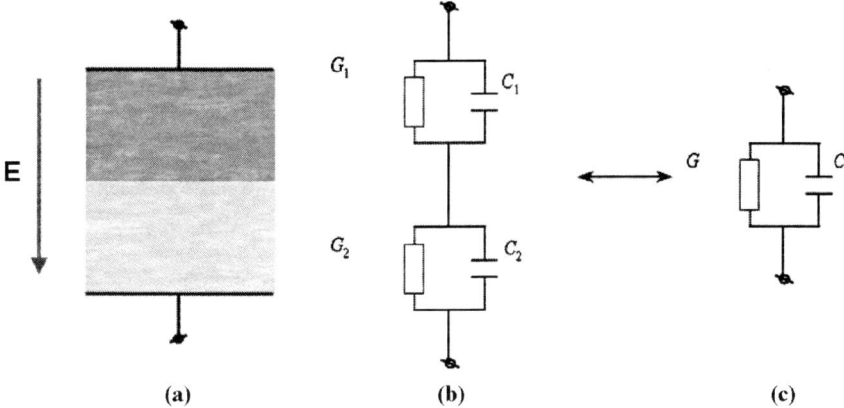

Fig. 3.6 Two stacked dielectrics placed between the plates of a parallel-plate capacitor (**a**), each being characterized by an electrical capacitance C_i and a conductance G_i ($i = 1, 2$) (**b**), present an equivalent capacitance C and conductance G (**c**), which are both frequency dependent, as described in the text.

(a parameter that has dimension of time), and

$$G_h = \frac{G_1 G_2}{G_1 + G_2}.$$

Further, if we divide the whole relationship (3.39) by the geometrical factor, S/d, to return to complex permittivity, we get the Debye dispersion function (Debye, 1945; Takashima, 1989),

$$\varepsilon^* = \varepsilon_h + \frac{\varepsilon_l - \varepsilon_h}{1 + j\omega\tau} - j\frac{\sigma_l}{\omega}, \qquad (3.40)$$

where τ is the *relaxation time*, ε_l and σ_l are the *limiting permittivity* and *conductivity at low frequencies* (i.e., for $\omega \to 0$), ε_h and σ_h the *limiting permittivity* and *conductivity at high frequencies* (i.e., for $\omega \to \infty$) and $\varepsilon_l - \varepsilon_h = \delta\varepsilon$ is called the *dielectric increment*.

> **Observation:** A *conductivity increment*, $\sigma_l - \sigma_h = \delta\sigma$, can also be defined, but it is rarely used in practice.

By defining τ in terms of a *characteristic frequency*, f_c ($= 1/2\pi\tau$), and using the frequency f instead of the angular frequency ω, we can re-write equation (3.40) as:

$$\varepsilon^* = \varepsilon_h + \frac{\varepsilon_l - \varepsilon_h}{1 + jf/f_c} - j\frac{\sigma_l}{2\pi f}. \qquad (3.41)$$

Finally, by using the Kramers-Krönig relationship between $\delta\varepsilon$ and $\delta\sigma$, which for the Debye case takes the simple form (Hanai, 1960; Takashima, 1989) $\delta\varepsilon = \tau \cdot \delta\sigma$, we obtain

$$\varepsilon = \varepsilon_h + \frac{\varepsilon_l - \varepsilon_h}{1 + (f/f_c)^2}, \qquad (3.42)$$

for the real part of the equation (3.41), and

$$\sigma = \sigma_l + \frac{\sigma_h - \sigma_l}{1 + (f/f_c)^2} \cdot (f/f_c)^2, \qquad (3.43)$$

for the imaginary part. These two quantities represent the *equivalent permittivity* and the *equivalent conductivity*, respectively, of the stack of two dielectric layers, and both depend on the frequency of the applied field as shown in Fig. 3.7. This frequency dependence is called dielectric dispersion.

Interestingly, a system formed by two dielectric layers stacked together behaves similarly to a pure system of permanent dipoles placed in an alternating electrical field, for which Debye equation has actually been derived initially. The role of electrical dipoles is played in this case by the accumulation of electrical charges at the interface between the two dielectric layers; the frequency dependence arises from the fact that the magnitude of the charge falls off with the increase in the frequency of the applied field. The mechanism of dielectric dispersion is known in the literature on dielectric spectroscopy as *Maxwell-Wagner interfacial polarization* or simply as *interfacial polarization*.

More insight can be gained into the relevance of this phenomenon to the study of biological systems by considering the case of dielectric particles suspended in electrolytic solutions. A mathematical treatment will be presented in section 3.3.3. Next, we focus our attention on the application of a model for stacked dielectric layers to probing the molecular organization of bilayer lipid membranes.

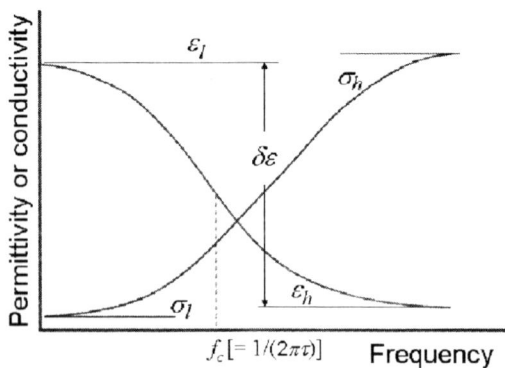

Fig. 3.7 Real [equation (3.42)] and imaginary [equation (3.4)] parts of the Debye dispersion function [equation (3.41)] and their parameterization.

3.3.2.2 The Dielectric Structure of Synthetic Lipid Bilayers and Plasma Membranes

It is possible to create experimentally bilayer lipid membranes that fill a small orifice in a slab of solid material immersed in an electrolyte, and to study the properties of the bilayer. Even before the fluid mosaic model of the membrane has been proposed (Singer and Nicolson, 1972), several researchers have used this method to study the dielectric behaviour of lipid bilayers as a function of the frequency of an alternating field applied between two electrodes each located on one side of the membrane (see, e.g., Ashcroft et al., 1981; White, 1973).

We will show here that, by using the lipid bilayer as a model for natural membranes and by regarding it as a dielectric multilayer, it is possible to validate the structural aspect of the Singer-Nicolson model by comparing its predicted dispersion curves to the experimental data. In doing so, we take into account not only the layered structure of the bilayer itself, but also an electrical *double layer capacitance* for each side of the *lipid bilayer*, as given by equation (3.27). This membrane model and its equivalent electrical circuit is shown in Fig. 3.8. By using this dielectric model, one can derive a formula for the real part of the admittance of the membrane (i.e., for equivalent conductance, G) and one for the imaginary part of the admittance divided by the frequency (which provides the equivalent capacitance, C), in terms of the specific electrical properties of each dielectric layer.

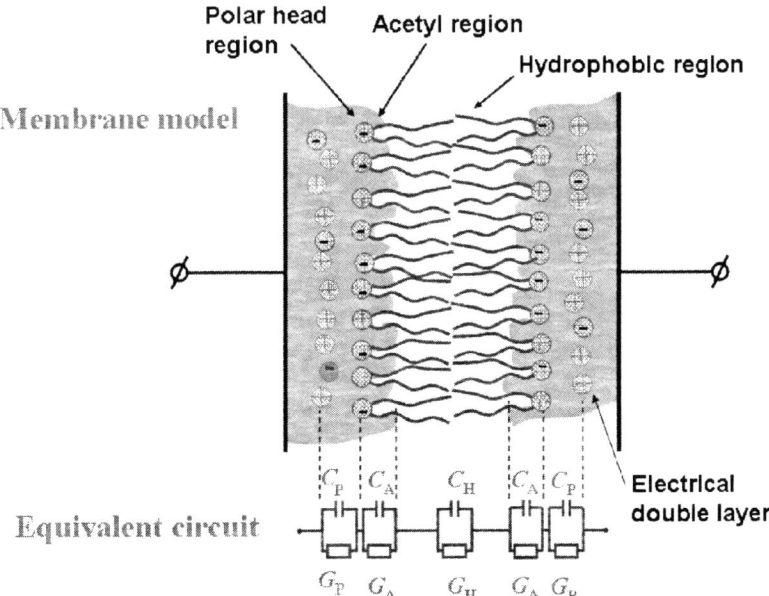

Fig. 3.8 Structural model of a bilayer lipid membrane placed in an electrolytic solution and its equivalent electrical circuit. Note that each lipid monolayer contributes an identical group of capacitance and conductance. C_P, C_A and C_H represent the capacitances, and G_P, G_A and G_H represent the conductances of the membrane layers.

Ashcroft and co-authors (1981) have performed measurements of capacitance and conductance of lipid bilayers (made from egg phosphatidyl-choline dissolved in n-tetradecane or n-decane solvents) in aqueous electrolyte solutions of KCl with various concentrations ranging from 1 to 1,000 mM. The results of their measurement of the electrical capacitance for the case of phosphatidyl-choline/tetradecane bilayer in 1 mM KCl solution are shown in Fig. 3.9, together with the curve predicted by the model from Fig. 3.8.

> **Quiz 2.** Derive expressions for the equivalent conductance and capacitance of the model membrane presented in Fig. 3.8. Use equation (3.33) to express the admittance of each dielectric layer in terms of its specific conductance and capacitance.

To test the dielectric model of the lipid membrane, Ashcroft et al. have adjusted the C and G values corresponding to each dielectric layer until the simulated capacitance curve (represented by lines in Fig. 3.9) accurately fitted the data (represented by filled circles or triangles). These *best-fit values* have then been taken as the actual electrical properties of each particular layer. (Conductance dispersion curve and conductances for each layer have also been obtained by Ashcroft et al., but these are not very relevant to our discussion here.)

The specific capacitances of each dielectric layer for the membrane for bilayers immersed in 10 mM KCl solution were: $43\,\mu F/cm^2$ for the polar head layer (which includes the electrical double layer), $45\,\mu F/cm^2$ for the acetyl region layer, and $0.54\,\mu F/cm^2$ for the hydrophobic layer. (Note that these values include a factor of two that comes from the double layered structure of the membrane). Of these three values, most important for our discussion here are the first value, which compares well to what is expected ($\sim 100\,\mu F/cm^2$) from electrical double layer considerations

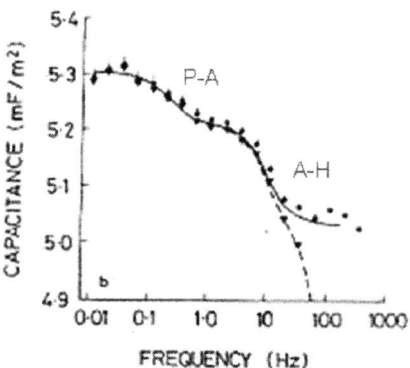

Fig. 3.9 Specific electrical capacitance of a model lipid bilayer in 1 mM aqueous KCl solution as a function of frequency of an applied electrical field. Points, experimental data; lines, simulations by using the model in Fig. 3.8. The filled circles and the solid line represent the data for the bilayer alone, while the dashed line and the triangles include uncorrected contribution from the bulk electrolyte. Note that there are two interfaces for each lipid monolayer, viz., one between the polar head layer and electrolyte and one between the hydrophobic layer and the acetyl layer, each of which contributes one dispersion (for a total of two sub-dispersions) represented by "P-A" and "A-H," respectively. Figure reprinted from Ashcroft et al. (1981) with permission from Elsevier.

(see section 3.2), and the one obtained for the hydrophobic layer. The latter, can be used to estimate the thickness of the hydrophobic layer (by using a value of 2.1 for the relative dielectric constant of that layer), which turns out to have a thickness of 3.5 nm. This value is very close to the one obtained by Fricke for the plasma membrane (mentioned in section 3.1.) from measurements in the radiofrequency range (Fricke, 1927) of plasma membrane, and for very good reasons.

To be meticulous, however, we should remark that slightly larger values ($0.73\,\mu F/cm^2$) have been obtained for the specific capacitance of the hydrophobic layer (Benz and Janko, 1976) in the case of bilayers that contained no dissolved solvent (alcohol) in their hydrophobic core (so-called *black lipid bilayers*). In this case, the membrane thickness is slightly smaller, since only the lipid tails are occupying now the hydrophobic layer volume.

3.3.3 Dielectric Properties of Random Suspensions of Particles with Particular Relevance to Biological Cells

Specific capacitance of the plasma membrane can be measured for instance by using micropipette techniques, in which portions of the membrane are attached to the tip of a micropipette and subjected to an applied electrical field (see, e.g., Asami and Takashima, 1994). However, to determine the dielectric properties of biological cells in a completely noninvasive manner, one typically performs measurements on suspensions.

Dielectric properties of systems of microscopic particles depend on a number of physical as well as geometrical factors:

1. Electrical properties of the particles and the suspending medium (see below)
2. The particles concentration (see below)
3. The particle shape (Takashima, 1989)
4. Whether or not the particle has a peculiar structure (i.e., it is itself heterogeneous, see below)
5. Whether the particles are individually dispersed to form random suspensions (Raicu et al., 1996, 1998a) or more orderly aggregates (Raicu et al., 1998b, 2001)

In this regard, one distinguishes three types of aggregates: (a) dilute random suspensions of particle aggregates; (b) concentrated suspensions of aggregates; and (c) large, self-similar aggregates, such as percolation and *Cantorian fractals* (see chapter 5 for a discussion of fractals and fractal structures in biophysics).

In this book we are only interested in the rather simple theory of Maxwell-Wagner relaxation as it applies to the plasma membrane.

3.3.3.1 Homogeneous Particles in Applied Homogeneous Electrical Field

In order to determine the equivalent dielectric constant of a suspension, one has to calculate, by solving the Laplace equation (Jackson, 1998) the electric potential at a

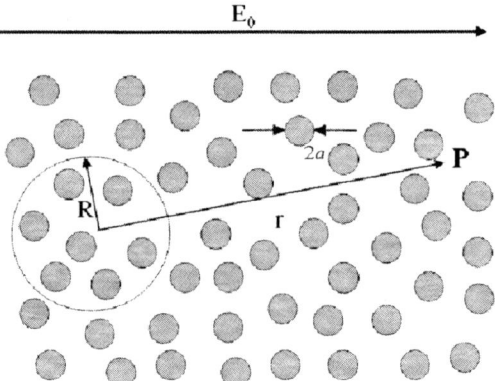

Fig. 3.10 The geometry parameters necessary for the determination of the potential of a suspension of spherules of radius a (at a point, P, within the suspension). The complex permittivity, ε_s^*, is assumed to be distributed at random in a uniform medium of complex permittivity, ε_e^*.

point, P, due to a spherule of radius a placed in an electrical field, E_0. Then, a region in space, containing n spherules, delimited by an imaginary sphere of radius, R, is considered as having dielectric properties similar to the whole suspension of interest (Fig. 3.10). The distances from each spherule to the point, P, are considered so large that the spherules can be regarded as being all located at the same distance r from the point P. The potential is then obtained as a superposition of all the potentials calculated for each spherule (Hanai, 1960; Takashima, 1989):

$$\Phi(r,\theta) = -E_0 r \cdot \cos\theta + n\frac{a^3}{r^2} \cdot \frac{\varepsilon_s^* - \varepsilon_e^*}{\varepsilon_s^* + 2\varepsilon_e^*} E_0 \cos\theta. \tag{3.44}$$

where n is the number of particles and θ represents the angle between $\vec{E}_0$ and $\vec{r}$.

Next, the potential due to the large sphere of radius, R, assumed to be homogeneous and having the complex permittivity, ε^*, is obtained as:

$$\Phi(r,\theta) = -E_0 r \cdot \cos\theta + \frac{R^3}{r^2} \cdot \frac{\varepsilon^* - \varepsilon_e^*}{\varepsilon^* + 2\varepsilon_e^*} E_0 \cos\theta. \tag{3.45}$$

Since the potentials determined by the two methods should be equal (they characterize the same system), we obtain:

$$\frac{\varepsilon^* - \varepsilon_e^*}{\varepsilon^* + 2\varepsilon_e^*} = p \cdot \frac{\varepsilon_s^* - \varepsilon_e^*}{\varepsilon_s^* + 2\varepsilon_e^*} \tag{3.46}$$

where $p = n\frac{a^3}{R^3}$ is the *volume fraction* of the suspended particles.

Upon some algebraic manipulation of equation (3.46), an equation formally equivalent to the Debye equation (see equation 3.41) is obtained for the equivalent complex permittivity of the suspension, ε^* (Hanai, 1960).

3.3.3.2 Particles Covered by a Thin Membrane: The Single-Shell Model

Mile and Robertson (1932) made use of the classical potential theory and calculated the potential outside of a shelled sphere suspended in a medium of permittivity ε_e^* as:

$$\Phi(r,\theta) = -E_0 r \cdot \cos\theta + n\frac{R^3}{r^2} \cdot \frac{(\varepsilon_m^* - \varepsilon_e^*)(\varepsilon_i^* + 2\varepsilon_m^*) + (\varepsilon_i^* - \varepsilon_m^*)(\varepsilon_e^* + 2\varepsilon_m^*)v}{(\varepsilon_m^* + 2\varepsilon_e^*)(\varepsilon_i^* + 2\varepsilon_m^*) + 2(\varepsilon_i^* - \varepsilon_m^*)(\varepsilon_e^* - \varepsilon_m^*)v} E_0 \cos\theta, \tag{3.47}$$

where ε_e^*, ε_i^*, ε_m^* represent the complex permittivities of the layers in Fig. 3.11. The equation for permittivity, obtained from equality of this potential to the one given by equation (3.45), is:

$$\frac{\varepsilon^* - \varepsilon_e^*}{\varepsilon^* + 2\varepsilon_e^*} = p \cdot \frac{(\varepsilon_m^* - \varepsilon_e^*)(\varepsilon_i^* + 2\varepsilon_m^*) + (\varepsilon_i^* - \varepsilon_m^*)(\varepsilon_e^* + 2\varepsilon_m^*)v}{(\varepsilon_m^* + 2\varepsilon_e^*)(\varepsilon_i^* + 2\varepsilon_m^*) + 2(\varepsilon_i^* - \varepsilon_m^*)(\varepsilon_e^* - \varepsilon_m^*)v}, \tag{3.48}$$

where $p = na^3/R_c^3$ and $v = (R_c - d)^3/R_c^3$ (see Fig. 3.11). This equation is sometimes improperly attributed to Pauly and Schwan (1959).

Dänzer (1993) and, later, Pauly and Schwan (1959) have undertaken detailed analyses of the equation for shelled spheres (equation 3.48) and found after somewhat cumbersome calculations that it is exactly decomposable into two terms of a Debye type, corresponding to the two interfaces of the particles.

Fricke (1955) had generalized the Mile and Robertson model to include multi-shelled particles, and obtained an equivalent admittance of the form of a continued fraction. Later, Irimajiri et al. (1979) have shown that a number of sub-dispersions equal to the number of interfaces should be expected for multi-shelled particle suspensions. For single-shelled particles, therefore, two sub-dispersions are expected to occur, but usually only one sub-dispersion is important – the one due to the polarization at the interface between the suspending medium and cell membrane. Pauly and Schwan (1959) have derived this result by considering the following reasonable approximations:

$$\frac{\sigma_m}{\sigma_e}, \frac{\sigma_m}{\sigma_i}, \frac{d}{R} \ll 1, \tag{3.49}$$

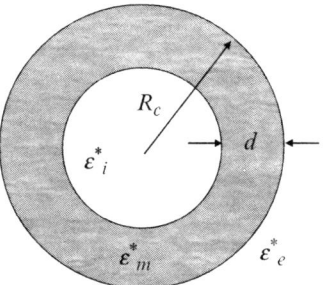

Fig. 3.11 The *single-shell dielectric model* of a particle covered with a thin shell.

which state that the cell membrane is insulating, compared to the internal and external electrolytes, and very thin. Thus, the parameters for the equivalent Debye equation are:

$$\sigma_l = \sigma_e \frac{2(1-p)}{2+p} \tag{3.50}$$

$$\delta\varepsilon \equiv \varepsilon_l - \varepsilon_h = \frac{9p}{(2+p)^2} \frac{RC_m}{\varepsilon_0} \tag{3.51}$$

$$\tau \equiv \frac{1}{2\pi f_c} = RC_m \left(\frac{1}{\sigma_i} + \frac{1-p}{2+p} \frac{1}{\sigma_e} \right), \tag{3.52}$$

where C_m is the specific plasma membrane capacitance, defined as

$$C_m = \frac{\text{Membrane capacitance}}{\text{Surface area}} \equiv \frac{\varepsilon_m}{d}. \tag{3.53}$$

C_m can be easily obtained from equation (3.52), upon knowledge of the volume fraction, p (derived from 3.50).

By performing dielectric measurements on suspensions of cells, it has been possible to determine the specific electrical capacitance of the plasma membrane for many different types of biological cells. For instance, a value of $0.72\,\mu F/cm^2$ has been obtained for erythrocyte ghosts (Asami et al., 1989), while for yeast cells values between 0.65–$0.75\,\mu F/cm^2$ have been reported (Raicu et al., 1996, 1998a).

For a better understanding of those results, in which a constant membrane capacitance is obtained at all frequencies above kHz-region, it is necessary to discuss them in light of the results presented above for lipid bilayers. Specifically, it was observed that the dielectric dispersion of the membrane itself takes place at low frequencies. After 1 kHz, the only contribution to the membrane capacitance is due to the hydrophobic layer alone; this is therefore the contribution that was expected to be obtained from measurements of cells in suspension, which are carried out at frequencies above 1 kHz. Therefore, one can conclude that the capacitance of biological cell membrane, measured in radiofrequency range provides a value for the thickness of the hydrophobic layer of the order of 3–4 nm, which supports the current membrane model, in particular the fact that a lipid bilayer provides a matrix for all membrane components.

References

Asami, K. and Takashima, S. (1994) Membrane admittance of cloned muscle cells in culture: use of a micropipette technique, *Biochim. Biophys. Acta*, **1190**: 129

Asami, K., Takahashi, T. and Takashima, S. (1989) Dielectric properties of mouse lymphocytes and erytrocytes, *Biochim. Biophys. Acta*, **1010**: 49

Ashcroft, R. G., Coster, H. G. L. and Smith, J. R. (1981) The molecular organisation of bimolecular lipid membranes. The dielectric structure of the hydrophilic/hydrophobic interface, *Biochim. Biophys. Acta*, **643**: 191

References

Atkins P. and de Paula J. (2002) *Atkins' Physical Chemistry*, 7th ed., Oxford University Press, New York

Berg, J. M., Tymoczo, J. L. and Stryer, L. (2002) *Biochemistry*, 5th ed., W. H. Freeman, New York

Brett, C. M. A. and Oliveira Brett, A. M. (1993) *Electrochemistry. Principles, Methods, and Applications*, Oxford University Press, Oxford/New York/Tokyo

Benz, R. and K. Janko (1976) Voltage-induced capacitance relaxation of lipid bilayer membranes. Effects of membrane composition, *Biochim. Biophys. Acta*, **455**: 721

Dänzer, H. (1938) In: E. B. Rajewsky (ed.) *Ergebnisse der Biophysikalischen Forschung*, Georg Thieme, Leipzig, p. 193

Debye, P. (1945) *Polar Molecules*, Dover, New York

Fricke, H. (1927) The electric capacity of suspensions of red corpuscles of a dog, *Phys. Rev.*, **26**: 682

Fricke, H. (1955) The complex conductivity of a suspension of stratified particles of spherical cylindrical form, *J. Phys. Chem.*, **59**: 168

Frye, C. D. and Edidin, M. (1970) The rapid intermixing of cell surface antigens after formation of mouse-human heterokaryons, *J. Cell. Sci.*, **7**: 319

Glaser, R. (2001) *Biophysics*, Springer, Berlin/Heidelberg/New York

Goodwin, J. S., Drake, K. R., Remmert, C. L. and Kenworthy, A. K. (2005) Ras diffusion is sensitive to plasma membrane viscosity, *Biophys. J.*, **89** (2): 1398

Hanai, T. (1960) Theory of the dielectric dispersion due to the interfacial polarization and its application to emulsions, *Kolloid. Z.* **171**: 3

Hille, B. (2001) *Ion Channels of Excitable Membranes*, Sinauer, Sunderland, MA

Hunter, R. J. (1987) *Foundations of Colloid Science*, Vol. 1, Oxford University Press, Oxford

Irimajiri A., Hanai T. and Inouye H. (1979), A dielectric theory of 'multi-stratified shell' model with its application to a lymphoma cell, *J. Theor. Biol.*, **78**: 251

Jackson, J. D. (1998) *Classical Electrodynamics*, Wiley, New York, p. 154

Lodish, H., Berk, A., Matsudaira, P., Kaiser, C. K., Krieger, M., Scott, M. P., Zipursky, S. L. and Darnell, J. (2004) *Molecular Cell Biology*, 5th ed., W. H. Freeman, New York

Mile, J. B. and Robertson, H. P. (1932) By assuming the material to be made up of elements of different relaxation times, interface may play a large role during the formation of a field-induced dipole moment, *Phys. Rev.*, **40**: 583

Movileanu, L., Popescu, D., Stelian, I. and Popescu, A. I. (2006) Transbilayer pores induced by thickness fluctuations, *Bull. Math. Biol.* **68**: 1231

Pauly, H. and Schwan, H. P. (1959) Über die Impedanz einer von kugelförmigenTeichen mit eine Schale *Z. Naturforsch.*, **14b**: 125

Popescu, D., Ion, S., Popescu, A. and Movileanu, L. (2003) Elastic properties of BLMs and pore formation, In: *Planar Lipid Bilayers (BLMs) and Their Applications*, Tien, H.T. and Ottova-Leitmannova, A. (Eds.), Elsevier Science, Amsterdam/New York, Chapter 5, p. 173

Raicu, V., Raicu, G. and Turcu, G. (1996) Dielectric properties of yeast cells as simulated by the two-shell model, *Biochim. Biophys. Acta*, **1274**: 143

Raicu, V., Gusbeth, C., Anghel, D. F. and Turcu, G. (1998a) Effects of cetyltrimethylammoniumbromide (CTAB) surfactant upon the dielectric properties of yeast cells, *Biochim. Biophys. Acta*, **1379**: 7

Raicu, V., Saibara, T. and Irimajiri, A. (1998b) Dielectric properties of rat liver in vivo: Analysis by modeling hepatocytes in the tissue architecture, *Bioelectrochem. Bioenerg.*, **47**: 333

Raicu, V., Sato, T. and Raicu, G. (2001) Non-Debye dielectric relaxation in biological structures arises from their fractal nature, *Phys. Rev. E*, **64**: 02191

Singer, S. J. and Nicolson, G. L. (1972) The fluid mosaic model of the structure of cell membranes, *Science*, **175**: 720

Stern, O (1924) Zur Theorie der Electrolytischen Doppelschicht, Z. Elektrochem., **30**: 508

Takashima, S. (1989) *Electrical Properties of Biopolymers and Cell Membranes*, Adam Hilger, Bristol

White, S. H. (1973) The surface charge and double layers of thin lipid films formed from neutral lipids, *Biochim. Biophys. Acta* **323**: 343

Chapter 4
Substance Transport Across Membranes

4.1 Brief Overview

The plasma membrane controls the traffic in both directions between a cell and its environment of small molecules, macromolecules, supra-molecular complexes or larger particles. The types of substance transport across the cell membrane may be classified, by taking into account the size of the particles, into two categories: (a) *macrotransport* of relatively large quantities of solution, molecular complexes, or even other cells, and (b) *microtransport* of ions, small molecules and macromolecules.

Macro-transport refers to three types of processes (Fig. 4.1): *endocytosis*, whereby various particles are internalized by the cells, *exocytosis*, in which macroparticles are externalized by the cell, and *transcytosis*, which consists of an endocytosis process followed by exocytosis of the same particle. During endocytosis and exocytosis, the target particles are wrapped in membranous material which is produced and/or destroyed by the cell as necessary.

A particular type of endocytotic process is *phagocytosis*, whereby *immunocompetent* cells (i.e., macrophages, involved in an organism's defense against viruses and microbes), are able to engulf and destroy viruses and microbes invading the organism, as well as dead cells and cellular remnants.

Figure 4.2 illustrates two different instances of exocytosis: *constitutive* or *unregulated secretion* and *regulated secretion*. In the first case the particles (proteins) are sorted and packed into vesicles by the Golgi aparatus and then delivered to the plasma membrane, which they will eventually cross by following the mechanism of membrane fusion sketched in Fig. 4.2. The *regulated* or *receptor-mediated secretion* is triggered by a signal (such as a hormone or a neurotransmitter) received by a receptor on the plasma membrane. Therefore, in this case, the proteins are only secreted as needed. More detailed information about this fascinating topic can be found in biology textbooks (Alberts et al., 2002, Lodish et al., 2004).

Microtransport across membrane is performed also in a variety of modalities. By taking into account the energy involved during the process, one can distinguish: *passive transport*, and *active transport*. The *passive transport* is always directed

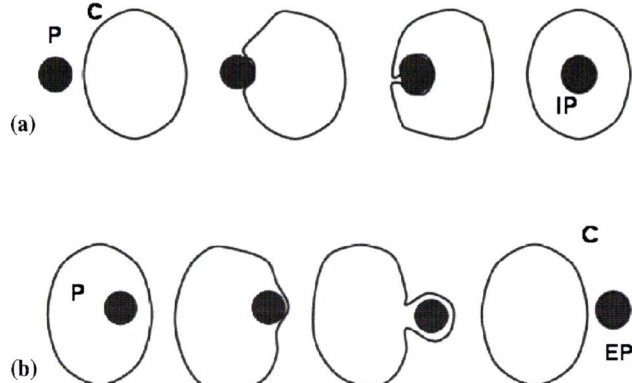

Fig. 4.1 Schematic representation of two modalities of biological macrotransport across membrane: (**a**) endocytosis; (**b**) exocytosis. The process involving both (**a**) and (**b**) is called transcytosis. Significance of the symbols: C – cell; P – particle; IP – internalized particle; EP – externalized particle.

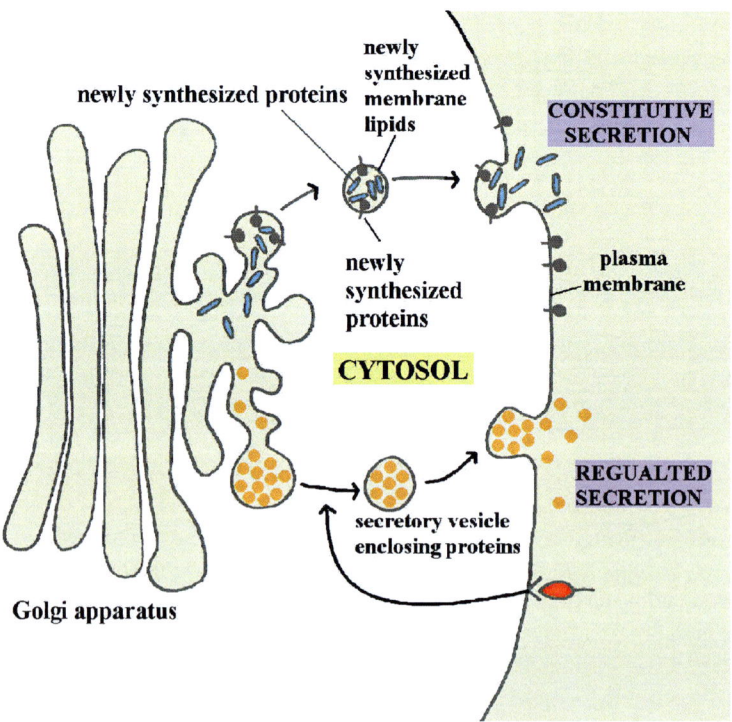

Fig. 4.2 Constitutive (unregulated) and receptor-mediated (regulated) exocytosis.

from higher concentrations of molecules to lower concentrations, that is, in the direction of the concentration gradient. This transport is due to diffusion of the particles and thus occurs spontaneously, without energy consumption (i.e., it is *exergonic*, with $\Delta G < 0$, and with entropy increase, $\Delta S_{ex} > 0$). Passive transport can be classified into: *simple diffusion*, which also occurs in non-biological systems, and *facilitated transport*, which is *specific* to biological systems.

In *active transport* the substance is always transferred *from lower to higher concentrations of molecules*, that is, in the opposite direction of the concentration gradients, and involves free energy consumption (i.e., the process is *endergonic*, with $\Delta G > 0$, and antientropic, $\Delta S_{en} < 0$).

4.2 Diffusion in Biological Systems

In biological systems, the *simple* (*passive*) *diffusion* may take place in bulk liquid phases as well as across membranes. As we have emphasized on several occasions, the interior of a cell is a highly *inhomogeneous* and *anisotropic milieu*. Therefore, the process of simple diffusion of cytosolic particles undergoes successive discontinuities as the particles encounter successive membrane interfaces, which give the whole diffusion process a complex character of chained serial processes. There is indication that diffusion processes in the cytosol may not follow classical pattern that diffusion in homogeneous media does, as we will discuss in the next chapter. At this stage in our discussion, however, we only consider the general properties of the medium as averages over large volumes, in particular when talking about transport between two reservoirs of substance. In the case of cellular membranes, the microtransport is performed across the *lipid matrix*, constituted by the lipid bilayer, through *short-lived pores* or through the cores of specialized proteins called *ionic channels* or *carriers*.

In the following, we shall assume, for simplicity, that the dimensions of the cellular compartments are much greater than the dimensions of the diffusing particles (rigorously, greater than *the mean free path* of the particles). Although rather crude, this approximation permits the use of well-known classical laws to mathematically describe the simple diffusion, as we shall see momentarily. One can regard the simple diffusion as a result of huge numbers of chaotic collisions (due to thermal agitation) both with other particles and with the solvent molecules.

4.2.1 Fick's Laws of Diffusion

In order to put the description of diffusion on a quantitative basis, we shall first review some indispensable concepts such as *flux*, *flux density*, and *gradient*.

The *flux*, J_Y, of a physical quantity, Y, across a surface area, A, is given by the *first derivative of Y with respect to time*:

$$J_Y = \frac{dY}{dt} \tag{4.1}$$

For instance, the flux of mass through a specified surface area is the mass flow, $J_m = dm/dt$. The flux divided by the *surface area* (i.e., flux through a unit surface area),

$$j_Y = \frac{1}{A}\frac{dY}{dt} \tag{4.2}$$

is called *flux density*.

By using the definition of the gradient of a conjugate scalar quantity, X_Y, as the vector quantity

$$\mathbf{grad}\, X_Y \equiv \nabla X_Y = \frac{\partial X_Y}{\partial x}\vec{i} + \frac{\partial X_Y}{\partial y}\vec{j} + \frac{\partial X_Y}{\partial x}\vec{k} \tag{4.3}$$

where $\vec{i}, \vec{j}, \vec{k}$ are unit vectors associated with the spatial coordinates of the Cartesian reference frame, and ∇ is the del operator (represented by the symbol "nabla"), we can introduce a *vectorial equation* describing the transport of the quantity, Y:

$$\vec{j}_Y = \frac{\vec{J}_Y}{A} = -C_Y \nabla X_Y \tag{4.4}$$

where C_Y is a specific coefficient describing the transport of Y. This is *Fick's first law* for transport across surfaces, which basically states that the flux of the quantity Y is directly proportional to the gradient of the conjugated quantity X_Y.

By setting $Y = m$ (mass), $X_Y = \rho$ (density), and $C_Y = D$ (translational diffusion coefficient), we obtain Fick's *first law of simple diffusion*:

$$\vec{J}_m = -AD\nabla\rho \tag{4.5}$$

with $[J_m]_{SI} = kg/s$.

In the case of one-dimensional diffusion of neutral particles across membranes, Fick's first law is written in the following scalar form:

$$J_m = -AD\frac{d\rho}{dx} \tag{4.6}$$

We emphasize that it is more common in the fields of biophysics and biochemistry to use instead of mass, m, the *quantity of substance*, ν, expressed in mol, and instead of *density*, ρ, the *molar concentration*, c_M, expressed in mol/l. In this case, an analog of (4.5) is obtained as:

$$\vec{J}_\nu = -AD\nabla c_M \tag{4.7}$$

where $J_\nu = d\nu/dt$, and $[J_\nu]_{SI} = mol/s$.

The diffusion coefficient, D, depends on the external medium temperature, T, and dynamic viscosity coefficient, η, as well as on the particle geometry. In the case of a spherical particle of radius, r, the diffusion coefficient is given by Einstein's formula:

$$D = \frac{k_B T}{6\pi\eta r} \tag{4.8}$$

where k_B is the Boltzmann constant.

4.2 Diffusion in Biological Systems

At a first glance, equation (4.8) seems to predict a linear dependence of the diffusion coefficient on temperature. In fact, the temperature dependence of D is more intricate, because the viscosity coefficient itself depends on temperature, being inversely proportional to T. Equation (4.8) also predicts that the bigger the particles the slower the diffusion. For instance, in the case of ions and small molecules, $D \approx 10^{-9}\, m^2 s^{-1}$, while in the case of macromolecules, $D \approx 10^{-11}\, m^2 s^{-1}$.

The temporal variation of the particle concentration is directly proportional to the spatial variation of the concentration gradient, according to *Fick's second law* of diffusion:

$$\frac{\partial c_M}{\partial t} = -\nabla(-D\nabla c_M) \tag{4.9}$$

which is easily obtained from equation (4.7) and the *continuity equation* (conservation of mass).

If D is the same in all space directions, Fick's second law becomes:

$$\frac{\partial c_M}{\partial t} = D\nabla^2 c_M \equiv D\Delta c_M \tag{4.10}$$

where Δ is the Laplacean. In one dimension, the second law of diffusion takes the form:

$$\frac{\partial c_M}{\partial t} = D\frac{\partial^2 c_M}{\partial x^2} \tag{4.10'}$$

which suggests a straightforward physical interpretation: the rate of concentration change at a given point is directly proportional to the second derivative, or the "curvature" of the concentration. This means that, if the local curvature of concentration is large (i.e., the concentration varies strongly in space), then the concentration changes rapidly with time. The diffusion equation (4.10') reflects the mathematical formulation of the "natural tendency for the wrinkles in a distribution to disappear. More succinctly: Nature abhors a wrinkle" (Atkins and de Paula, 2002). However, we note that living bodies also generate and exploit all kinds of wrinkles (e.g., concentration gradients, electrical potential gradients). It is therefore the interplay between the mechanisms of creating curvatures and the above mechanisms of actually removing them that distinguishes inanimate from animate objects.

4.2.2 Simple Diffusion Through Membranes

If the molecules present in the cytosol or in the outer medium are soluble in the cell membrane, they can diffuse from one side to another across the membrane. For instance, lipid-soluble molecules (e.g., fatty acids, alcohols, etc.) can diffuse through biological membranes, while lipid-insoluble charged (e.g., amino acids) and uncharged molecules (e.g., glucose) cannot diffuse easily. We shall consider a homogenous but anisotropic membrane of thickness, d_m (Fig. 4.3) separating two homogenous solutions of the same solute, but with different solute concentrations: $c_{M1} > c_{M2}$. In order to characterize mathematically the *simple diffusion through*

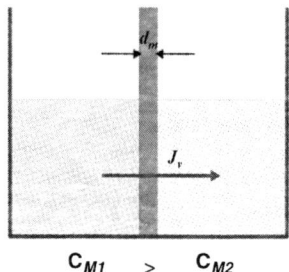
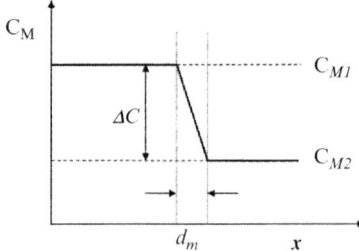

Fig. 4.3 Simple diffusion through a membrane of thickness d_m. Note that the concentration falls linearly across the membrane, with the slope $\delta c/\delta x = \delta c/d_m$. See text for the significance of the symbols.

membrane (also called *permeation*), we write Fick's first law, for a short, but finite time interval, δt, as follows:

$$J_v = \frac{\delta v}{\delta t} = -AD\frac{\delta c_M}{\delta x} = -AD\frac{c_{M2} - c_{M1}}{d_m} \quad (4.11)$$

where we have substituted the membrane thickness, d_m, with δx. This equation indicates that the flux through the membrane is directly proportional to the solute concentration difference. It assumes also that the concentrations in the media on each side of the membrane are independent of time. This is usually a good approximation in the case of biological cells, where the diffusion through the membrane is much slower than diffusion within each of the two media, which means that the membrane is the only major barrier to the motion of solute molecules. (There is another implicit approximation made in deriving equation (4.11), which will be discussed in the next section.)

By replacing the ratio D/d_m with a parameter, P, called *permeability coefficient*, equation (4.11) becomes:

$$J_v = \frac{\delta v}{\delta t} = -AP(c_{M2} - c_{M1}) \quad (4.12)$$

If there are no molecules in the second compartment (or if they are rapidly consumed as a result of some biochemical reaction), equation (4.12) gives:

$$J_v = APc_{M1} \propto c_{M1} \quad (4.13)$$

which suggests that the flux of simple (passive) diffusion is directly proportional to the concentration of the particle transported and, therefore, does not present saturation.

Observation: In reality, significant discontinuities in the concentrations are to be expected in biological membranes due to the absorption of the diffusing substances at the membrane surfaces, the change of ion concentrations in the electric double layers associated to membrane surfaces, and variations in the diffusing particles mobilities inside the membrane caused by dielectric inhomogeneities (Glaser, 2001).

Membranes of different cells present different values of the permeability coefficients. For instance, in the case of *erythrocyte membrane*, P is 3×10^{-6} m s^{-1} for water, and 2×10^{-9} m s^{-1} for glucose. In general the permeability of the lipid bilayer of the cell membranes spans about 10 orders of magnitude, ranging from 10^{-14} m s^{-1}, for Na$^+$, to $\sim 5 \times 10^{-5}$ m s^{-1} for water molecules (Stryer, 1988; Alberts et al., 2002).

4.2.3 Determination of Membrane Permeability from Membrane Potential Energy Profile

The method used above for deriving the expression for the flux of substance passing through the membrane [equation (4.12)] has the distinct advantage of simplicity. However, as already seen, that derivation relied on the validity of Fick's first law, and on the assumption that the mean free path of diffusing molecules is smaller than the spatial dimensions of the media, which cannot be *a priori* tested for the case of very thin membranes. Also, as Davson and Danielli (1970) have pointed out, the major difficulty with applying Fick's laws to media separated by biological membranes is the assumption that the diffusing particles are much larger than the membrane molecules which resist the motion of the solute. Actually, most passively diffusing molecules in biological membranes are comparable in size to the phospholipids that make up the lipid bilayer. In this section, we introduce a method for deriving a formula equivalent to equation (4.12) along the lines presented by Danielli (see Davson and Danielli, 1970).

A solute particle which diffuses freely within an aqueous medium (such as the cytosol of a cell) encounters two types of resistance when it crosses a membrane interface that separates the first medium from another with lower concentration of solute: the potential energy barriers at the interfaces between the two media and the membrane polar layer (μ_a), and the energy barrier at crossing the polar layer of the membrane into the "bulk" of the membrane hydrophobic layer (μ_i). There are also smaller energy barriers associated with the particle jumping from one site to another after entering the membrane (μ_m).

Figure 4.4 shows two potential energy profiles that correspond to solute molecules with different polarities (relative to water). Depending on the nature of the molecule, the first and last barriers (i.e., those located at the two interfaces of the membrane) may or may not be the highest of all the barriers that the particle will meet. The concentrations across the membrane will therefore vary accordingly.

As regards the particular motion of the diffusing molecules, in general, each molecule can oscillate about a mean position between barriers or, if it possesses enough kinetic energy, can jump over the barrier to reach the next minimum of potential.

The concentrations of solute in the internal and external media are denoted by c_1 and c_2, respectively, while the concentrations at the internal minima of potential energy by $c'_1, c'_2, \ldots, c'_n$. We start from the medium 1 and write the rate of transfer

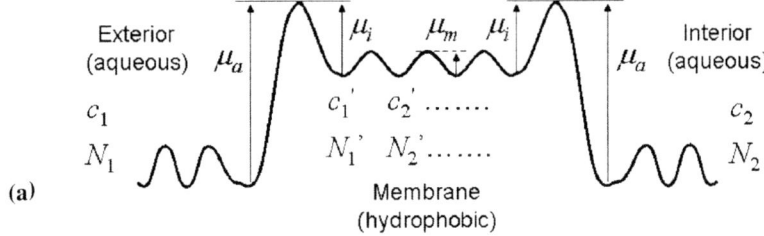

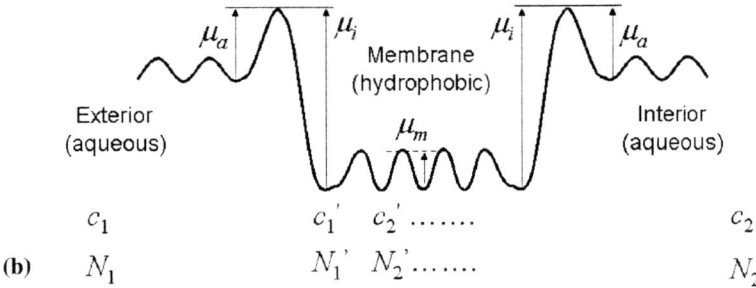

Fig. 4.4 Potential energy profiles of (**a**) a polar solute molecule (i.e., less soluble in the membrane hydrophobic layer) and (**b**) a solute molecule that is less polar than water (i.e., more soluble in membrane).

of the number of molecules, N_1, from the left to the right side of the membrane,

$$\frac{dN_1}{dt} = ac_1 \qquad (4.14)$$

and the rate of transfer of molecules, N'_1, in the reverse direction,

$$\frac{dN'_1}{dt} = -bc'_1 \qquad (4.15)$$

where a and b are two rate constants. Similarly, we can write the forward and reverse rates of transfer of solute at the interface with medium 2 as:

$$\frac{dN'_n}{dt} = bc'_n \qquad (4.16)$$

$$\frac{dN_2}{dt} = -ac_2 \qquad (4.17)$$

When a steady flow from left to right is reached, the flux of substance dS/dt across the membrane can be written as:

$$J = \frac{dS}{dt} = \frac{dN_1}{dt} + \frac{dN'_1}{dt} = ac_1 - bc'_1, \qquad (4.18)$$

4.2 Diffusion in Biological Systems

and:
$$J = \frac{dS}{dt} = \frac{dN'_n}{dt} + \frac{dN_2}{dt} = bc'_n - ac_2 \tag{4.19}$$

By taking the semisum of (4.18) with (4.19), we obtain:
$$J = \frac{dS}{dt} = \frac{a}{2}(c_1 - c_2) + \frac{b}{2}(c'_n - c'_1) \tag{4.20}$$

A similar treatment for the steady flow within the membrane will give:
$$J = \frac{dS}{dt} = e(c'_1 - c'_2) \tag{4.21}$$

$$J = \frac{dS}{dt} = e(c'_2 - c'_3), \tag{4.22}$$

$$\vdots$$

$$J = \frac{dS}{dt} = e(c'_{n-1} - c'_n) \tag{4.23}$$

where n is the number of (identical) barriers in the membrane and e is the rate constant inside the membrane.

By summing up equations (4.21) through (4.23) and dividing by n, we obtain:
$$J = \frac{dS}{dt} = \frac{e}{n}(c'_1 - c'_n) \tag{4.24}$$

Finally, by imposing the condition that the right-hand side of equation (4.20) equals the right-hand side of equation (4.24), an expression for $c'_1 - c'_n$ is obtained, which is substituted into equation (4.24) to obtain:
$$\frac{dS}{dt} = -PA(c_1 - c_2) \tag{4.25}$$

where
$$P = \frac{ea}{A(2e + bn)} \tag{4.26}$$

represents the permeability coefficient of the membrane.

As seen, equation (4.25) is of the same type as equation (4.12). In either case, the membrane permeability can be determined experimentally. It can be therefore concluded that both, Fick's laws as well as the above general considerations of potential energy barriers and rate of change in the number of molecules (or moles) between energy barrier maxima give the same results for the flux of substance across the membrane as a result of free diffusion.

Quiz 1. Plot the concentration profile across the membrane and from compartment 1 to compartment 2, corresponding to both cases shown in Fig. 4.4.

4.3 Osmosis and Osmotic Pressure

A particular case of *simple diffusion* is *osmosis*, which takes place when two solutions with different concentrations of solutes are separated by a *selective membrane*, that is, a membrane that permits the diffusion of solvent molecules but not of the solute (Fig. 4.5). Since we are exclusively interested in describing the process of osmosis in living matter, we shall refer here only to water as a solvent.

The fact that the solute concentration is lower in one compartment than in the other one can be reinterpreted as the *"concentration of water"* is lower in the compartment on the left-hand-side. Due to thermal agitation, the water molecules will spontaneously (i.e., with $\Delta G < 0$) diffuse from the compartment with more "concentrated" water towards the one with a lower water "concentration," where the level of the solution will increase as a result (Fig. 4.5). This process of solvent passive diffusion through a selective membrane is called *direct osmosis*. The direct osmosis is an *entropy-generating process* ($\Delta S > 0$), having the tendency to equalize the solute (or solvent) concentrations among the two compartments.

Observation: One may indeed speak of the solvent (i.e., water) concentration. For instance, at $t = 4\,°C$, one mol ($\nu = 1$) of pure water has a mass, $m = 18\,g$, occupying a volume, $V = 18\,cm^3$. The molar concentration of the pure water, $C_M = \nu/V$, is $C_M^{H_2O} \approx 55.55\,M$.

We can write the expressions for the chemical potentials of water "dissolved" in the two compartments, also by taking pressure into account:

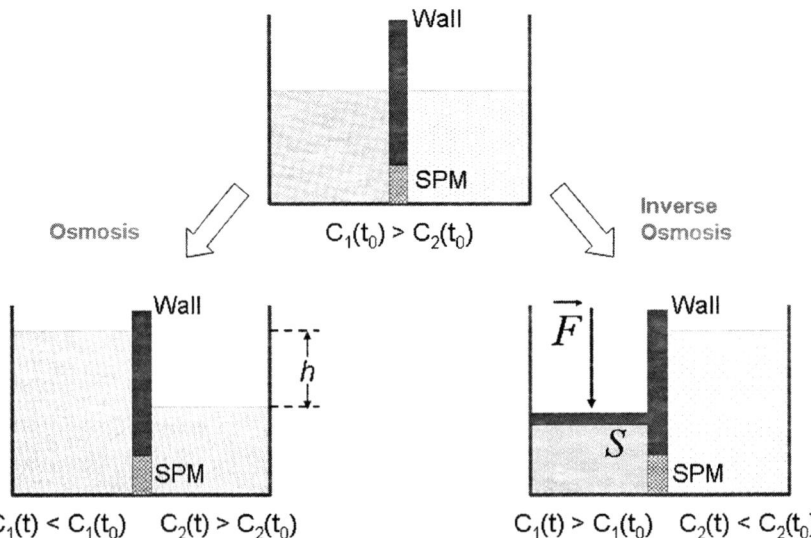

Fig. 4.5 Direct and inverse osmosis through a selective permeable membrane (SPM) separating two solutions with different initial solute concentrations, $C_1(t_0) > C_2(t_0)$.

4.3 Osmosis and Osmotic Pressure

$$\mu_1 = \mu_0 + RT \ln X_1^w + v^w(p_a + \rho^w g h_1) \qquad (4.27)$$

$$\mu_2 = \mu_0 + RT \ln X_2^w + v^w(p_a + \rho^w g h_2) \qquad (4.28)$$

where R is the ideal gas constant, T is the absolute temperature, $X^1{}_w$ and $X^2{}_w$ are the concentrations of water in mole fraction units, μ_0 is the standard chemical potential of water *dissolved* in pure water (i.e., the chemical potential of the pure solute in the same state of aggregation as the solvent), v^w is the partial molar volume of water, p_a is the atmospheric pressure, and $\rho^w g h_i (i = 1, 2)$ is the hydrostatic pressure (where the density has been assumed equal between the two compartments).

In a hypothetical situation of imponderability (i.e., $g = 0$), the equality between the chemical potentials given by equations (4.27) and (4.28) gives at equilibrium:

$$X_1^w = X_2^w \qquad (4.29)$$

However, in the presence of *gravity*, equal concentrations of solute between the two compartments cannot be achieved, since a *hydrostatic pressure* would counterbalance the process of water ascension in the compartment on the left-hand side, as given by:

$$X_1^w = X_2^w e^{v^w \rho^w g h / RT} \qquad (4.30)$$

The hydrostatic pressure in this equation, $p_h = \rho^w g h$, is a measure of the "suction pressure," known as *osmotic pressure*, π, and which is exerted by the solution containing larger amount of solute onto the one containing less.

If one applies *external pressure*, $p_e = F/S$, on the more concentrated solution (Fig. 4.5) some of the water molecules will be forced to traverse the membrane towards the left (because the solute cannot pass through the membrane), reaching the first compartment. As a result, the concentrated solution will become more concentrated, while the dilute one will become even more diluted. At equilibrium, the following equation can be easily derived by including the external pressure term in equation (4.30):

$$X_1^w = X_2^w e^{v^w(-p_e + \rho^w g h)/RT} \qquad (4.31)$$

This energy-consuming process ($\Delta G > 0$), called *reverse osmosis*, can be encountered in, e.g., some special glands of marine animals, where the salted water of seas and oceans is purified to be used as drinking water. This reverse-osmosis process is also used by humans to obtain *desalinated* water from salted sea water.

4.3.1 van't Hoff's Laws

As has been shown by van't Hoff, there exists an interesting and productive analogy between *chemical solutions* and *gases* (van't Hoff, 1887; also reprinted in Kepner, 1979). Osmotic pressure of *ideal* (dilute) solutions is described by *van't Hoff's laws* which are equivalent to the laws of the ideal gas; in the case of very concentrated solutions and protein solutions (the correspondent of real gases), deviations from van't Hoff's laws occur, similar to the case of nonideal gases.

Let us consider a volume, V, of solution obtained by dissolving a mass, m_s, of solute with molecular mass, μ. The percentile concentration is $C_S = \frac{m_s}{V}$ at the absolute temperature, T. The osmotic pressure, π, of the solution satisfies a relation, formally identical with the *general equation* of state of the ideal gas:

$$\pi V = \frac{m_s}{\mu} RT \tag{4.32}$$

Using the definition of C_s it follows that:

$$\pi = \frac{C_S}{\mu} RT \tag{4.33a}$$

or, because $m_s/\mu = v$ and $v/V = C_M$ (molar concentration), one can write an equivalent relation:

$$\pi = C_M RT \tag{4.33b}$$

This last expression includes van't Hoff's laws for ideal solutions, which state that *the osmotic pressure*, π, of an *ideal solution* is: (a) directly proportional to the percentile solute concentration, C_S or molar concentration, C_M; (b) directly proportional to the absolute temperature, T, of the solute (an expected result, because the average thermal agitation energy of the molecules is proportional to the absolute temperature).

Osmotic pressure is an important parameter for biological fluids (cytosol, interstitial liquid, blood plasma, etc.). All these liquids must be compatible from the point of view of their osmotic pressures. The osmotic pressure of biological fluids is equal to the osmotic pressure of a *standard solution*, called *physiological saline solution*: an aqueous solution containing 0.145 M NaCl (or, approximately 9% NaCl) at pH=7 and room temperature.

All solutions having *the same osmotic pressure* as the *physiological saline* solution are called *isotonic* solutions. Other solutions are either *hypotonic*, when their osmotic pressure *is lower* than that of an isotonic solution, or *hypertonic*, in the opposite case.

When the cells are suspended in a *hypotonic aqueous solution*, the water molecules will enter the cells, due to osmosis (see above), increasing the cellular volume and provoking thus cellular *swelling* (Fig. 4.6). If the process is not stopped, the swelling will be followed by *cytolysis*, that is, by membrane rupture. A particular case of cytolysis is *hemolysis* (i.e., the lysis of erythrocytes). By using this phenomenon of hemolysis of erythrocytes suspended in water, for example, one can obtain *hemoglobin*.

If, on the contrary, the cells are suspended in *hypertonic solutions*, they will begin to *shrink* (Fig. 4.6), and, if not removed from the hypotonic solution, *plasmolysis* will occur, that is, the cytosol will be completely lost. In the case of plant cells or bacteria, the cell membranes can detach from the rigid cell walls, as a result of severe hypertonic treatment.

4.3 Osmosis and Osmotic Pressure

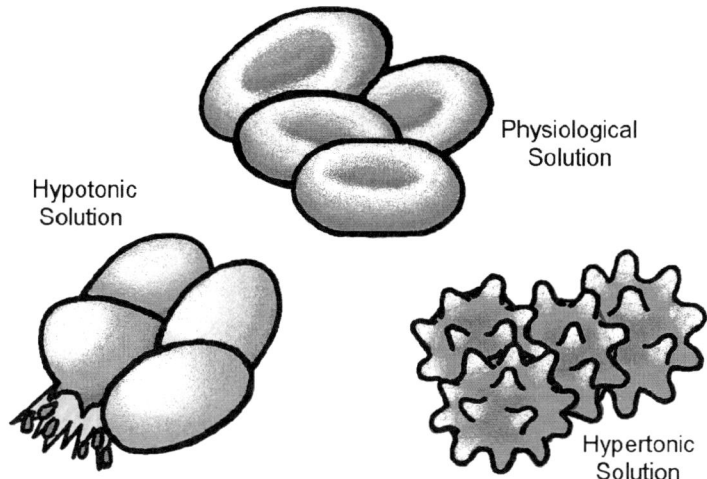

Fig. 4.6 The effect of hypotonic and hypertonic solutions on red blood cells.

Observation: In the case of electrolyte solutions, it is necessary to use in equation (4.33b) a corrected concentration, C_{MS}^* of the solute, which takes into account the *dissociation degree*, α, of the solute molecules, defined as the ratio between the dissociated molecules, N_D, and the total number, N, of initial solute molecules ($\alpha = N_D/N$). Due to the dissociation of the solute molecules, the number of osmotically active particles increases, and the osmotic pressure increases as a result. Considering the most frequent case of molecular dissociation in only two fragments, one can easily find that $C_{MS}^* = (1+\alpha)C_M$. In the case of complete dissociation ($\alpha = 1$, as in the case of NaCl), it results that $C_{MS}^* = 2\,C_M$. Therefore, the osmotic pressure of the *physiological saline solution* ($C_M = 0.145\,\text{M}$; pH $= 7$; $T = 298.15\,\text{K}$), $\pi^* = 2\pi$, is equivalent with the osmotic pressure of a solution of an undissociated solute (e.g., glucose solution) that is two times more concentrated.

Quiz 2. Show that $C_{MS}^* = (1+\alpha)C_M$.

Quiz 3. Compute the osmotic pressure of the standard solution of physiological saline solution at room temperature, $t = 25\,°\text{C}$.

4.3.2 Deviations from van't Hoff Laws

The *analogy* between solutions and gases *holds* even in the case of *real solutions and real gases*. For instance, one can adopt the *virial equation of state* from the theory of *real gases*, in order to describe the behavior of real solutions of macromolecules (e.g., proteins); in fact, there are several known virial equations of state, one of which is the following (Atkins and de Paula, 2002):

$$\frac{pV_M}{RT} = 1 + \frac{B}{V_M} + \frac{C}{V_M^2} + \ldots \tag{4.34}$$

where p is the gas pressure, V_M, the molar volume, R the gas constant, T, the absolute temperature, B and C constants. By using only the first terms of the right-hand side of equation (4.34), and using the notation $V_M = V/\nu$ (where $\nu = m_s/\mu$, and m_s is the solute mass) the virial equation of state for real macromolecular solutions is:

$$\pi V \approx \nu RT + \frac{\nu^2 RT B}{V} \qquad (4.35)$$

Dividing both terms of equation (4.35) by V, and taking into account that $m_S/V = C_s$ (i.e., mass percentile concentration), one can write:

$$\pi \approx RT\left(\frac{C_s}{\mu}\right) + RTB\left(\frac{C_s}{\mu}\right)^2 \qquad (4.36)$$

This last expression for the osmotic pressure of the real solutions can be used to estimate experimentally the *molecular mass* of a macromolecule by simply measuring the osmotic pressure (using an *osmometer*) of a series of solutions with different solute concentrations.

Quiz 4. Using $\pi/C_s \approx RT/\mu + aC_s$ (where $RTB/\mu^2 = constant = a$), show graphically how the molar mass, μ, of a macromolecule, can be determined by measuring the osmotic pressure of a series of solutions with different macromolecular concentrations (assume that the temperature is equal to the room temperature of 298.15 K).

4.3.3 Osmotic Pressure of Biological Liquids

The osmotic pressure of biological fluids (such as cytosol, blood, interstitial liquid, cerebrospinal liquid, synovial liquid, etc.) can be calculated according to "Dalton's law," i.e., π is equal to the sum of the *partial pressures* of the different components (ions, i, small molecules, m, and macromolecules, M):

$$\pi = \sum_{k=1}^{N_i} \pi_{ik} + \sum_{k=1}^{N_m} \pi_{mk} + \sum_{k=1}^{N_M} \pi_{Mk} \qquad (4.37)$$

where N_i, N_m, N_M represent the number of ionic (Na^+, K^+, Cl^-, HCO_3^-, etc.), micro-molecular and macromolecular species in solution. The contribution of macromolecules is generally much less than that of ions and micro-molecules. In the case of the blood plasma, for example, the total osmotic pressure of about 7.6 bar comes mostly from ions and micro molecules (representing about 1% of its mass), while the macromolecules (representing 9% of the plasma mass) are exerting an osmotic pressure of only 0.037 bar (≈ 28 mmHg).

The small pressure exerted by the macromolecules is called *colloid osmotic pressure*. The *colloid osmotic pressure* of the *interstitial liquid*, π_{IL} (i.e., the liquid bathing the tissue cells) is smaller than the plasma colloid osmotic pressure π_P.

Any tissue is irrigated by a complex network of capillary vessels through which blood flows, having practically the same colloid osmotic pressure as plasma. The wall of the blood capillaries (called *endothelial wall*) is impermeable to proteins but is permeable to water molecules, ions, and micro molecules. Therefore there is a difference between the colloid osmotic pressures across the capillary wall, called the *effective colloid osmotic pressure*, π_{ECO}, given by the relation:

$$\pi_{ECO} = \pi_P - \pi_{IL} \approx 22\,\text{mm Hg} > 0 \tag{4.38}$$

It is this "positive" effective colloid osmotic pressure that drives water molecules out of the interstitial space, diffusing through the capillary walls in order to equalize the global osmotic pressures of the two liquids.

4.3.4 The Cellular "Osmotic Pressure Menace"

The cell membrane is permeable to water molecules, ions and small molecules, but impermeable to macromolecules. It is interesting to note that the cell membrane is more permeable to water than to smaller ions, due to the existence of the specific *water channels*, called *aquaporins*, that prevent even protons from passing through (Lodish et al., 2004). Although the cell membrane is essential for proper functioning of the cell, it exposes the cell to a permanent danger: due to the selective transfer of water inside the cells, imposed by the difference in *the colloid osmotic pressure*, the cell is under a permanent threat to *swell* and to finally *burst*. This imminent "osmotic pressure menace" is fortunately actively avoided using an energy-consuming process of ion expulsion from cells, in order to maintain isotonicity of the cytosol. We shall see later that this active transport of ions is produced by certain molecular machines called *ionic pumps*, which use energy liberated from ATP hydrolysis.

> **Observation:** In plant cells as well as in bacteria, fungi, and algae, this danger of cellular burst, although present, is counteracted by a special cell morphology: their plasma membrane is surrounded by *rigid walls*, so that, even when suspended in *hypotonic solutions*, these cells maintain their volume, due to the mechanical pressure exerted by their walls. This mechanical pressure that pushes the plasma membrane towards the rigid wall is called *turgor pressure*. Moreover, plant cells are endowed with vacuoles which contain concentrated electrolyte and are able to "absorb" water from the cytosol through osmosis.

One can also speak of "osmotic stress" induced by addition of inert osmolyte (e.g., polyethylene glycol) to any aqueous system, which generates compacting forces on membranes and other interesting biochemical and biophysical perturbations, as shown by Cohen and Highsmith (1997) (see also: http://*aqueous.labs.brocku.ca:osfile.html* and http://www. *mgsl.dcrt.nih.gov/docs/osmdata/osmdata.html*).

4.4 Facilitated Transport

As discussed in section 4.2, small essential ions (e.g., Na^+, K^+ and Cl^-) and small molecules (e.g., glucose) are not easily dissolved into the lipid bilayer and therefore present very small permeability coefficients. However, in reality they do permeate the membrane very efficiently, which suggests that an alternative modality of permeation is at work. Indeed, the flow of small ions and molecules into or out of the cells is *facilitated* by the existence of a special type of *transporters*. This process of transport through the membrane is called *facilitated transport* or *facilitated diffusion* and occurs spontaneously *down* the particles concentration gradients. Similar to the simple (passive) diffusion, facilitated diffusion is an exergonic process and a generator of entropy (i.e., $\Delta G < 0$, $\Delta S > 0$). Unlike the simple diffusion, however, facilitated diffusion is aided by specific *channel proteins* or by specific *carrier proteins* inserted in the cell membrane, and it is only encountered in biological systems.

4.4.1 Channel-Mediated Transport

Protein channels allow selective passage of a specific ion or molecule (e.g., Na^+, K^+, Cl^-, Ca^{2+}, H_2O, etc.), and are essentially *two-state structures* whose opening or closing are activated *chemically, electrically, mechanically*, or by *light*. There are *endogenous* ion channels, which are secreted by the cells in whose membranes they reside, as well as *exogenous* channels, which are formed in the plasma membrane by insertion of molecules secreted by other cells. A common example is the secretion of *gramicidin A* channel by the bacteria *Bacillus brevis* (Fig. 4.7). The gramicidin channel is a dimer formed by association across the membrane of two β-*helix* polypeptide strands (i.e., parallel β-strands, folded into a large helix). These exogenous channels are non-gated and permanently open; they allow passage of Na ions at a very high rate of $10^7 \, Na^+/s$. Because of this, gramicidin channel inserted into the membranes of the target cells kills those cells by "erasing" their electrochemical gradients, which are vital for proper cell functioning. Therefore, gramicidin acts like an antibiotic, constituting a *chemical weapon* against the nutrient competitors of *Bacillus brevis*. The antibiotic activity of pore-forming molecules, such as gramicidin A, is exploited in pharmacology; several other pore-forming molecules (e.g., *nistatin, amfotericin B*) are used as *antifungal drugs*.

4.4.2 Carrier-Mediated Transport

Certain small molecules including sugars (glucose, fructose, mannose) and amino acids cannot diffuse through the cell membranes, because they are insoluble in the lipid matrix of the membrane, and can not be translocated through ion channels

4.4 Facilitated Transport

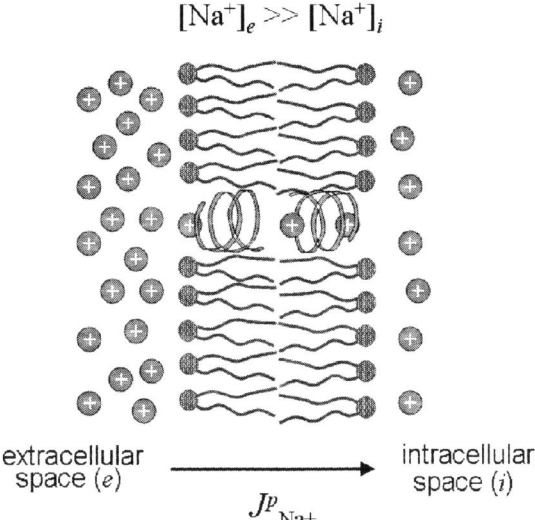

Fig. 4.7 The pore formed by two gramicidin molecules inserted into the membrane bilayer of the aggressed fungi cell. J_{Na}^p = passive flux of Na$^+$. ⊕ – sodium ions; Blue filled circles – polar heads of lipid molecules. Only Na ions are depicted, for the sake of simplicity.

either. These important molecules can still enter the cell through the plasma membrane via some specific carriers, called *uniporter* carriers. The process is driven by concentration gradients across the membrane, caused by consumption of the transported molecules in various metabolic reactions.

Observation: Although highly specific, carriers cannot function against concentration gradients.

As in the case of channels, there are *endogenous* as well as *exogenous* carriers. The latter type is secreted by certain bacteria and inserted into the membrane of a host cell. An example is *valinomycin*, an antibiotic secreted by *Streptomyces fulvissimus*. Due to its appreciable hydrophobicity, this carrier is able to insert itself into the membrane of target cells, inducing uncontrolled K$^+$ leakage from the cells. Therefore valinomycin, like gramicidin A, represents another natural *chemical weapon* in the struggle of cellular species for survival.

Unlike ionic channels, carriers undergo cyclic *conformational changes*, as they bind molecules from one side of the membrane and release them to the other side. This process of repeated binding–release of small molecules is time-consuming, which confers carrier-facilitated diffusion (10^2–10^4 ions/s) a diffusion rate which is several orders of magnitude lower than the channel-facilitated diffusion (10^7–10^8 ions/s).

A well-studied example of endogenous carrier is the *glucose carrier* of the erythrocyte (or red blood cell), which is called GLUT1 (Lodish et al., 2004), and has two distinct conformational states: one associated with the binding of glucose on the extracellular side of the membrane and another when the glucose is released on

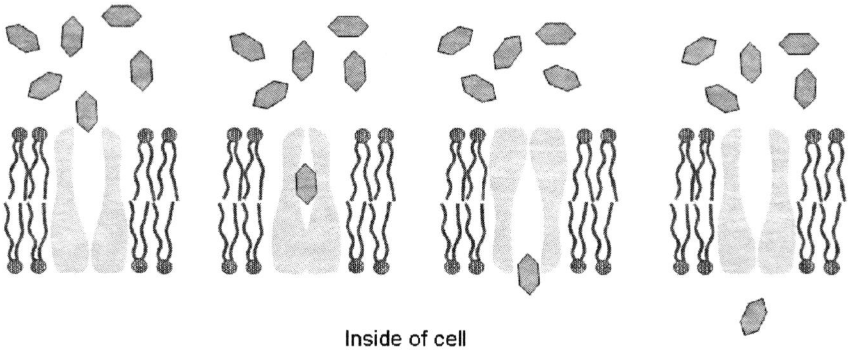

Outside of cell

Inside of cell

Fig. 4.8 Schematic representation of the mode of operation of a glucose carrier embedded in a phospholipid matrix. The hexagonal shapes stand for sugar molecules.

the intracellular side (Fig. 4.8). Although glucose is continuously transported across the membrane into the erythrocyte, it is not accumulated inside the cell because it is rapidly transformed into *glucose 6-phosphate* by the glucose metabolism. Therefore the GLUT1 carrier functions ceaselessly. It is interesting to note that there are two other competitors for GLUT1 binding site: D-galactose and D-mannose, which, however, bind to GLUT 1 with lesser affinities compared to glucose.

By analogy to enzymatic reactions, the process of facilitated diffusion of a sugar, S, mediated by a carrier, C, is described by:

$$S + C \underset{k_{-1}}{\overset{k_1}{\rightleftarrows}} SC \overset{k_{dis}}{\longrightarrow} S \quad (4.39)$$

where k_1 and k_{-1} are reaction rate coefficients for the external side of the membrane, and k_{dis} represents the rate constant of complex dissociation at the intracellular face of the membrane.

From the reaction (4.39) one can derive the rate of SC complex formation,

$$J_{C \to SC} = \frac{d[SC]}{dt} = k_1[S][C] \quad (4.40)$$

as well as the rate of complex dissociation,

$$J_{SC \to C} = \frac{d[SC]}{dt} = (k_{-1} + k_{dis})[SC] \quad (4.41)$$

where $[X]$ = molar concentration (measured in $\text{mol}/\text{dm}^3 = \text{M}$) of the X species.

In a steady-state, $J_{C \to SC} = J_{SC \to C}$ and one can write:

$$k_1[S][C] = (k_{-1} + k_{dis})[SC] \quad (4.42)$$

4.4 Facilitated Transport

or

$$\frac{[S][C]}{[SC]} = \frac{k_{-1}+k_{dis}}{k_1} = K_M \qquad (4.43)$$

where K_M represents the *Michaelis constant*.

The concentration, $[C]$, of the *free carrier* molecules is given by the obvious relation:

$$[C] = [C]_T - [SC] \qquad (4.44)$$

where $[C]_T$ represents the total concentration of the carrier molecules. Substituting $[C]$ from (4.44) into (4.43), and after rearranging, one obtains:

$$[SC] = [C]_T \frac{[S]}{[S]+K_M} \qquad (4.45)$$

From (4.45) it follows immediately that the rate of S release on the intracellular side of the membrane into the cell, or the *flux of facilitated diffusion* is given by:

$$J_{FD} \equiv \frac{d[S]}{dt} = k_{dis}[SC] = k_{dis}[C]_T \frac{[S]}{[S]+K_M} \text{(mol/s)} \qquad (4.46)$$

From the last expression it follows that: (i) if $[S] = 0$, then $J_{FD} = 0$; (ii) if $[S] \ll K_M$, then, J_{FD} is directly proportional to $[S]$; (iii) if $[S] \to \infty$, that is, when sugar is in excess, then $J_{FD} \to k_{dis}[C]_T = J_{FD}^{Max}$, in this case all the carriers being saturated with sugar molecules; (iv) finally, if $[S] = K_M$, then $J_{FD} = J_{FD}^{Max}/2$. Therefore, K_M represents that sugar concentration for which the flux of the facilitated diffusion is *half of the maximum flux*.

The flux dependence on the concentration of solute (sugar, in this case) is given comparatively in Fig. 4.9 for carrier-facilitated diffusion and simple diffusion.

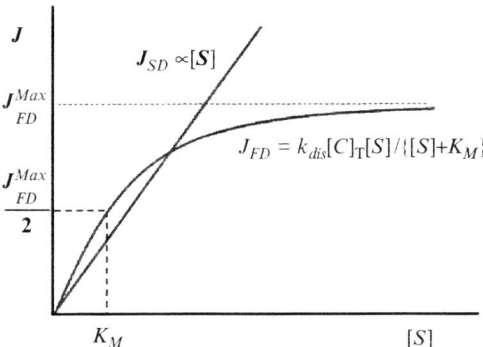

Fig. 4.9 Qualitative dependence of the simple diffusion flux (J_{SD}) and the facilitated diffusion flux (J_{FD}) on the concentration $[S]$ of the transported molecules.

4.4.3 Main Characteristics of Facilitated Transport

The two types of facilitated transport described above present the following features:

1. They are both *performed passively*, without energy consumption and driven by a concentration and/or electrical gradient
2. The *rate* of facilitated diffusion is *much higher* than the simple diffusion through lipid bilayers for a given substance
3. Both processes are *highly specific*, in the sense that a certain type of molecule can be translocated only by a given type of channel or carrier
4. Their rate of transport presents *saturation* as the small molecule concentration is increased, due the limited number of channels and/or carriers, within the cell membrane
5. Both types present *competition*, that is, the presence of other types of molecules capable to bind to the channel or carrier decreases the transport rate of the molecule of interest
6. They may be blocked by *inhibitors* (for instance, the Na^+ channels of the axon membranes can be blocked by the toxins *tetrodotoxin*, TTX, and *saxitoxin*, STX)

   Observation: Under certain conditions, the saturation plateau can be increased or decreased by inserting transporters into or eliminating them from the membranes, as it is done for instance in the treatment of certain diseases.

4.5 Active Ion Transport

Living cells maintain a huge ion concentration disparity between their intracellular and extracellular faces of the membrane. As a general though not absolute rule, the concentration of Na^+, Cl^- and Ca^{2+} is much higher *outside* the cells, while the K^+ concentration is much higher *inside* the cells. Passive diffusion processes discussed above have the tendency to equalize this strongly antientropic ion distribution. However, with some minor fluctuations, this asymmetry is actively maintained by the cell, this being a *sine qua non* condition for cellular viability.

There are specific biological mechanisms by which ions are transported against their electrical and chemical gradients (collectively termed *electrochemical gradients*), thereby ensuring ionic concentration asymmetry between both sides of the cell membrane; these mechanisms are called *active transport*. This type of solute transport (Fig. 4.10) against concentration gradients is performed with consumption of cellular energy (i.e., $\Delta G > 0$ for the solute) and generation of entropy ($\Delta S < 0$ for the solute). Most cells are using up to one third of their ATP reserve in order to maintain active transport, while *excitable cells* such as neurons, which will be discussed in chapter 7, consume even more – approximately *three quarters* of their ATP reserve. This kind of active transport is usually termed *primary* active transport; there is also a type of transport called *secondary* active transport, which is driven by concentration gradients of a solute that has been created by primary active transport.

4.5 Active Ion Transport

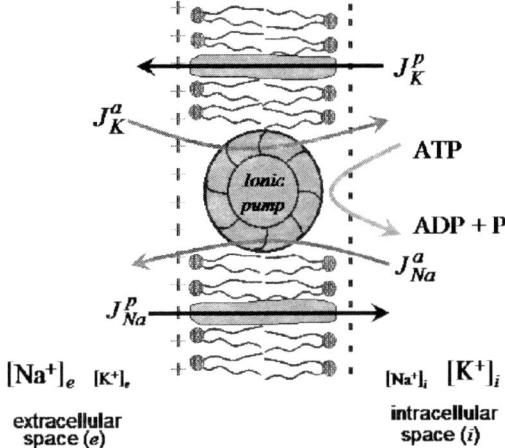

Fig. 4.10 Schematic representation of a membrane ionic pump and its function. Significance of symbols: J^a_{Na}, J^a_K, J^p_{Na}, and J^p_K, active (superscript a) and passive (superscript p) fluxes of Na$^+$ and K$^+$ through protein specific channels, depicted in green; [Na$^+$] and [K$^+$], ion concentrations (the different font sizes signify that the concentration is comparatively large or small). Purple filled circles are the polar heads of the membrane lipid bilayers. Electric polarization of the membrane is produced by net positive (outside the cell) and negative (inside the cell) charges.

The main characteristics of the primary active transport can be summarized as follows:

1. It is a highly endergonic process ($\Delta G > 0$)
2. It is absent from dead cells
3. It depends on temperature to a much higher degree than it does simple passive diffusion

   Observation: In order to characterize the influence of the temperature on the rate of a process (physical, chemical, biological), occasionally one uses the so-called *temperature quotient*, Q_{10}, given by:
   $$Q_{10} = J_{t+10}/J_t \qquad (4.47)$$
   where J_t is the rate of the process at a temperature, t, while J_{t+10} is its rate at $t + 10\,°C$. Interestingly, for most physical and chemical processes, this quotient varies over the interval [2, 3], while for biological processes, $Q_{10} > 3$.

4. It may be indirectly diminished by certain *metabolic inhibitors* (such as cyanides, azides, dinitrophenols) and by hypoxia (i.e., reduced oxygen supply), which impede ATP synthesis
5. It may be blocked by specific inhibitors including *cardiotonic steroids* (e.g., ouabain and digitoxigenin) which have a high affinity for ionic pumps (Berg et al., 2002)

   Observation: It has been known for a very long time that the cardiotonic steroids ouabain and digitoxigenin (Stryer, 1988) exert pharmacological effects on the heart by increasing the contraction force of the cardiac muscle. *Cardiotonic steroids* are used clinically in the treatment of congestive heart failure.

Although the specific biological process responsible for actively "pumping" Na^+ into and K^+ out of the erythrocyte ghosts was first observed long time ago (Gardos cited by Stein, 1990), detailed knowledge of the "machines" (called later *ionic pumps*) performing such functions, in all the living cells, is still being accumulated.

The study of structure and mechanisms of action of ionic pumps continues to be an active area of research to these days, as will be discussed in more details in chapter 7.

References

Alberts, B., Johnson, A., Lewis, J., Raff, M., Roberts, K. and Walter, P. (2002) *Molecular Biology of the Cell*, 4th ed., Garland Science/Taylor & Francis, New York

Atkins, P. W. and de Paula, J. (2002) *Atkins' Physical Chemistry*, 7th ed., Oxford University Press, New York

Berg, J. M., Tymoczo, J. L. and Stryer, L. (2002) *Biochemistry*, 5th ed., W. H. Freeman, New York

Cohen, J. A. and Highsmith, S. (1997) An improved fit to website osmotic pressure data, *Biophys. J.*, **73**: 1689–1694

Davson, H. and Danielli, J. F. (1970) *The Permeability of Natural Membranes*, Hafner, Darien, CT

Glaser, R. (2001) *Biophysics*, Springer, Berlin/Heidelberg/New York

Kepner, G. R. (ed.) (1979) *Benchmark Papers in Human Physiology, Cell Membrane Permeability and Transport*, Dowden, Hutchinson & Ross, Stroudsburg, PA

Lodish, H., Berk, A., Matsudaira, P., Kaiser, C. K., Krieger, M., Scott, M. P., Zipursky, S. L. and Darnell, J. (2004) *Molecular Cell Biology*, 5th ed., W. H. Freeman, New York

Stein, W. D. (1990) *Channels, Carriers, and Pumps: An Introduction to Membrane Transport*, Academic/Harcourt Brace Jovanovich, San Diego, CA/New York/Boston/London/Sydney/Tokyo/Toronto

Stryer, L. (1988) *Biochemistry*, 3rd ed., W. H. Freeman, New York

van't Hoff, J. H. (1887) The role of osmotic pressure in the analogy between solutions and gases, *Z. Physik. Chemie*, **1**: 481

Chapter 5
Reaction, Diffusion and Dimensionality

As we have seen so far, biological cells are complex systems containing many different molecular species that interact with one another to form molecular complexes or entirely different molecular species. Biomolecular interactions may be conveniently described as chemical reactions, and, in fact, the cell itself can be regarded as a complex biochemical reactor, in which many reactions occur simultaneously. Some examples have already been introduced in previous chapters (see, e.g., the self-association of amphiphiles into micelles and membranes), with others yet to follow.

In the next section, we will lay down the classical framework for describing reaction kinetics. We will first consider that biochemical reactions take place in an aqueous solution (e.g., the cell cytosol), assumed to be homogenous, and that the chemical reaction of interest does not interfere with others taking place simultaneously in the same cellular volume. Many of these approximations do seem to break down under most circumstances in biological cells. In the second part of this chapter, therefore, we will relax some of the approximations, and will make use of fractal concepts to incorporate deviations of biological systems from the Euclidian geometry of smooth objects, which may impinge on the reaction kinetics inside the cell.

5.1 Equilibrium and the Law of Mass Action

5.1.1 Molecular Association

Let us consider the simplest (bio)chemical reaction between two reactants: a macromolecule, M (e.g., an *enzyme*), and a partner of interaction, hereafter called *ligand*, L (e.g., specific *substrate*). Due to permanent thermal agitation, reactant molecules are subject to random collisions, some of which may result in formation of molecular complexes, ML; this is an antientropic process. On the other hand, ML complexes may collide with individual reactant molecules, other ML complexes, and solvent (water) molecules, which may lead to dissociation into M and L – an entropy-generating process.

Considering the simplest process, with a 1:1:1 stoichiometry (i.e., one of each, M, L, and ML), the complex formation or dissociation may be described by the following simple reaction:

$$\text{M} + \text{L} \underset{k_{dis}}{\overset{k_{as}}{\rightleftarrows}} \text{ML} \quad (5.1)$$

where k_{as} and k_{dis} are the rate constants of association and, respectively, dissociation.

To a first approximation, the rate of complex formation (i.e., the scalar flux J_{as}) is proportional to the reactants concentrations, namely:

$$J_{as} = \frac{d[ML]}{dt} = k_{as}[M][L] \quad (5.2)$$

where $[X]$ represents the molar concentration of the X species ($X = M, L, ML$). The rate of complex dissociation (i.e., the scalar flux, J_{dis}) is proportional to the concentration of complexes, as expressed by the equation:

$$J_{dis} = -\frac{d[ML]}{dt} = k_{dis}[ML] \quad (5.3)$$

If the system is isolated, certain physical and chemical parameters (e.g., T, total reactants concentrations, pH, etc.) are constant, and a *chemical equilibrium* will be reached between all the chemical species after a short transitory period. At equilibrium, the net flux is zero, i.e., the reaction does not advance in any direction, and the rate of complex formation thus equals the rate of complex dissociation. This gives:

$$K_{eq} \equiv k_{as}/k_{dis} = [ML]/\{[M][L]\} \quad (5.4)$$

where K_{eq} is called the *equilibrium constant* and has units of inverse of concentration.

The last relation is one of the mathematical expressions of the *Law of Mass Action* (LMA), which states that, at equilibrium, the ratio between the complex concentration and the product of the reactants concentrations is constant in time, if the total concentrations of M and L are constant. Note that we have already used LMA in deriving equation (1.48) in chapter 1, without getting into many details.

Observation: In the general case of $m : l$ stoichiometry (with m, l natural numbers), i.e.,

$$m\text{M} + l\text{L} \underset{k_{dis}}{\overset{k_{as}}{\rightleftarrows}} \text{M}_m\text{L}_l \quad (5.5)$$

the LMA gives the following relation for the equilibrium constant:

$$K_{eq} = [M_mL_l]/\left\{[M]^m[L]^l\right\} \quad (5.6)$$

5.1 Equilibrium and the Law of Mass Action

Quiz 1. Compute the equilibrium constant of the interaction of a macromolecule M($[M(t = 0)] = 100\,\mu M$) with the ligand L($[L(t = 0)] = 200\,\mu M$), assuming that M presents two identical independent binding sites. Assume that $[ML(t \to \infty)] = 10, 50, 90, 99, 100\,\mu M$. Compare the results to the case of 1:1 stoichiometry.

The equilibrium constant is often called *association* or *affinity* constant, K_a. A higher association constant means stronger binding, or higher *affinity* of the reactants for one another. More commonly, one uses the dissociation constant, which is equal to the inverse of the association constant (i.e., $K_d = 1/K_a = 1/K_{eq}$), which has units of concentration. In this way, one may easily assess whether binding or dissociation dominate in the system, depending on whether the concentrations are above or below K_d.

Depending on the specific properties of M and L, the association constant for specific interactions, can vary over many orders of magnitude (usually, from $\sim 10^4$ to 10^{10}). For example, the Xeroderma pigmentosum group C protein binds with a K_a of $\sim 10^8 \, M^{-1}$ to human centrin 2 (a calcium-binding protein) (Popescu et al., 2003), while the highly specific interaction between some IgG *antibodies* (immunoglobulin G) and their *antigen* (Decker, 2007) is characterized by an association constant of $10^9 \, M^{-1}$.

The association and dissociation constants of a (bio)chemical reaction connect the formalism described above to the thermodynamics underlying the process through the well-known expression of the Gibbs free energy variation associated with the binding process:

$$\Delta G = -RT \ln K_a = +RT \ln K_d \quad (5.7)$$

For reactions with high affinity ($K_a \gg 1$), equation (5.7) predicts a large negative change in the Gibbs free energy, the process being *exergonic* and therefore taking place spontaneously. On the contrary, in the case of reactants with very low affinity constant ($K_a \ll 1$), a positive variation of the Gibbs free energy results, the process being *endergonic* and requiring external energy in order to take place. In biological systems, an endergonic process is always coupled to some exergonic process, which supplies the necessary energy. For example, in the biosynthesis of proteins, nucleic acids, etc., energy is supplied by ATP, which is dissociated into ADP and P, in order to produce energy.

5.1.2 Determination of Affinity Constant by Equilibrium Dialysis

Dialysis is a physical process of diffusion through a selective permeable membrane of ions and small molecules, driven by diffusion and osmosis (see chapter 4). It is used by chemists and bio-chemists for purifying proteins and polymers and by medical doctors for removing organic waste products (such as urea, a waste product from protein metabolism) from the blood in case of kidney failure. In this section,

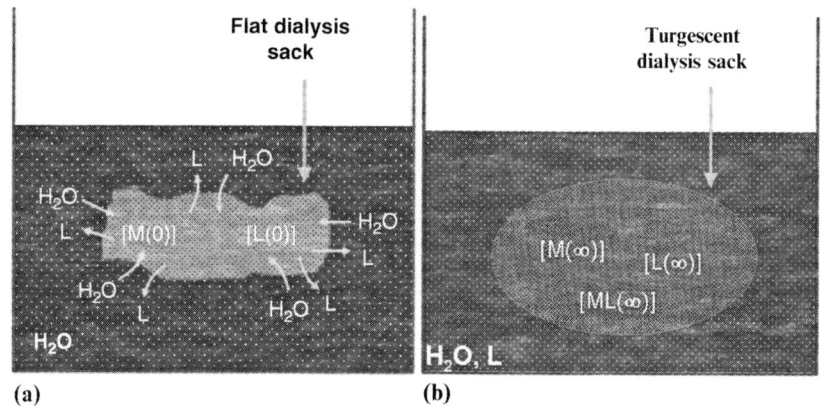

Fig. 5.1 The process of dialysis in (**a**) the initial state and (**b**) final state. [X(0)] is the initial concentration while [X(∞)] is the final concentration ($t \to \infty$) of species X.

we will show how equilibrium dialysis can be used in quantitative studies of the binding of a macromolecule to its ligand.

Let us consider that a species of macromolecules, M, and its ligand, L, are dissolved in water and placed into a dialysis sack, which is made of a semi-permeable membrane that allows diffusion of L and water molecules but not of M (Fig. 5.1). For simplicity, we assume that binding of L to M inside the sack is governed by equation (5.1), i.e., that there is only one binding site per macromolecule. The initial concentrations (at $t = 0$) of macromolecule and ligand are known and are denoted by [M(0)] and [L(0)], respectively.

Because L diffuses freely through the membrane, at equilibrium (which is attained theoretically for $t \to \infty$) its concentration inside the sack, $[L(\infty)]_{in}$, equals the one outside the sack, $[L(\infty)]_{out}$. Since $[L(\infty)]_{out}$ may be determined experimentally at equilibrium, as for example from spectrophotometric measurements, $[L(\infty)]_{in}$ is also known. In addition, the concentrations of bound macromolecules, bound ligands, and complexes, [ML], are equal to one another, for any time t. Therefore, the concentration of free macromolecules in the sack at equilibrium is given by:

$$[M(\infty)] = [M(0)] - [ML(\infty)] \tag{5.8}$$

where the concentration of ML complexes is:

$$[ML(\infty)] = [L(0)] - [L(\infty)] \tag{5.9}$$

To express the amount of liganded (i.e., bound) macromolecules relative to the total amount of macromolecules, one may employ the *fractional saturation*, or the fraction of bound (liganded) macromolecule, as expressed by:

$$f \equiv \frac{[ML(\infty)]}{[M(0)]} = \frac{[L(0)] - [L(\infty)]}{[M(0)]} \tag{5.10}$$

5.1 Equilibrium and the Law of Mass Action

The second equality in the above equation may be used for experimental determination of f, while the first equality may be used to relate f to the affinity constant by using equations (5.8) and (5.4) to obtain:

$$f = \frac{[ML(\infty)]}{[M(\infty)] + [ML(\infty)]} = \frac{K_a[L(\infty)]}{1 + K_a[L(\infty)]} \tag{5.11}$$

Rearranging equation (5.11), the following simple linear relationship is obtained:

$$f/[L(\infty)] = K_a - K_a f \tag{5.12}$$

By performing a series of dialysis experiments for various initial concentrations of ligand, and plotting the $f/[L(\infty)]$ determined experimentally against the measured f, one obtains a *Scatchard plot*, from which the affinity constant is obtained as the slope of the line.

If, instead of one binding site per molecule, the macromolecule, M, possesses n independent sites to which the ligand, L, binds with the same affinity, equation (5.10) needs to be modified by replacing the concentration M(0) with $n[M(0)]$:

$$f' = [ML(\infty)]/\{n[M(0)]\} = f/n \tag{5.10'}$$

By taking this into account and if the affinity constant is the same for all sites, equation (5.12) is replaced by (Atkins and de Paula, 2002):

$$f/[L(\infty)] = nK_a - K_a f \tag{5.12'}$$

As in the case of a single binding site, by performing dialysis experiments for various initial concentrations of ligand, a Scatchard plot may be obtained from which the affinity constant is extracted as the slope of the line. Once K_a has been determined, one may compute the number of binding sites for a given macromolecule by extrapolating the straight line to $f = 0$.

5.1.3 Competitive Binding

If a ligand, L, competes with another ligand, I, for the same binding site of a macromolecule, M, the ligand I inhibits the binding of the ligand L to M. Two separate reactions, analogous to equation (5.1), may be written for this situation, as:

$$M + L \underset{k_{disL}}{\overset{k_{asL}}{\rightleftarrows}} ML \tag{5.13a}$$

$$M + I \underset{k_{disI}}{\overset{k_{asI}}{\rightleftarrows}} MI \tag{5.13b}$$

which are characterized by the following equilibrium constants:

$$K_{aL} \equiv k_{asL}/k_{disL} = [ML]/\{[M][L]\} \tag{5.14a}$$

$$K_{aI} \equiv k_{asI}/k_{disI} = [MI]/\{[M][I]\} \tag{5.14b}$$

In determining the fractional saturation for the ligand, L, we now also need to take into account the concentration of macromolecules ligated by I. In this case, the fraction of macromolecules liganded by L is obtained immediately by making use of equations (5.14):

$$f = \frac{[ML(\infty)]}{[M(\infty)] + [ML(\infty)] + [MI(\infty)]} = \frac{K_{aL}[L(\infty)]}{1 + K_{aL}[L(\infty)] + K_{aI}[I(\infty)]} \tag{5.15}$$

Rearranging (5.15), we obtain:

$$f/[L(\infty)] = K_{aL} - K_{aL}\,f - K_{aI}\frac{[I(\infty)]}{[L(\infty)]} \tag{5.16}$$

which reduces to equation (5.12) in the absence of the inhibitor, I (as it should). Obviously, it is possible to write a similar equation for the inhibitor. One should also notice that although our choice to call one of the ligands "inhibitor" may appear arbitrary, it has an analog in biological processes, for instance in the case of binding of CO to the heme of myoglobin, which inhibits the physiologically normal function of myoglobin to bind O_2 or CO_2 (with potentially lethal consequences for the human body).

5.1.4 Allosteric Activation and Inhibition of Binding

Another biologically relevant type of molecular binding occurs when, for a macromolecule M, binding of ligand I to one site changes (increases or decreases) the affinity for binding L to another site on the same molecule. This is called *allosteric binding* or simply *allostery*; this word originates from the Greek words *allos* (meaning "other") and *stereos* (meaning "space"), and signifies changes induced in binding affinity of a protein as a result of binding of a ligand to a regulatory site. The concept has been introduced originally to account for changes in the catalytic activity of an enzyme due to conformational changes induced by binding of certain ligands called *effectors*. When the first binding event stimulates the next event, the effect is called *positive allostery*, while the effect of reducing the binding affinity for subsequent binding events is called *negative allostery*. *Allostery* is tightly connected to another process called *cooperativity*, which occurs when ligands bind to multiple binding sites on M (Helmstaedt et al., 2001).

Allosteric binding may be formally described by a number of *sequential* reactions equal to the number of binding subunits of M. In this model, due to Koshland, Nemethy and Filmer (Koshland et al., 1966) and also known as the *induced fit* model, ligand binding to one subunit leads to conformational changes of other

5.1 Equilibrium and the Law of Mass Action

subunits, which modify their binding affinity. For $n = 2$, the sequential model describes two serial reactions,

$$M + 2L \underset{k_{dis1}}{\overset{k_{as1}}{\rightleftarrows}} ML + L \underset{k_{dis2}}{\overset{k_{as2}}{\rightleftarrows}} ML_2 \qquad (5.17)$$

which are characterized by separate affinity constants:

$$K_{a1} \equiv k_{as1}/k_{dis1} = [ML][L]/\{[M][L]^2\} = [ML]/\{[M][L]\} \qquad (5.18a)$$
$$K_{a2} \equiv k_{as2}/k_{dis2} = [ML_2]/\{[ML][L]\} \qquad (5.18b)$$

Using equation (5.17), the fractional saturation of ML_2, at equilibrium may be written as:

$$f = \frac{[ML_2(\infty)]}{[M(\infty)] + [ML(\infty)] + [ML_2(\infty)]} = \frac{K_{a1}K_{a2}[L(\infty)]^2}{1 + K_{a1}[L(\infty)] + K_{a1}K_{a2}[L(\infty)]^2} \qquad (5.19)$$

which depends on the second power of the concentration of free ligand, and reduces to equation (5.11) when the second and the third term in the denominator are much greater than one. Note that the *fractional saturation* of ML_2, as defined by equation (5.19) differs from the fractional binding of ligand, which is also used in the literature on the subject.

Quiz 2. Derive the following equation for the fractional saturation in the case of sequential binding to a macromolecule with an arbitrary number of sites, n:

$$f = \frac{[ML_n]}{\text{Total}[M]} = \frac{K_{a1}K_{a2}\ldots K_{an}[L]^n}{1 + K_{a1}[L] + K_{a1}K_{a2}[L]^2 + K_{a1}K_{a2}\ldots K_{an}[L]^n} \qquad (5.19')$$

Depending on the relation between the affinity constants of the two reactions, three different cases may be distinguished for cooperative binding, as follows:

- *Positive cooperativity*, for $K_{a1} \ll K_{a2}$ – in this case, binding of the first ligand makes it easier for the second ligand to bind
- *No cooperativity*, for $K_{a1} = K_{a2}$ – binding of the first ligand does not affect the binding affinity of M
- *Negative cooperativity*, for $K_{a1} \gg K_{a2}$ – binding of the first ligand makes it more difficult for the second ligand to bind

These cases are illustrated by Fig. 5.2, in which the fractional saturation as a function of the free monomer concentration is shown comparatively for different types of binding.

Observation: While allosteric binding is explained in some cases of biological interest (e.g., for the Na-K pump) by the sequential model, some other cases (e.g., G-protein-coupled receptors in the membrane) are better explained by the *independent binding* model of Monod, Wyman and Changeux, MWC (Monod et al., 1965). The MWC model basically assumes that the n binding sites are all available to the ligand to bind, and that ligand binding to any one of the sites increases (or decreases) the affinity of the remaining sites. A comparative discussion of the sequential and the independent binding schemes is presented in a paper by Weiss (Weiss, 1997).

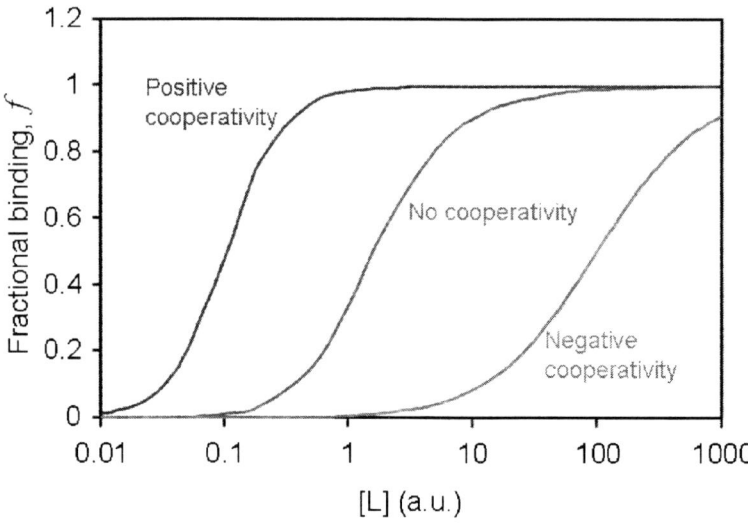

Fig. 5.2 Fractional saturation vs. concentration of free monomers in the case of positive cooperativity for $K_{a1} = 1$, and $K_{a2} = 100$ (in arbitrary units, a.u.), no cooperativity for $K_{a1} = K_{a2} = 1$ (a.u.), and negative cooperativity for $K_{a1} = 100$, and $K_{a2} = 1$ (a.u.). Note that the slope of positive cooperativity is steeper than the one for negative cooperativity.

If the binding of the first ligand increases the binding affinity of adjoining sites very strongly (i.e., $K_{a2} \gg K_{a1}$), at equilibrium most macromolecules will be either totally unliganded or totally liganded, in which case a global reaction may be written as:

$$M + 2L \underset{k_{dis}}{\overset{k_{as}}{\rightleftarrows}} ML_2 \tag{5.17'}$$

The affinity constant in this case is:

$$K_a = [ML_2]/\{[M][L]^2\} \tag{5.20}$$

which leads to the following expression for the fractional saturation:

$$f = \frac{[ML_2(\infty)]}{[M(\infty)] + [ML_2(\infty)]} = \frac{K_a[L(\infty)]^2}{1 + K_a[L(\infty)]^2} \tag{5.21}$$

In the general case of arbitrary n, the fractional saturation becomes:

$$f = \frac{[ML_n(\infty)]}{[M(\infty)] + [ML_n(\infty)]} = \frac{K_a[L(\infty)]^n}{1 + K_a[L(\infty)]^n} \tag{5.22}$$

which is known as the *Hill equation*, and n is the *Hill coefficient*. This equation has been first used by Hill (1910) to explain the sigmoidal binding curve of O_2 to hemoglobin, which is due to (positive) *cooperativity*.

In a very concisely written paper, Weiss (1997) warns of the potential dangers and pitfalls in using Hill equation for data fitting. In particular, one needs to be aware that the Hill coefficient is only equal to the number of binding sites in the macromolecule in the very particular, and physically unlikely, case of strong cooperativity, in which ligands bind simultaneously to all binding sites. In the case of hemoglobin, for instance, the Hill coefficient ranges between 1.7 and 3.2, which is significantly different from the number of available binding sites ($n = 4$). In fact, the Hill coefficient is usually considered only to provide a lower limit for the number of binding sites. For the above as well as other reasons, Weiss proposes that one avoids altogether the use of Hill equation in data fitting, since Hill's choice of this simplified equation (compared to, e.g., a sequential binding scheme) rests on it being conveniently handled in the process of data analysis in an era without computers.

Quiz 3. Name an advantage conferred by the fact that binding of oxygen to hemoglobin presents cooperativity.

To conclude section 5.1, we would like to mention that, while our attempt to introduce the reader to the rich problem of ligand binding has been limited to a few simple binding schemes, we hope that it has given the reader a glimpse of the nature and scope of this kind of studies. A very accessible presentation of the MWC model and some other binding schemes is presented in a recent book by Jackson (2006), while a very rigorous and comprehensive treatment of a wide array of ligand-binding problems, including numerical methods and error analysis, is given in an extensive review by Wells (1992).

5.2 Introduction to Fractals

Standard textbooks on Euclidian geometry teach that a point has dimension zero, a line has dimension one, a square has dimension two, a cube has dimension three, and so on. Even though our ability to picture a multidimensional object ends at three, special relativity added a fourth dimension – the time – while some modern physical theories even toy with the idea of more than 10 spatial dimensions. These examples constitute a particular direction of departure of science from our intuitive understanding of the geometry of Nature.

Many revolutions in science have started by redefining the process of measurement of space and/or time or redefining the geometry altogether. Most modern theories of physics reinterpret the concept of measurement or geometry. Relatively recent mathematical work, notably, that of Mandelbrot, started a new revolution in thinking by redefining the concept of dimension (Mandelbrot, 1999). Mandelbrot's ideas, which constitute another departure from the classical path and will be discussed in this chapter, stretch our imagination in a different direction: fractional number of dimensions.

5.2.1 "...The Measure of Everything"

Measurements of lengths have been usually expressed throughout human history in units that are of the same order of magnitude as the human size (foot, yard, meter, etc.). Not surprisingly, given the fact that spatial dimensions have been among the very first to be measured by ancient humans, which could not avail themselves of sophisticated tools beyond their immediate body parts, sticks, etc. Technological progress has brought the possibility of measuring very small dimensions, after magnifying them with the help of optical and other types of microscopes, so that the number of macroscopic yardsticks that fit within the magnified images can be counted. Once this number is found and divided by the magnification factor, the real size of a magnified object may be expressed as a fraction of a yardstick, i.e., as submultiples of a fundamental unit (the meter, in SI). We shall illustrate next this process of measurement.

5.2.1.1 Euclidian Versus Fractal Objects

Let us consider a square with sides a_0, perimeter $P_0 = 4a_0$, and area $A_0 = a_0^2$, which is magnified by a scale factor of, let us say, 3, to reveal its microscopic features. This measurement procedure is shown in Fig. 5.3, together with the perimeter and area after each magnification step. The general formulae for the side, perimeter, and area of the object obtained after n steps of $\times 3$ magnification are $a_n = 3^n a_0$, $P_n = 3^n P_0$, and $A_n = 3^{2n} A_0$, respectively. A relationship between a_n and A_n may be derived by taking the logarithm of each expression,

$$\ln a_n = n \ln 3 + {}^1\!/_2 \ln A_0 \qquad (5.23)$$

$$\ln A_n = 2n \ln 3 + \ln A_0 \qquad (5.24)$$

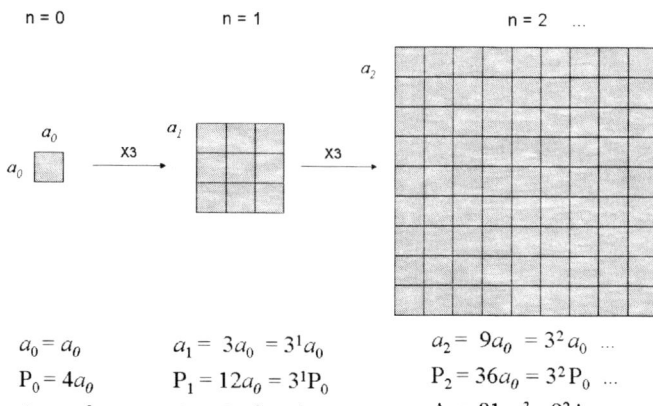

Fig. 5.3 Determination of the size of microscopic features of a simple structure by magnification and counting of the length and area of the features.

5.2 Introduction to Fractals

and then eliminating n to obtain:

$$a_n = A_n^{1/2} \tag{5.25}$$

Analysis of equation (5.25) reveals that the exponent is equal to the inverse of the dimension of the elementary square (which is equal to two). If instead of length and area we considered the surface area of a cube and its volume, a similar relationship would have been obtained, in which the exponent equals the inverse of the dimension of the cube, which is three. Further, if we used different geometrical shapes, e.g., circle or sphere instead of square and cube, similar power laws would have been obtained. All these observations lead to the conclusion that the denominator of the exponent in the power law relating the linear parameter (or scale) to the n-dimensional manifold represents the dimension of the space considered. This idea can be generalized further, as we will show momentarily.

Quiz 4. Derive the general relationship between surface area and volume of a sphere of radius r_0 after n steps of magnification.

An important observation has to be made at this point: the uniform coverage of the above structure with "microscopic" elements (i.e., squares) would have been very likely the product of human activity. This is because humans have the capacity to analyze and correct "imperfections" caused by missing squares. Now, suppose that such a structure is created following a more natural process: small square rocks are dropped one at a time in a rapid stream of water that drags the rocks slowly downstream. If some obstacle blocks the "flow" of the rocks but not the water (which accumulates and then rises above the barrier level), the squares will accumulate by bumping into one another, to cover more or less randomly the surface at the bottom of the stream of water. One would expect certain defects in the area coverage, as represented by the missing squares in the structure shown in Fig. 5.4.

Note that, although 5.2.2 is still an idealized picture of a natural process (no natural process relies on perfect squares of uniform size), it generalizes the geometry in the desired direction to capture the idea of structural "defects."

The general formulae for the side length and perimeter of the object in Fig. 5.4 obtained after n steps of $\times 3$ magnification are identical to those derived for Fig. 5.3, i.e., $a_n = 3^n a_0$ and $P_n = 3^n P_0$, while the formula for the area is $A_n = 8^n A_0$. A relationship between side length and area may be derived as above, by taking the logarithm of a_n and A_n,

$$\ln a_n = n\ln 3 + \ln a_0 \tag{5.26}$$

$$\ln A_n = n\ln 8 + \ln A_0 \tag{5.27}$$

and then eliminating n. The result is:

$$a_n = a_0^{1-2\ln 3/\ln 8} A_n^{\ln 3/\ln 8} \tag{5.28}$$

which, simplifies to

$$a_n = A_n^{\ln 3/\ln 8} \tag{5.28'}$$

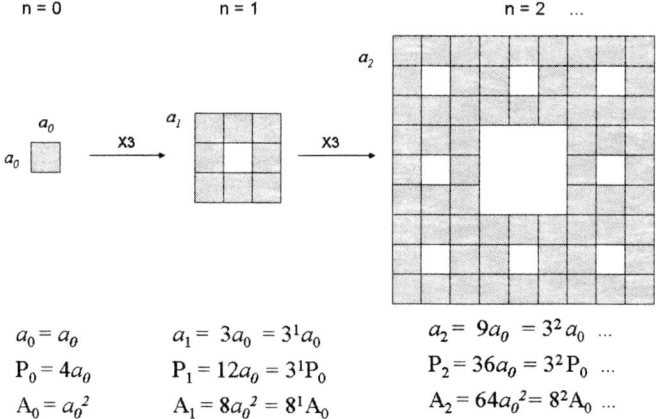

Fig. 5.4 Measuring the perimeter and area of a structure with defects – *Sierpinski's carpet*. Note that defects (i.e., missing squares) are revealed after magnification of what it seems to be a square area without defects.

if we take a_0 as our "yardstick," i.e., the unit of length. This choice may be justified by the fact that our method of measurement of the size of microscopic features magnifies the features so that they can be compared to the size of the initial square of side a_0, which is of an order of magnitude that can be detected by our senses. In other words, the act of measurement is inevitably related to the sensitivity of the human senses (vision, in this case), which work best for distances comparable to our own size.

Now, returning to equation (5.28′) and applying our above interpretation of the exponent of A_n, the dimension of the structure in Fig. 5.4,

$$D_f = \frac{\ln 8}{\ln 3} = 1.893 < 2,$$

is found, which represents a clear departure from the dimension of the Euclidian objects within which the structure is embedded. This structure is known as the *Sierpinski carpet* (Sierpinski, 1916; Mandelbrot, 1999, p. 144) and, in addition to taking non-integer values, it has the remarkable property that its area tends to zero for $n \to \infty$, while its perimeter remains finite and equal to P_0. To see this, we first need to scale back down the perimeter and the area of the nth magnification step, by dividing the length of the magnified image by 3^n. The two parameters thus obtained, $P'_n = 3^n 4a_n/3^n = P_0$ and $A'_n = 8^n(a_n/3^n)^2 = (8/9)^n a_0^2$, respectively, lead immediately to the results stated above for $n \to \infty$.

In Euclidian geometry, something that has zero area should be a curve, which Sierpinski's carpet indeed is; however, unlike lines in Euclidian geometry, our line has a dimension significantly greater than 1, although not quite equal to 2. Structures, such as Sierpinski's carpet, which present fractional dimension, are called *fractal structures* or simply *fractals*, while D_f is called the *fractal dimension*.

5.2 Introduction to Fractals

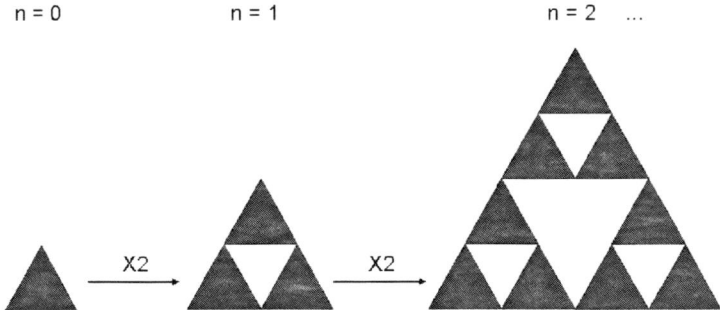

Fig. 5.5 Sierpinski's gasket, referred to in Quiz 5.

Quiz 5. Using the procedure described above, compute the fractal dimension of *Sierpinki's gasket* shown in Fig. 5.5 (Mandelbrot, 1999). Compare it to the fractal dimension of the Euclidian gasket.

We will see in the next sections that Nature prefers to use fractal geometry instead of Euclidian geometry. This is not accidental, as we will discuss in section 5.2.3.

5.2.1.2 General Expression for Estimation of the Fractal Dimension

The length, perimeter and the area of a fractal structure after n magnification steps may be expressed more generally as:

$$a_n = s^n a_0 \tag{5.29}$$
$$P_n = s^n P_0 \tag{5.30}$$
$$A_n = r^n A_0 \tag{5.31}$$

where s is the *linear scaling* (or magnification) *factor*, and r is the *scaling factor for the high-dimensional manifold* (the area, for Sierpinski's carpet). By taking the logarithm of equations (5.29) and (5.31), and substituting for n between them, the following relation is obtained between the length and area:

$$a_n = a_0^{1-2\ln s/\ln r} A_n^{\ln s/\ln r} \tag{5.32}$$

which, simplifies to:

$$a_n = A_n^{\ln s/\ln r} \tag{5.32'}$$

if we again consider $a_0 = 1$ (unit of length). The last equation suggests a general definition for the fractal dimension of any fractal shape, as:

$$D_f = \frac{\ln r}{\ln s} \tag{5.33}$$

For Sierpinski's carpet, this formula gives immediately $\ln 8/\ln 3 = 1.893$ for the fractal dimension, while for Sierpinski's gasket mentioned in Quiz 5, it gives $\ln 3/\ln 2 = 1.585$. Finally, for a Euclidian (i.e., non-fractal) structure similar to Sierpinski's gasket (except for the missing triangles), the fractal dimension is two (i.e., $\ln 4/\ln 2$), as it should be. It is also possible to apply this method of calculating the fractal dimension to fractals that extend in three-dimensional Euclidian space, as we will show briefly below.

5.2.2 Examples of Fractals of Biological Interest

An extension of Sierpinski's carpet to three-dimensional Euclidian space is the so-called *Menger's triadic sponge* (Mandelbrot, 1999). This fractal starts from a single cube, which, after magnification by a linear scaling factor, $s = 3$, and removal of the middle cubes, gives $r = 20$ small cubes (Fig. 5.6). The fractal dimension of this structure is then $\ln 20/\ln 3 = 2.727$. If the magnification and removal of the middle cubes continues for $n \to \infty$, it is found that the structure is in fact a surface packed into a three dimensional Euclidian space.

Mengers's sponge with a finite number of generations of cubes, n, constitutes a typical example of what came to be known in condensed matter physics as a *percolation*, or *percolation cluster* (Sahimi, 1994). In biology, Menger's sponge constitutes a competent model for the structure of certain body organs. A typical example is the liver (Weibel et al., 1969; Weibel and Paumgartner, 1978; Raicu et al., 1998) whose functions include intermediary metabolism and detoxification of the blood. The liver cells, or hepatocytes, which are the main cellular species in the liver, are aggregated into one-cell-thick clusters (the hepatic plates) that extend for several cells in two dimensions and are highly twisted and interconnected to form large *percolation clusters* in three dimensions (see Fig. 5.7). The space between plates, called hepatic sinusoids, exposes maximally the flowing biological fluids to the surface of each hepatocyte (Elias, 1949; Forker, 1989), which actively controls the composition of the fluid by diffusion through its plasma membrane. Although Menger's

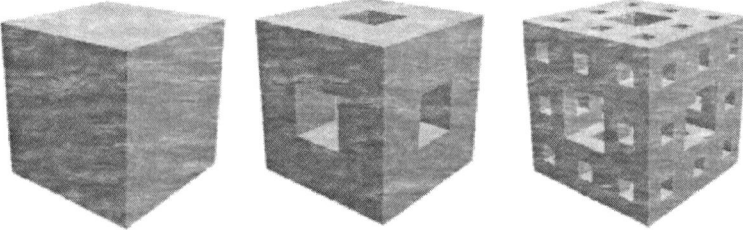

Fig. 5.6 Menger's triadic sponge. Figure obtained from http://local.wasp.uwa.edu.au/~pbourke/fractals/gasket, with permission from Paul Bourke.

5.2 Introduction to Fractals

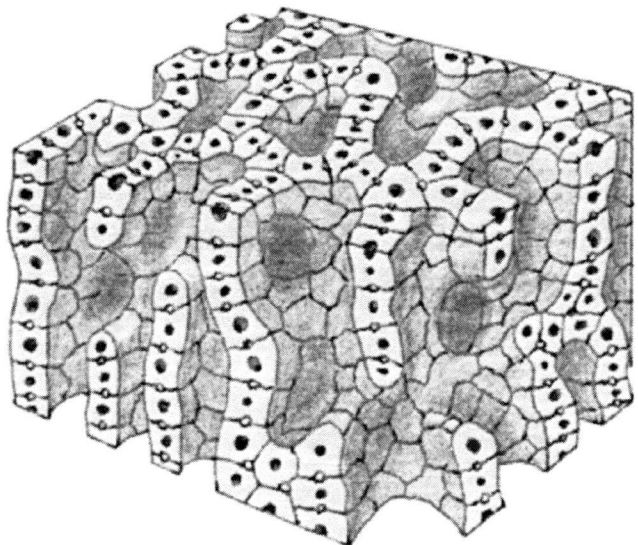

Fig. 5.7 Schematic representation of liver three-dimensional architecture after removal of bile ducts, blood vessels and connective tissue. Figure reproduced from Raicu et al. (1998), with permission from Elsevier.

sponge does not capture the random distribution of the hepatocytes within the liver, it is a good model for the percolation of fluids through the liver.

The plasma membrane of the hapatocyte itself is highly folded (Blouin et al., 1977; Raicu et al., 1998), and falls into a different category of fractals – Cantorian fractals –, which will be described below.

Another case of percolation may arise with regard to the cellular cytoplasm. The cytoplasm is characterized by macromolecular crowding, which is caused by large numbers of molecular species, each of which being at relatively low concentrations (Kopelman, 1988; Schnell and Turner, 2004), and may generate zones of heterogeneity in the cytoplasm caused by molecular *exclusion volume*. It is also possible that reduced individual concentrations of molecular species result in rapid consumption of certain reactants, leading to creation of *depletion regions* inside the cytoplasm. In both of these cases, if such small regions may exceed a critical concentration, they can form continuous percolation clusters within the cytoplasm. These in turn may induce deviations of reaction kinetics in the cytoplasm from the law of mass action (Kopelman, 1988; Dewey, 1997; Schnell and Turner, 2004). Diffusion of ligands in the cytoplasm towards a carrier or a receptor on the membrane may be also affected, with significant implications for the transport across the membrane. In the next section, we will deal with this problem in some significant details.

Another fractal of biological interest is *Cantor's bar* (or set), which is generated by ×3 magnification of a line segment followed by removal of the middle segment (Fig. 5.8a). Its fractal dimension of 0.631 is greater than the dimension of a Euclidian point and less than that of a line, while its length tends to zero for $n \to \infty$.

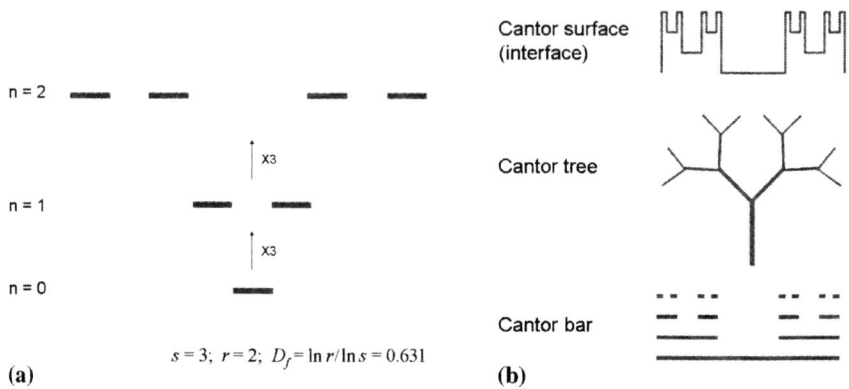

Fig. 5.8 (**a**) Generation of Cantor bar by x3 magnification and removal of the middle segment. (**b**) Equivalent forms of the Cantor bar.

The Cantor bar lends its name to an entire class of fractals – called Cantorian fractals – which have found widespread use as models of tree-like structures and rough interfaces in physics (Liu, 1985), geology and geophysics (Turcotte and Newman, 1996; McNamee, 1991), biophysics and physiology (Raicu et al, 2001; Dewey, 1997; Shlesinger and West, 1991; McNamee, 1991), etc. The archetypal example of Cantorian trees in biophysics is represented by the vascular and bronchial trees in lungs (Raicu et al, 2001; Shlesinger and West, 1991; Lefevre, 1983), while the rough, highly folded plasma membrane of the hepatocyte could be given as an example of Cantorian interface in two dimensions. The highly folded hepatocyte membrane maximizes the area of the interface between the cell and its environment for increased flux of substance across the membrane. This feature is likely to be met in other cells whose roles are to rapidly exchange substance with their environment, and may also explain the mitochondrial membrane folding. It is also interesting to remark that mitochondria in the cytoplasm of cardiac myocytes (i.e., heart cells) form a complex percolation network with precise physiological role (Aon et al., 2004).

The list of examples of fractal structures could be extended significantly beyond the few examples mentioned above, and more examples may be found throughout the references cited in this chapter. However, we hope that these few examples have already convinced the reader that fractal geometry deserves closer attention.

5.2.3 Practical Considerations and Limitations of the Above Theory

The fractal theory summarized herein is in essence a theory of measurement, which arises from applying the method developed by humans for measuring man-made objects to measuring naturally occurring objects. As such, its role is not necessarily

5.2 Introduction to Fractals

explanatory, but rather descriptive. This statement should not be read in any other way than as an attempt to draw a (flexible) border for the area of applicability of fractal geometry. The recognition of the reality of this unconventional geometry, at the time when it was introduced by Mandelbrot, constituted a big leap in our way of thinking about natural phenomena.

To cite but a few of the explanations attempted in the literature, the fractal geometry is adopted by natural structures as a necessity for keeping to a minimum the amount of genetic information to be transmitted in the process of organism growing (Weibel, 1991; Mandelbrot, 1999), for optimal functioning of a system with minimal consumption of energy (West et al., 1997, 1999), or for an increased length/volume or surface/volume ratio in systems requiring rapid and efficient transport of substance through tree-like structures or across folded membranes (Weibel, 1991; Turcotte et al., 1996). Attempts have been also made to explain the mechanisms that govern the formation of such structures, which include the theory of *diffusion-limited aggregation* (Witten and Sander, 1981) and the *constructal* theory (Bejan, 2006).

To measure the fractal dimension of a structure, such as a rough surface, one could, for example, perform measurements of the surface area, A, using a box (i.e., a square or a circle) of varying sizes, and then plot the logarithm of the area against the logarithm of the box size, l. This method follows immediately from the procedure described above for defining fractal dimensions, since making the box smaller is the same as increasing the magnification in the above discussion. If the log-log plot is well described by the straight line, the slope of the line gives the fractal dimension, D_f, according to the equation

$$\ln A = -D_f \ln l \tag{5.34}$$

which follows immediately from equation (5.32'). This method is called the *box counting method* (Barabási and Stanley, 1995), and represents just one of the many possible methods described in the literature for determination of the fractal dimension.

Naturally occurring fractals may depart from the ideal cases described above in two main ways. First, the number of generations, n, in natural structures presents a cutoff, after which the structure may follow a different type of fractal pattern. The value of n at which the cutoff occurs could occasionally be quite large, but always finite. Such systems, called *multifractals*, may no longer be described by a unique fractal dimension; in other words, the log-log representation suggested above does not give a straight line. Secondly, and as we have already mentioned with regard to the supra-cellular architecture of liver, natural fractals are often less regular than our idealized versions discussed above. Specifically, the feature size within one generation may not follow rigorously a particular rule, in that the scaling factor, s, may not be exactly the same for all generations, but only statistically so. Such structures are called *stochastic fractals*, and are to be distinguished from the *deterministic fractals* presented above.

5.3 Fractal Diffusion and the Law of Mass Action

The fractal geometry has been used for physical modeling along two main directions. The first approach is to associate fractal geometry with a new 'breed' of calculus – *fractional calculus* (Kilbas et al., 2006) – in order to derive and solve equations pertaining to the phenomena of interest. Although its use in science is not yet widespread, fractional calculus dates back to Leibnitz, one of the fathers of standard calculus (Kulish and Lage, 2002). In fractional calculus, the usual integral of a function, $f(x)$, is replaced by the (more general) *Riemann-Liouville integral* of order α,

$$J^\alpha f(x) = \frac{1}{\Gamma(\alpha)} \int_0^x \frac{f(\zeta)}{(x-\zeta)^{1-\alpha}} d\zeta,$$

where Γ is the Gamma function and α is a fractional number. The usual derivative is replaced by the *Riemann-Liouville fractional derivative* of fractional order α, $\frac{d^\alpha f(x)}{dx^\alpha} \equiv D^\alpha f(x)$, defined as

$$D^\alpha f(x) = \frac{d}{dx}\left[J^{1-\alpha} f(x)\right] = \frac{1}{\Gamma(1-\alpha)} \frac{d}{dx} \int_0^x \frac{f(\zeta)}{(x-\zeta)^\alpha} d\zeta,$$

or by the Caputo fractional derivative of fractional order α, defined as

$$D^\alpha f(x) = J^{1-\alpha}\left[\frac{df(x)}{dx}\right] = \frac{1}{\Gamma(1-\alpha)} \int_0^x \frac{1}{(x-\zeta)^\alpha} \frac{df(\zeta)}{d\zeta} d\zeta.$$

Examples of fractional differential equations include the fractional diffusion equation (Chechkin et al., 2005) and the fractional viscous-diffusion equation (Kulish and Lage, 2002). The fractional calculus method is perhaps the most rigorous way to deal with fractal processes, although its use in practical problems still poses significant challenges.

The second approach to theoretical modeling relies on classical calculus and uses scaling concepts to relate the differential equations to fractal geometry of the system modeled. This appears to be primarily a physicist's way of tackling transport phenomena on fractal lattices. Although they are less rigorous than the first method, scaling approaches are relatively easy to implement and provide significant insight into the processes investigated (O'Shaughnessy and Procaccia, 1985; Dewey, 1997).

In the next sections, we will summarize the application of the second method to the problem of diffusion of particles on fractal structures, and will then use it to investigate the changes that the theory predicts for the kinetics of carrier-mediated glucose transport across a membrane, which has been discussed first in section 4.4.2.

5.3.1 Diffusion on Fractal Lattices

It has been known for a long time that reduced dimensionality impacts diffusion processes. For example, diffusion of proteins within the membrane plane leads to a high probability for protein interactions, compared to diffusion in three dimensions. On the other hand, the rate of binding of the *lac repressor* to a specific site on the DNA is larger than expected for three-dimensional diffusion. The lac repressor's role is to bind to the DNA and inhibit the expression of genes coding for lac proteins involved in metabolism of lactose in yeast. In order to rapidly find its specific binding site on the DNA strand, the lac repressor first binds non-specifically to DNA and then diffuses in one dimension on the DNA chain (Dewey, 1997). Fractal structures introduce an entirely new flavor of reduced dimensionality to the diffusion problem, by casting diffusion pathways into complicated surfaces packed into a three-dimensional space; as we have seen in the previous section, this leads to fractional dimensions.

O'Shaughnessy and Procaccia (1985) used scaling arguments to modify classical diffusion equation to incorporate diffusion on fractal objects. Their derivation starts by considering a fractal structure of dimension, D_f, and writing the probability, $M(r,t)$, of finding, at a time t, a particle within a hyperspherical shell (on the fractal) between r and $r + dr$. $M(r,t)$ should obey the continuity equation:

$$\frac{\partial M(r,t)}{\partial t} = -\frac{\partial J(r,t)}{\partial r} \tag{5.35}$$

where $J(r,t)$ is the flux (or current) density of probability (in units of s^{-1}) flowing across the hyperspherical shell.

Further, it is reasonable to assume that the probability $M(r,t)$ scales with the *average probability*, $p(r,t)$, of finding a particle at a particular site in the volume, as:

$$M(r,t) \sim r^{D_f - 1} p(r,t) \tag{5.36}$$

where $r^{D_f - 1}$ is the number of sites. With this in mind, an analog to Fick's first law of simple diffusion, given by equation (4.5), may be postulated for the fractal structure as:

$$J(r,t) = -r^{D_f - 1} D(r) \frac{\partial p(r,t)}{\partial r} \tag{5.37}$$

where $D(r)$ is the position-dependent diffusion coefficient.

Plugging equations (5.36) and (5.37) into equation (5.35) yields:

$$\frac{\partial p(r,t)}{\partial t} = \frac{1}{r^{D_f - 1}} \frac{\partial}{\partial r} \left[D(r) r^{D_f - 1} \frac{\partial p(r,t)}{\partial r} \right] \tag{5.38}$$

which is the diffusion equation on a fractal lattice.

Observation: Note that it has been tacitly assumed above that $r^{D_f - 1}$ is independent of time.

Additional scaling arguments have lead O'Shaughnessy and Procaccia to the following relationship between the position dependent diffusion coefficient and the diffusion coefficient in Euclidian space:

$$D(r) = Dr^{-\theta} \quad (5.39)$$

where $\theta = D_f + \alpha - 2$ is called the *exponent of anomalous diffusion* and α is a scaling exponent for conductivity. With this relationship, the solution to the diffusion equation for fractal objects is found to be:

$$p(r,t) = \frac{2+\theta}{D_f \Gamma[D_f/(2+\theta)]} \left[\frac{1}{(2+\theta)^2 Dt}\right]^{D_f/(2+\theta)} \exp\left[-\frac{r^{2+\theta}}{(2+\theta)^2 Dt}\right] \quad (5.40)$$

which can be recast into the following form, which emphasizes the time-dependence:

$$p(r,t) = p_0 \, t^{-h} \exp\left(-\frac{\tau}{t}\right) \quad (5.41)$$

where $p_0 = [(2+\theta)^2 D]^{-D_f/(2+\theta)} / \{D_f \Gamma[D_f/(2+\theta)]\}$, $h = D_f/(2+\theta)$ and $\tau = r^{2+\theta}/[D(2+\theta)^2]$.

For $D_f = 2$ and $\alpha = 0$ (i.e., $\theta = 0$), equation (5.40) reduces itself to the solution of the three-dimensional Euclidian diffusion equation,

$$p(r,t) = \frac{1}{4Dt} \exp\left(-\frac{r^2}{4Dt}\right) \quad (5.42)$$

Quiz 6. Show that equation (5.41) is a solution to the fractional equation of diffusion on a fractal lattice (5.38).

It is interesting to note that equation (5.40) leads to an expression for the mean square displacement, i.e., the *average* distance traveled by a diffusing particle (O'Shaughnessy and Procaccia, 1985):

$$\langle r^2(t) \rangle = \frac{\Gamma[(2+D_f)/(2+\theta)]}{\Gamma[D_f/(2+\theta)]} \left[D(2+\theta)^2 t\right]^{2/(2+\theta)} \quad (5.43)$$

that depends nonlinearly on time, unlike the classical Einstein equation (for $D_f = 2$),

$$\langle r^2(t) \rangle = Dt \quad (5.43')$$

Finally, equation (5.37) together with equations (5.39) and (5.41) lead to the following relation for the flux of substance diffusing on the fractal lattice:

$$J(r,t) = \varphi \, t^{-h-1} \exp\left(-\frac{\tau}{t}\right) \quad (5.44)$$

where $\varphi = r^{D_f} p_0$ and the other notations were defined above.

5.3.2 The Carrier-Mediated Transport of Glucose Revisited

In chapter 4, we derived an expression for the flux of glucose diffusion across the erythrocyte membrane facilitated by the glucose carrier GLUT1 in red blood cells. In that case, we considered continuous supply of glucose at the external side of the membrane. In this section, we will generalize that treatment by assuming that the transport of sugar, S, across the membrane, facilitated by a carrier, C, is coupled to a process of diffusion on a fractal lattice, which brings sugar molecules from the source to the extracellular side of the plasma membrane. These processes are described by the following reaction:

$$S \xrightarrow{\text{fractal diff.}} S+C \underset{k_{-1}}{\overset{k_1}{\rightleftharpoons}} SC \xrightarrow{k_{dis}} S \qquad (5.45)$$

where, as before, k_1 and k_{-1} are reaction rate constants for the binding of S to C on the external side of the membrane, and k_{dis} represents the rate constant of complex dissociation at the intracellular face of the membrane.

Here again, the rates of SC complex formation and dissociation are given by

$$J_{C \to SC} = \frac{d[SC]}{dt} = k_1[S][C] \qquad (5.46)$$

and,

$$J_{SC \to S} = -\frac{d[SC]}{dt} = (k_{-1}+k_{dis})[SC] \qquad (5.47)$$

where $[S]$ and $[C]$ are the molar concentrations of unbound S and C on the outer side of the membrane. Assuming a steady state in the membrane, i.e., $J_{C \to SC} = J_{SC \to C}$, and using the total concentration of carrier, $[C]_T$, equations (5.46) and (5.47) yield:

$$[SC] = [C]_T \frac{[S]}{[S]+K_M} \qquad (5.48)$$

where, as before, $K_M = (k_{-1}+k_{dis})/k_1$ is the *Michaelis constant*. The last equation gives immediately the rate of S release on the intracellular side of the membrane into the cell:

$$J_{FD} = k_{dis}[SC] = k_{dis}[C]_T \frac{[S]}{[S]+K_M} \qquad (5.49)$$

in which $[S]$ should be now related to the diffusion on the fractal lattice.

At this point, we need to take the first part of the reaction (5.45) into account, which means to acknowledge the fact that $[S]$ depends on time. This is the main point of departure from the mathematical treatment of the facilitated diffusion in chapter 4. We use for this purpose equation (5.41), which upon a simple renotation becomes:

$$[S] = [S]_0 t^{-h} \exp\left(-\frac{\tau}{t}\right) \qquad (5.50)$$

where $[S]_0$ is the concentration at the point where S is added to the system (which includes the cell of interest), and $h < 1$. Taking this equation into account, equation (5.49) becomes:

$$J_{FD} = k_{dis}[SC] = k_{dis}[C]_T \frac{[S]_0 t^{-h} \exp\left(-\frac{\tau}{t}\right)}{[S]_0 t^{-h} \exp\left(-\frac{\tau}{t}\right) + K_M}, \quad h < 1 \qquad (5.51)$$

which states that the flux of facilitated diffusion depends on time nonlinearly.

For very short times, since the power function increases slowly with decreasing t, while the exponential decreases dramatically, the exponential term in equation (5.51) dominates over the fractional power function (see the numerical estimates in Table 5.1), leading to the following asymptotic behavior of the flux:

$$J_{FD} \propto \frac{k_{dis}}{K_M} [C]_T [S]_0 \exp\left(-\frac{\tau}{t}\right) \qquad (5.52)$$

which is reduces to zero for $t \to 0$. In this case, the glucose transport is *limited* by the fractal diffusion, which limits the amount of glucose available on the outer side of the membrane.

For very long times, the power function dominates the temporal behavior $[\exp(-\tau/t) \to 1]$ of equation (5.51), and J_{FD} presents the following asymptotic behavior:

$$J_{FD} \propto k_{dis}[C]_T \frac{[S]_0 t^{-h}}{[S]_0 t^{-h} + K_M} \qquad (5.53)$$

Here, much of the glucose diffused on the fractal lattice nearby the membrane. This case is somewhat similar to the case of facilitated diffusion treated in chapter 4, since the overall transport is *limited* by the facilitated diffusion through the membrane and depends only moderately on time.

Observation: The exponent h is subunitary only for a fractal lattice, for Euclidian space of any dimension being equal to unity, as may be seen from the definition of the symbols used in equation (5.41). In the latter case, the power function in equation (5.53) depends more strongly on time than in the former.

Table 5.1 Numerical estimates for the exponential and the fractional power function used for determining the asymptotic behavior of equation (5.51). Assumed values: $h = 0.5$, $\tau = 1$

Time (a.u.)	$\exp\left(-\frac{\tau}{t}\right)$	t^{-h}	$\exp\left(-\frac{\tau}{t}\right) t^{-h}$
0.01	3.72×10^{-44}	10	3.72×10^{-43}
0.1	4.54×10^{-05}	3.2	0.0001
1	0.37	1	0.37
10	0.90	0.32	0.29
100	0.99	0.1	0.099
1,000	0.999	0.032	0.032
10,000	0.9999	0.01	0.00999

Let us note that the facilitated diffusion of glucose across the RBC membrane is certainly not the only process of biological relevance that may couple to a process of diffusion on a fractal lattice; in fact, it may not even be the most important one in this regard. We chose it here for reasons of consistency with chapter 4. Diffusion on fractals may also be involved, at least in principle, in the self-association (or homo-oligomerization) of membrane protein receptors. In this case, heterogeneity in the membrane composition may lead to percolation-cluster-like structures, which will affect the mobility of the individual proteins, and thereby the rate of encounter between them.

Reaction constants between biological macromolecules may also become time-dependent as a result of fractal geometry (Schnell and Turner, 2004; Dewey, 1997; Kopelman, 1988); this may be even more relevant to biophysics, as reaction-limited kinetics seem to be more common in biological cells than diffusion-limited process are. It goes of course without saying that the time dependence of a reaction constant *in vivo* amounts to violations of the law of mass action, which we have assumed as valid at the beginning of this chapter. A review of available experimental evidence and the theoretical approaches in this regard is presented in a very accessible paper by Schnell and Turner (Schnell and Turner, 2004), which also includes a novel fractal-like approach to modeling biochemical reactions.

References

Atkins, P. and de Paula, J. (2002) *Atkins' Physical Chemistry*, 7th ed., Oxford University Press, New York

Aon, M. A., Cortassa S. and O'Rourke, B. (2004) Percolation and criticality in a mitochondrial network, *Proc. Natl. Acad. Sci. USA*, **101**: 4447

Barabási, A.-L. and Stanley, H. E. (1995) *Fractal Concepts in Surface Growth*, Cambridge University Press, New York

Bejan, A. (2006) *Advanced Engineering Thermodynamics*, 3rd ed., Wiley-Interscience, Hoboken, NJ

Blouin, A., Bolender, R. P. and Weibel, E. R. (1977) Distribution of organelles and membranes between hepatocytes and nonhepatocytes in the rat liver parenchyma. A stereological study, *J. Cell Biol.*, **72**: 441

Chechkin, A. V. Gorenflo, R. and Sokolov, I. M. (2005) Fractional diffusion in inhomogeneous media, *J. Phys. A: Math. Gen.*, **38**: L679

Decker, J. M. (2007) http://microvet.arizona.edu/Courses/MIC419/Tutorials/antibody.html

Dewey, G. T. (1997) *Fractals in Molecular Biophysics*, Oxford University Press, New York

Elias, H. (1949) A re-examination of the structure of the mammalian liver: II. The hepatic lobule and its relation to the vascular and biliary systems, *Am. J. Anat.*, **85**: 379

Forker, E. L. (1989) Hepatic transport of organic solutes, in: S. G. Schultz (ed.) *Handbook of Physiology. A Critical, Comprehensive Presentation of Physiological Knowledge and Concepts*, Section 6, Vol. III, American Physiological Society, Bethesda, MD, p. 693

Helmstaedt, K., Krappmann, S. and Braus, G.H. (2001) Allosteric regulation of catalytic activity: Escherichia coli aspartate transcarbamoylase versus yeast chorismate mutase, *Microbiol. Mol. Biol. Rev.*, **65**: 404

Hill, A. V. (1910) The possible effects of the aggregation of the molecules of haemoglobin on its dissociation curves, *J. Physiol. (Lond.)*, **40**: iv

Jackson, M. B. (2006) *Molecular and cellular biophysics*, Cambridge University Press, Cambridge

Kilbas, A. A., Srivastava, H. M. and Trujillo, J. J. (2006) *Theory and Applications of Fractional Differential Equations*, Elsevier, Amsterdam

Kopelman, R. (1988) Fractal reaction kinetics, *Science*, **241**: 1620

Koshland, D. E., Némethy, D. E. and Filmer, D. (1966) Comparison of the experimental binding data and theoretical models in proteins containing subunits, *Biochemistry*, **5**: 365

Kulish, V. V. and Lage, J. L. (2002) Application of fractional calculus to fluid mechanics, *J. Fluids Eng.*, **124**: 803

Lefevre, J. (1983) Teleonomical optimisation of a fractal model of the pulmonary arterial bed, *J. Theor. Biol.*, **102**: 225

Liu, S. H. (1985) Fractal model for the ac response of a rough interface, *Phys. Rev. Lett.* **55**: 529

Mandelbrot, B. B. (1999) *The Fractal Geometry of Nature*, W. H. Freeman, New York

McNamee, J. E. (1991) Fractal perspectives in pulmonary physiology, *J. Appl. Physiol.*, **71**: 1

Monod, J., Wyman, J. and Changeux, J. (1965) On the nature of allosteric transitions: a plausible model, *J. Mol. Biol.*, **12**: 88

O'Shaughnessy, B. and Procaccia, I. (1985) Analytical solutions for diffusion on fractal objects, *Phys. Rev. Lett.*, **54**: 455

Popescu, A., Miron, S., Blouquit, Y., Duchambon, P., Christova, P. and Craescu, C. T. (2003) Xeroderma pigmentosum group C protein possesses a high affinity binding site to human centrin 2 and calmodulin, *J. Biol. Chem.*, **278**: 40252

Raicu, V., Saibara, T., Enzan, H. and Irimajiri, A. (1998) Dielectric properties of rat liver in vivo: analysis by modeling hepatocytes in the tissue architecture, *Bioelectrochem. Bioenerg.*, **47**: 333

Raicu, V., Sato, T. and Raicu, G. (2001) Non-Debye dielectric relaxation in biological structures arises from their fractal nature, *Phys. Rev. E*, **64**: 021916

Sahimi, M. (1994) *Applications of Percolation Theory*, Taylor & Francis, London

Schnell, S. and Turner, T. E. (2004) Reaction kinetics in intracellular environments with macromolecular crowding: simulations and rate laws, *Prog. Biophys. Mol. Biol.*, **85**: 235

Shlesinger, M. F. and West, B. J. (1991) Complex fractal dimension of the bronchial tree, *Phys. Rev. Lett.*, **67**: 2106

Sierpinski, W. (1916) Sur une courbe cantorienne qui contient une image biunivoque et continue de tout courbe donnée, *Comptes Rendus (Paris)*, **162**: 629

Turcotte, D. L. and Newman, W. I. (1996) Symmetries in geology and geophysics, *Proc. Natl. Acad. Sci. USA*, **93**: 14295

Turcotte, D. L. Pelletier, J. D. and Newman W. I. (1998) Networks with side branching in biology, *J. Theor. Biol.*, **193**: 577

Weibel, E. R. (1991) Fractal geometry: a design principle for living organisms, *Am. J. Physiol.*, **261**: L361

Weibel, E. R. and Paumgartner, D. (1978) Integrated stereological and biochemical studies on hepatocytic membranes: II. Correction of section thickness effect on volume and surface density estimates, *J. Cell Biol.*, **77**: 584

Weibel, E. R., Staubli, W., Gnagi, H. R. and Hess, F. A. (1969) Correlated morphometric and biochemical studies on the liver cell: I. Morphometric model, stereologic methods, and normal morphometric data for rat liver, *J. Cell Biol.*, **42**: 68

Weiss, J. N. (1997) The Hill equation revisited: uses and misuses, *FASEB J.*, **11**: 835

Wells, J. W. (1992) Analysis and interpretation of binding at equilibrium, in: E. C. Hulme (ed.) *Receptor-Ligand Interactions: A Practical Approach*, Oxford University Press, Oxford, p. 289

West, G. B., Brown, J. H. and Enquist, B. J. (1997) A general model for the origin of allometric scaling in biology, *Science*, **276**: 122

West, G. B., Brown, J. H. and Enquist, B. J. (1999) A general model for the structure and of allometry of plant vascular systems, *Nature*, **400**: 664

Witten, T. A. and Sander, L. M. (1981) Diffusion-limited aggregation, a kinetic critical phenomenon, *Phys. Rev. Lett.*, **47**: 1400

Chapter 6
Electrophysiology and Excitability

6.1 Electric Charges and the Transmembrane Potential

Although the phospholipid matrix of the membrane presents symmetry (relative to a middle plane between the ends of the hydrophobic tails of the phospholipids), active transport of ions across the membrane by ionic pumps leads to a pronounced asymmetric distribution of ions between the inner and outer sides of the membrane (Fig. 6.1). Common ions that can be found on both sides of most cellular membranes are sodium (Na^+), potassium (K^+), and chlorine (Cl^-) ions; negatively charged proteins (A_P^-, i.e., anionic proteins) are found inside the cells and may not diffuse through the membrane. As a general rule, Na^+ ions are more concentrated (by ~ 10 times) in the extra cellular milieu (e.g., in the interstitial liquid, in the case of a tissue), K^+ ions are more concentrated (by 30–50 times) inside the cell, Cl^- ions are more concentrated outside the cell, while A_P^- ions are about 15 times more concentrated inside the cell (e.g., in the case of vertebrate cells at pH = 7). This large ionic disparity among the two sides of the membrane leaves the membrane electrically polarized, with the outer face being positively charged while the inner face negatively charged. The membrane polarization in turn gives a *transmembrane potential difference* or, simply, a *transmembrane potential*, which is defined, by convention, as:

$$\varphi_{TM} = \phi_i - \phi_e \qquad (6.1)$$

where ϕ_i and ϕ_e are the intracellular and extracellular electrical potentials, respectively. A viable cell has always a non-zero transmembrane potential, and since the net charge on the inner face of the membrane is negative, it follows from our definition above that $\varphi_{TM} < 0$.

Observation: In the earlier literature the transmembrane potential is defined as $\varphi_{TM}^* = -(\phi_i - \phi_e)$.

The membrane potential difference varies widely among different cellular species. For example, in the case of erythrocytes it is $-10\,mV$, in the case of excitable animal cells (e.g., neurons and muscle cells) it ranges from -60 to $-100\,mV$, while for

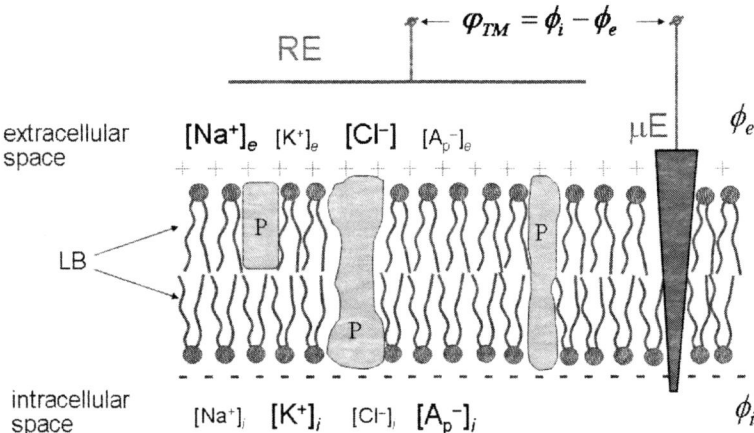

Fig. 6.1 Ionic distribution across the squid giant axon membrane and electrode configuration for transmembrane potential measurements. Significance of the symbols: e – extra cellular medium; i – intracellular medium; RE – reference electrode; μE – microelectrode; [X] – concentration (in mM) of the ionic species, X; P – proteins; LB – lipid bilayer. Other symbols are as defined in the text.

freshwater giant algae (*Nitella* and *Chara*) and for some higher plants (e.g., *Mimosa pudica*) it varies between -150 and $-230\,\text{mV}$. In the latter case the chlorine gradient is reversed compared to that in animal cells (Hille, 2001).

The transmembrane potential is sometimes called *resting membrane potential*, although a viable cell constitutes an 'arena' of many complex biochemical transformations, and, thereby, is never at rest. The membrane itself is continually traversed (both passively and actively) by electrical ionic currents. However, the resting membrane potential has a special significance in the case of *excitable cells*, and it is to be distinguished from *local* and *action potentials*, discussed later on in this chapter.

The most common method of determining the transmembrane potential (Fig. 6.1) relies on measuring the potential difference between a large reference electrode (RE) immersed in the suspending medium of the cell and a microelectrode (μE) inserted through the membrane into the interior of the cell (Koester, 1991; Guyton, 1992; Bear et al., 2001). It can also be measured by quantifying the fluorescence intensity and/or spectral shift of voltage-sensitive fluorescent dyes as they approach the membrane (Emaus et al., 1986; Hibino et al., 1993). More recently, an interesting variant of the dye-based method has been developed, which relies on dyes that generate *second harmonics* when hit by intense ultrashort (i.e., femtosecond to nanosecond) laser pulses (Sacconi et al., 2005).

Observation: *Second harmonic generation* (SHG) is a nonlinear optical process of generation of photons with a given frequency (wavelength) from two photons with half-frequency (double wavelength), upon interaction in a material with special properties (Boyd, 2003). SHG is commonly used in optics laboratories to efficiently produce light of certain wavelength from laser light with double the wavelength. More recently, it has been used in biophysical applications by relying either on tags that present the ability to generate second harmonics (such as in the example above) or on intrinsic SHG of certain biological molecules (Barzda et al., 2005).

6.1.1 Models for Calculating the Transmembrane Potential

6.1.1.1 The Nernst Equation

The transmembrane potential in the steady state can be related to the concentrations of ions on both sides of the membrane. Let us begin by considering a membrane that is permeable to a single ionic species, k, whose electrochemical potential in the cell interior is

$$\mu^k_i = \mu_0^k + RT \ln[X_k]_i + z_k F \phi_i^k \tag{6.2}$$

and for the cell exterior is

$$\mu^k_e = \mu_0^k + RT \ln[X_k]_e + z_k F \phi_e^k \tag{6.3}$$

In these equations, μ_0^k is the standard chemical potential (as defined in chapter 1), $[X_k]$ is the concentration of the species k, ϕ_i^k, and ϕ_e^k stand for intracellular and extracellular electrical potentials, respectively, z_k is the electrovalence and F is Faraday's number ($= eN_A = 96,500 \, C/mol$, where e is the elementary charge).

While ions continuously flow across the membrane as a result of active and passive transport, a steady state can be reached, for which the electrochemical potentials between the internal and external sides of the membrane are equal, i.e.:

$$RT \ln[X_k]_e + F z_k \phi_e = RT \ln[X_k]_i + F z_k \phi_i,$$

which gives immediately

$$\Delta \phi^k = \phi_i^k - \phi_e^k = -\frac{RT}{F z_k} \ln \frac{[X_k]_i}{[X_k]_e} \tag{6.4}$$

for the potential difference generated by ionic species k across the membrane. Equation (6.4) is the Nernst equation and relates the transmembrane potential of ionic species k to its concentrations on both sides of the membrane. For the room temperature of 20 °C, and after using $\ln = 2.3 \lg$, a practical form of equation (6.4) is obtained:

$$\Delta \phi^k = -\frac{58}{z_k} \lg \frac{[X_k]_i}{[X_k]_e} \, (\text{mV}) \tag{6.5}$$

To illustrate the magnitude of the transmembrane potential generated by ionic concentrations commensurate with those of biological cells, consider the case of the erythrocyte membrane, which contains the following ionic concentrations under normal physiological conditions: $[Na^+]_e = 155$ mM, $[Na^+]_i = 19$ mM, $[K^+]_e = 5$ mM, $[K^+]_i = 136$ mM, $[Cl^-]_e = 112$ mM and $[Cl^-]_i = 78$ mM. The transmembrane potential differences corresponding to each of these ions are: $\Delta\phi^{Na^+} = +53\,\text{mV}$, $\Delta\phi^{K^+} = -83\,\text{mV}$ and $\Delta\phi^{Cl^-} = -9\,\text{mV}$, respectively. However, one may not simply sum up the Nernst potential differences in order to determine the total transmembrane potential, because diffusing anions and cations must obey an additional restriction imposed by the so-called *Donnan equilibrium*, which will be described in the next section. Indeed, the measured transmembrane potential of the erythrocyte membrane is $\varphi_{TM} = -10\,\text{mV}$, which also incorporates contributions from A_P^-, which cannot diffuse across the membrane.

6.1.1.2 Donnan Equilibrium

Let us consider a simplified model of a cell represented by two aqueous compartments containing small ions and separated by a semipermeable membrane (Fig. 6.2), which allows only small ions to diffuse through it but not protein anions (A_P^-). To simplify the treatment, we consider only K^+ and Cl^-. In addition, we regard anionic proteins as $A_P^- K^+$ (or "proteinates" of K^+).

At $t = 0$ (Fig. 6.2), the number of positively charged ions equals the total number of negatively charged ions in each compartment (note that all ions are univalent in this example), i.e., $[K^+]_i^0 = [Ap^-]_i^0 \equiv C_i^0$, and $[K^+]_e^0 = [Cl^-]_e^0 \equiv C_e^0$. At the same time, Cl^- has a greater electrochemical potential in the external compartment, while the electrochemical potential of Ap^- is greater in the internal compartment. Since the membrane is permeable to Cl^-, some $\delta[Cl^-]$ ions will diffuse to the internal compartment to cancel its electrochemical potential imbalance. In order to maintain

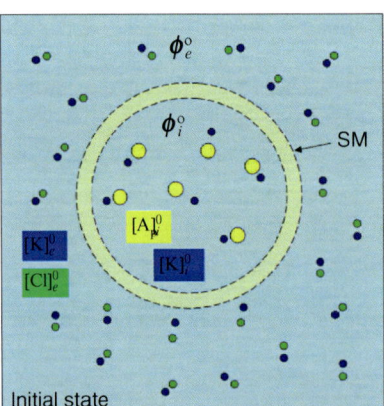
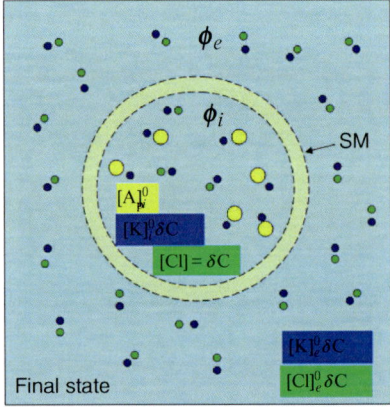

Fig. 6.2 Initial and final states of an aqueous solution of small ions (K^+ and Cl^-) and protein ions (A_P^-), separated by a semipermeable membrane (SM) into internal (i) and external (e) compartments. In the initial state, small ion concentrations are equal between compartments. The final state is governed by Donnan equilibrium. Symbols used: ϕ_i^0, ϕ_e^0, ϕ_i, and ϕ_e are electrical potentials in each compartment in the initial (superscript "0") and final (no superscript) states. Nondiffusing protein anions are drawn in yellow, while potassium and chlorine ions are drawn in blue and green, respectively.

the electroneutrality of the solutions on both sides of the membrane, it is necessary that an amount $\delta[K^+]$ of potassium ions also diffuse to the inner compartment. In the final state, when equilibrium is reached, the changes in the small ion concentrations should be equal, i.e., $\delta[Cl^-] = \delta[K^+] \equiv \delta C$. In that state, the transmembrane potential is given by the double identity:

$$\phi_i - \phi_e = \Delta\phi^{K^+} = \Delta\phi^{Cl^-} \tag{6.6}$$

By using equations (6.4) and (6.6) and taking into account the electrovalence of the two ions, $z_{K^+} = +1$, $z_{Cl^-} = -1$, the following relation is obtained:

$$\frac{[K^+]_i}{[K^+]_e} = \frac{[Cl^-]_e}{[Cl^-]_i} \tag{6.7}$$

where $[K^+]_i$, $[Cl^-]_i$, $[K^+]_e$, and $[Cl^-]_e$ are the total concentrations in the "interior" and, respectively, the "exterior" of the cell. Relation (6.7) is the mathematical formulation of the Donnan equilibrium.

Further, by using the notations introduced above for the concentrations, equation (6.7) becomes:

$$\delta C = \frac{(C_e^0)^2}{C_i^0 + 2C_e^0} \tag{6.8}$$

This equation gives the concentration of small ions transported from the exterior to the interior of the cell as a result of the impossibility of protein ions to diffuse across the membrane.

In conclusion, the presence of nondiffusing protein ions into the cell imposes an unequal repartition of small ions between the two sides of a semipermeable membrane.

Observation: If the Na^+-K^+-ATP pumps are inhibited (e.g., by chemical blockers – see below) then a living cell will shift its state towards a Donnan equilibrium (Glaser, 2001).

6.1.1.3 Equivalent Electrical Circuit of the Membrane

In chapter 3, we have introduced an electrical model for the plasma membrane in terms of the dielectric constant and conductivity of various membrane layers. Herein we will introduce a different type of electrical model, from the standpoint of electric currents flowing through the membrane and potential differences caused by ionic asymmetry of the extracellular and intracellular milieus. This will allow us to directly connect the transmembrane potential and, later on, its temporal evolution to the ionic distributions and their temporal changes. In this model, the main ions involved in the *electrogenesis* (i.e., the generation of the membrane potential difference) are Na^+ and K^+, all other ions being globally treated as *leakage ions*, L. In addition, instead of *diffusion coefficients* and *permeabilities* we will use *ionic current densities*, defined as *current per unit area*, J_k ($k = Na^+, K^+$), with units of $A\ m^{-2}$.

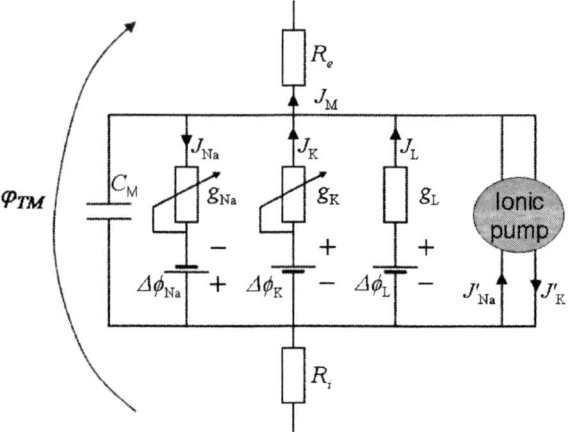

Fig. 6.3 The equivalent electrical circuit of a patch. g_k (k = Na, K, L) is the membrane specific conductivity (i.e., per unit area) for Na$^+$, K$^+$, and L (leakage) ions, respectively, $\Delta\phi^{Na}$, $\Delta\phi^K$, and $\Delta\phi^L$ are electrical potential differences across the membrane caused by gradients of Na$^+$, K$^+$ and L, respectively, φ_{TM} is the transmembrane potential, C_M membrane capacitance per unit area of membrane, R_e and R_i are resistances per unit area for the external and internal media, and J_{Na}, J_K, and J_L are Na$^+$, K$^+$ and L current densities (per unit area), respectively. J'_{Na}, J'_K are Na$^+$ and K$^+$ current densities injected by ionic pumps.

An equivalent electrical circuit of a patch of a cell membrane is presented in Fig. 6.3, which takes into account the current densities and membrane conductances for each ionic species. Because a membrane presents *distributed* (as opposed to lumped) passive electrical properties, specific quantities (i.e., per unit area) are used, viz. *specific ionic conductances*, g_k (in $\Omega^{-1}\text{m}^{-2}$ or $\text{S}\,\text{m}^{-2}$), *specific membrane capacitance*, C_M (in $\text{F}\,\text{m}^{-2}$) (see chapter 3), specific resistances of the internal (R_e) and external (R_i) media.

The *passive ionic currents* that enter or exit the cell are produced by the concentration gradients of these ions, which are represented by *electrical sources* in Fig. 6.3. The passive ionic currents are counterbalanced by the active current densities, J_{Na}' and J_K', produced by *Na-K-ionic pumps* (see chapters 4 and 7). The equivalent electrical circuit of a membrane was designed under the assumption of the existence of *specific channels* within cellular membrane for each ionic species, long before ionic channels where actually discovered. Also, ionic currents were assumed to flow independently through these channels (Hodgkin and Huxley, 1952). This is known as the *independence principle* (of ionic channels operation).

According to Fig. 6.3, the total passive current flowing through the membrane obeys the following equation:

$$J_M = J_{Na} + J_K + J_L \tag{6.9}$$

6.1 Electric Charges and the Transmembrane Potential

which may be rearranged, using potential differences across the membrane, as:

$$\varphi_{TM} = \frac{1}{g_M}(\Delta\phi_{Na}g_{Na} + \Delta\phi_K g_K + \Delta\phi_L g_L) \qquad (6.9')$$

where g_M is the membrane equivalent conductance. The potential differences may be related to the ionic concentrations on both sides of the membrane by using the Nernst formula, given by equation (6.4), for each ionic species k ($k = Na^+$, K^+, L).

A difficulty with equation (6.9') is the fact that the membrane equivalent conductance is not always known. Furthermore, in the resting state J_M is equal to zero (because the passive diffusion currents are perfectly cancelled out by those due to active transport), and the membrane potential may not be computed from equation (6.9').

6.1.1.4 The Goldman-Hodgkin-Katz Model of the Membrane

A different approach to deriving an equation for the transmembrane potential makes use of explicit expressions for the current densities and relates the membrane to concentrations via permeabilities to individual ionic species. The model makes the following assumptions:

(i) The membrane is homogeneous and electrically neutral
(ii) The properties of intracellular and extracellular media are uniform and unchanging in time
(iii) The membrane portion under investigation is planar and infinite in its lateral extent
(iv) The field intensity is constant across the membrane, i.e., $d\phi/dx \cong \varphi_{TM}/\delta_M$, where δ_M is the membrane thickness
(v) Fick's law applies to the passive diffusion through the membrane (in chapter 4, we have seen that an approach avoiding the use of Fick's law leads to the same expression for the flux of substance across the membrane as the one using Fick's law)

Let us write the flux density of ions of species k across the membrane (not an ionic current!) as:

$$\vec{j}_k = -D_k \nabla[X_k] - u_k \frac{z_k}{|z_k|}[X_k]\nabla\phi \qquad (6.10)$$

where the first term is the contribution of simple diffusion [assumption (v)], while the second term is due to the electrical potential gradient (i.e., electric field). By noticing that the diffusion coefficient, D_k, depends on the ion mobility, u_k, according to the equation

$$D_k = \frac{u_k RT}{|z_k|F} \qquad (6.11)$$

the ionic flux density may be rewritten as:

$$\vec{j}_k = -D_k\left\{\nabla[X_k] + \frac{z_k F}{RT}[X_k]\nabla\phi\right\} \qquad (6.12)$$

Because, according to the assumptions made above, only one spatial dimension is important (in the direction across the membrane and denoted herein by x), the concentration gradient may be expressed simply as $d[X_k]/dx$. Using this approximation and the assumption (iv) above, equation (6.12) can be rearranged as:

$$dx = -\frac{RT}{z_k F} \frac{\delta_M}{\varphi_{TM}} \frac{d\left\{\frac{j_k}{D_k} + \frac{z_k F}{RT}[X_k]\frac{\varphi_{TM}}{\delta_M}\right\}}{\frac{j_k}{D_k} + \frac{z_k F}{RT}[X_k]\frac{\varphi_{TM}}{\delta_M}} \tag{6.13}$$

which may be integrated from 0 to δ_M and from $[X_k]_{me}$ to $[X_k]_{mi}$ to give:

$$j_k = -\frac{z_k F}{RT}\frac{D_k \varphi_{TM}}{\delta_M}\frac{[X_k]_{mi} - [X_k]_{me} e^{-\frac{z_k F}{RT}\varphi_{TM}}}{1 - e^{-\frac{z_k F}{RT}\varphi_{TM}}} \tag{6.14}$$

Quiz 1. Prove equations (6.13) and (6.14).

At this point, the ion concentrations close to the interfaces with the internal and external media (but inside the membrane) may be related to the concentrations in the internal, $[X_k]_i$, and external, $[X_k]_e$, media (i.e., outside the membrane) according to the following simple relations: $[X_k]_{mi} = \beta_k [X_k]_i$, and $[X_k]_{me} = \beta_k [X_k]_e$ (where β_k is the partition coefficient of the ionic species k and the new concentrations are now measurable). With these equations, and after defining the permeability by

$$P_k = \frac{D_k \beta_k}{\delta_M} \tag{6.15}$$

the electrical current density, $J_k = z_k F j_k$, may be expressed as

$$J_k = -\frac{z_k^2 F^2}{RT} P_k \varphi_{TM} \frac{[X_k]_i - [X_k]_e e^{-\frac{z_k F}{RT}\varphi_{TM}}}{1 - e^{-\frac{z_k F}{RT}\varphi_{TM}}} \tag{6.16}$$

which holds for each of the three ionic species, Na^+, K^+ and Cl^-.

For the resting state of the membrane, the current density obeys the obvious relationship $J_K + J_{Na} + J_{Cl} = 0$, which, combined with equation (6.16), gives the *Goldman-Hodgkin-Katz* equation (Goldman, 1943; Hodgkin and Katz, 1949):

$$\varphi_{TM} = -\frac{RT}{F}\ln\frac{P_{Na}[Na^+]_i + P_K[K^+]_i + P_{Cl^-}[Cl^-]_e}{P_{Na}[Na^+]_e + P_K[K^+]_e + P_{Cl}[Cl^-]_i} \tag{6.17}$$

which, at 20 °C, becomes:

$$\varphi_{TM} = -58\lg\frac{P_{Na}[Na^+]_i + P_K[K^+]_i + P_{Cl^-}[Cl^-]_e}{P_{Na}[Na^+]_e + P_K[K^+]_e + P_{Cl}[Cl^-]_i}\,(\text{mV}) \tag{6.18}$$

Recalling the example of the erythrocyte's transmembrane potential discussed in the previous section, we note that φ_{TM} ($= -10\,\text{mV}$) is approximately equal to the

Nernst potential of the chlorine ions. In light of equation (6.18), this implies that the permeability of the erythrocyte membrane to the sodium and potassium ions is much lower than that to the chlorine ion.

6.2 Excitable Membranes

As discussed above, at their resting state, cell membranes are more or less electrically polarized, depending on the type of cells. Usually, *excitable cells* (i.e., neuronal, muscular and glandular cells) have higher resting potentials than non excitable cells. In the following sections, we will focus our attention onto the electrical behavior of the *neuron* (depicted in Fig. 6.4) – the brain's most important cell type which is involved in *memory* and *perception* (Kelly, 1991; Kupfermann, 1991; Guyton, 1992; Bear et al., 2001) – and in particular of its axon.

A cell is not an isolated system, but one that permanently experiences interactions (e.g., chemical, mechanical, thermal, electrical, and optical) with its environment. As a result of these interactions, a membrane's physical properties are continually changing; this includes changes in the transmembrane potential due to biological activity. According to Bullock (1959), membrane potentials fall into two main categories: *resting potentials* and *potential changes due to activity* or *action potentials*. Transmembrane potential changes are usually associated with changes in electrical conductances of the ionic channels present in membranes.

> **Observation:** As it was already mentioned before, depending on the type of natural agent (called *excitants* or *stimuli*) that triggers changes in their conductances, three main types of channels are known to be present in biological membranes: *chemical* (encountered especially in postsynaptic membranes), *electrical* (also called *voltage dependent* channels, and found in axons and muscle cells) and *mechanical* channels (found in hair cells). Although all three types of external agents may induce changes in membrane ionic conductivities, in experimental biophysics and physiology, one typically uses electrical stimuli.

6.2.1 Electrotonic Versus Action Potentials

Depending on the intensity of the stimuli, changes in the potential of a membrane region either vanish rapidly without leaving any influences on adjacent zones, or persist and eventually influence adjacent zones that will, in their turn, be structurally and functionally altered. These two situations will be discussed separately in the following sections.

Let us consider the simple electrical setup presented in Fig. 6.5, in which a current with a rectangular temporal profile, whose amplitudes are chosen by varying the resistance, R_x, is applied to the surface of the membrane via two electrodes.

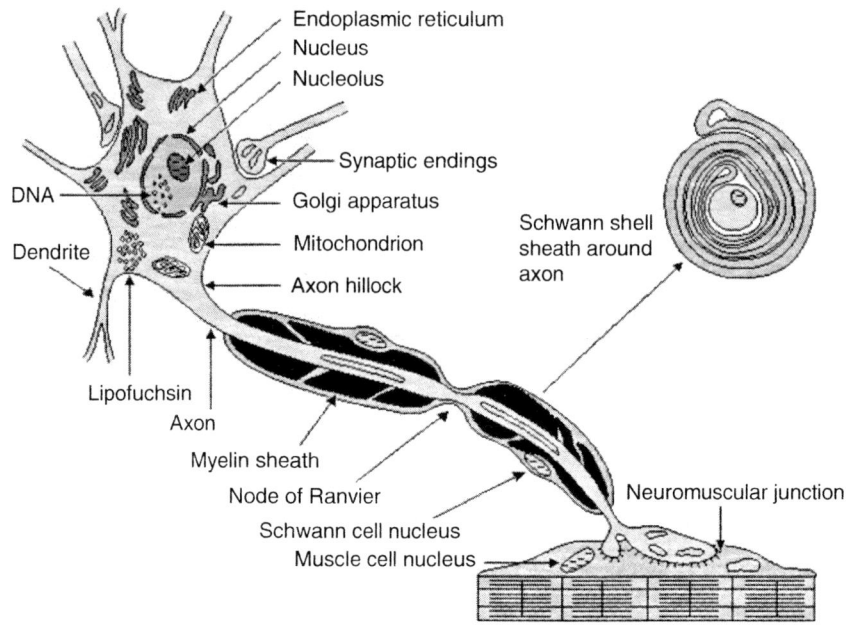

(a)

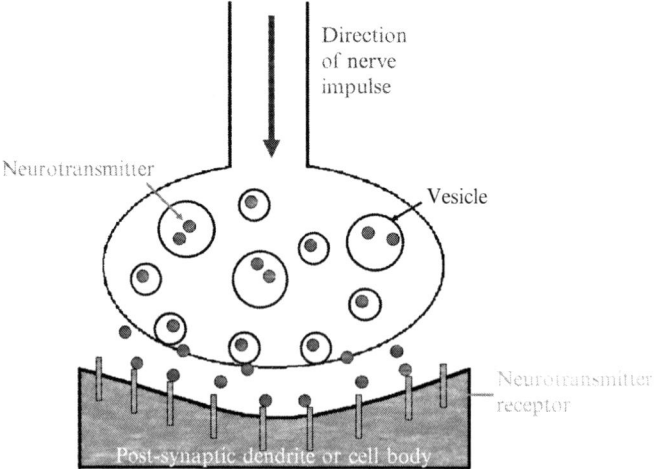

(b)

Fig. 6.4 Schematic representations of (**a**) a neuron and (**b**) its synapse, i.e., the terminal part of the axon interacting with the end of a dendrite via chemical neurotransmitters. Panel (**a**), reproduced from Malmivuo and Plonsey (1995) and Malmivuo and Plonsey (http://butler.cc.tut.fi/∼malmivuo/bem/bembook/) by permission of Jaakko Malmivuo and Oxford University Press, Inc.

6.2 Excitable Membranes

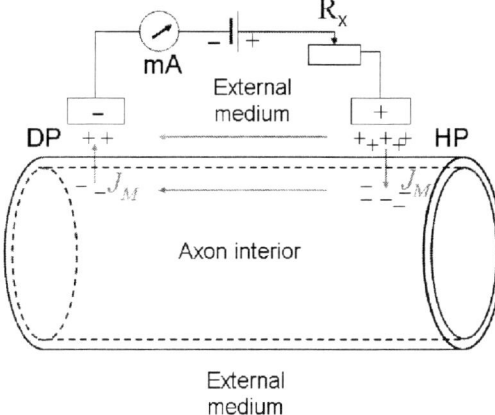

Fig. 6.5 Schematic of a simplified electrical setup used to electrically stimulate an axon. Symbols used: R_x – variable resistance; mA – milliamperemeter; DP – membrane depolarization; HP – membrane hyperpolarization.

If the stimulus (current) is relatively weak or, more precisely, its intensity does not exceed a specific *threshold* value, the membrane responds *electrotonically* (i.e., passively) inducing membrane *depolarization* (*cathelectrotonus*) in the vicinity of the cathode, as well as membrane *hyperpolarization* (*anelectrotonus*) in the vicinity of the anode. These two states can be evidenced using implanted microelectrodes.

Observation: The amplitudes of the membrane hyperpolarizations are always greater than the corresponding amplitudes of membrane depolarizations (see Fig. 6.6), due to the rectifier property of the membrane.

Immediately after the application of the electrical current (at time t_{ON}), the membrane potential increases exponentially, and then decays exponentially following the "OFF" part of the rectangular stimulus. The exponential behavior is due to charging/discharging of the membrane capacitance, C_M, via the membrane specific resistance, R_M (with a time constant $\tau_M = R_M C_M$).

If the amplitude of the stimulating current is increased further but it is kept under a threshold value (provoking, for instance, membrane depolarization smaller than 15 mV, in the case of squid giant axon) an additional hump appears in the potential curve, which overlaps with the electrotonic response. This additional transient response is called *local potential* and presents the following main features (Hille, 2001) (see Fig. 6.6):

- It is *a graded change*, its initial amplitude increasing with the stimulus intensity
- It is an *active response* of the axon, which may be suppressed by using anesthetics and metabolic inhibitors
- Its amplitude, at a given point, x, decreases exponentially in time following cessation of stimulation, according to the formula

$$\varphi_{TM}(x,t) = \varphi_{TM}(x, t_{OFF}) e^{-\frac{t}{\tau_M}} \tag{6.19}$$

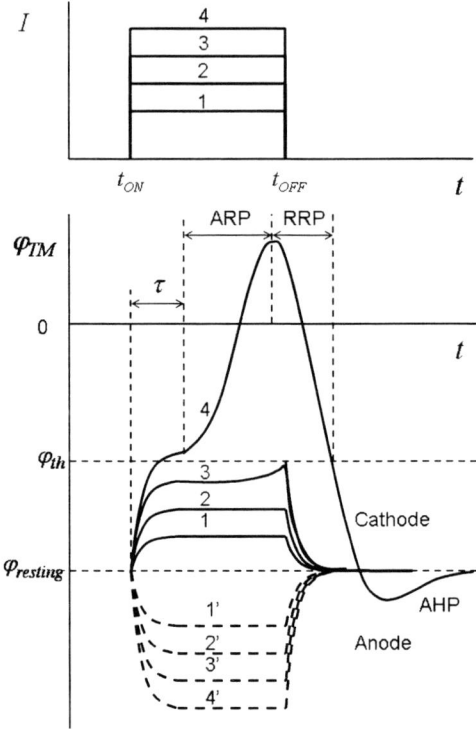

Fig. 6.6 Electrical responses of a membrane following stimulation by a rectangular current. 1, 2, 1′, 2′, 3′ and 4′ – electrotonic (passive) response; 3 – local potential; 4 – action potential (see text for details).

- It decreases exponentially in space at an arbitrary time, t, according to

$$\varphi_{TM}(x,t) = \varphi_{TM}(x_0, t)e^{-\frac{x}{\lambda}} \quad (6.20)$$

where λ is the *space attenuation constant* given by,

$$\lambda = \sqrt{\frac{R_M}{R_i + R_e}} \approx \sqrt{\frac{R_M}{R_i}} \quad (6.21)$$

where R_M is the membrane resistance per unit area, while R_i and R_e ($\ll R_i$), are resistances of the inner and outer media per unit area of membrane, as shown in Fig. 6.3.

Observation: The attenuation constant, λ, is proportional to the square root of the axonal diameter, d. Thus, in the case of an unmyelinated vertebrate axon (i.e., axon without myelin sheath) with $d = 2\mu m$, $\lambda \approx 0.33$ mm, while in the case of the squid giant axon with $d = 0.5$ mm, $\lambda \approx 5$ mm (Vasilescu and Margineanu, 1982) This potential perturbation is thus localized spatially, even in the case of very conducting unmyelinated axons. Therefore the local potentials cannot be used to convey information at a long distance. For that, a different mechanism is needed, as discussed next.

6.2 Excitable Membranes

If the intensity of the stimulating current exceeds a value that leads to membrane depolarization over the threshold value, φ_{th} (which is usually $\sim -50\,\text{mV}$), the membrane potential undergoes a rapid evolution, first increasing steeply, and then decreasing to the extent that it reverses sign briefly before recovery of the initial resting state. This type of response, called *action potential* (AP), is attributable to the highly nonlinear behavior of membrane ionic currents, which is due to the interplay between a number of different factors, which will be addressed in the next section.

Observation: In technical terms, the stimuli that evoke membrane potential responses with amplitudes that are lower than the threshold value are called *subliminal*, while those for which the response exceeds the threshold are called *supraliminal* stimuli.

One can distinguish the following successive phases in the evolution of action potentials (Fig. 6.6): (i) a period of latency, τ, (ii) a very short ascending phase, called *absolute refractory period* (ARP) during which another action potential cannot be generated, regardless of the applied stimulus intensity, (iii) a descending phase called *relative refractory period* (RRP) when another impulse can be generated, provided that the second stimulus is much higher than the previous one, (iv) an *after-hyperpolarization phase* (AHP) when the interior side of the membrane is more negative than in the resting state, and (v) after AHP, the resting state is attained, in which the axonal membrane can be excited again. Note that, unlike local potentials, action potentials can only be generated in *excitable cells*.

Observation: The action potential can be generated only in the vicinity of the cathode, which induces membrane depolarization, and never in the vicinity of anode which induces membrane hyperpolarization.

Action potentials present certain qualitative features that distinguish them from local membrane potentials, as listed below (see curve 4 in Fig. 6.6).

1. AP's are triggered only by *supraliminal stimuli*, which are able to induce potential depolarizations exceeding the threshold, φ_{th}.

   Observation: The threshold value varies among different cell types, the most sensitive cells presenting the lowest threshold values. In molecular terms, the threshold is correlated with the minimum number of sodium channels required for membrane depolarization (*vide infra*).

2. AP obeys an "*all-or-nothing*" law, in the sense that their time course and amplitude are always the same irrespective of the amplitude of the supraliminal stimuli.

   Observations: (a) The strength o the supraliminal stimuli must not exceed however, a reasonable value. Otherwise, the excitable cells themselves could be destroyed. (b) Only the axonal part of a neuron obeys the all-or-nothing law, the body of the neuronal cell (i.e., the *soma*) responding gradually to increasing stimuli amplitudes.

3. AP's are *rapid* and *self-maintaining* processes, with steeply ascending and descending slopes. The whole AP evolution lasts only a few milliseconds.

4. AP's are *irreversible* (that is, they cannot be stopped once triggered) but regenerative processes: once the membrane recovers its resting state they may be triggered again.

5. AP's are *not localized* depolarizations, *but self-propagating* perturbations, which may travel without attenuation over long distances (in some cases for up to 1 m). The propagation of the *action potential*, which is also called *nerve impulse*, will be analyzed in more details in the last section of this chapter.

6.2.2 The Voltage-Clamp Technique and Transmembrane Ionic Currents

The time course of the action potential can be understood by taking into consideration the transient *transmembrane ionic currents* flowing *into* and *out* of the excitable cell. The evolution of ionic currents is a natural consequence of the temporal variation of the membrane *conductances*, g_i (or, equivalently, of the membrane *permeabilities*) following a membrane excitation that exceeds the threshold potential, φ_{th}. This will be explained next, first by analyzing the temporal behavior of individual currents for fixed transmembrane potential.

Using the electrical circuit presented in Fig. 6.3 as an electrical model of the membrane, the total transient current traversing the membrane at any time, $J_M(t)$, may be written as the algebraic sum of all different currents flowing through the membrane, i.e.,

$$\begin{aligned} J_M(t) &= J_C(t) + J_{Na}(t) + J_K(t) + J_L(t) \\ &= C_M \frac{d\varphi_{TM}}{dt} + (\varphi_{TM} - \Delta\phi_{Na})g_{Na} + (\varphi_{TM} - \Delta\phi_K)g_K \\ &\quad + (\varphi_{TM} - \Delta\phi_L)g_L \end{aligned} \quad (6.22)$$

where the first term is the current for charging the membrane capacitance, $\Delta\phi_k$ are the Nernst potentials, given by equation (6.4), for each ionic species k ($k =$ Na$^+$, K$^+$, L), and the other symbols have been defined above.

Understanding the mechanisms responsible for the nerve impulse has dramatically benefited from the *voltage clamp technique* developed independently by Cole (1949) and Marmont (1949). This technique permits one to disconnect the normal feedback between transmembrane potential and ionic current during the generation of the nerve impulse by constraining the membrane potential to accept desired fixed values. Due to its scientific and also historical importance, this technique will be described below in its essential details (see Fig. 6.7).

Two wire electrodes are inserted inside the axon along its axis, and the whole system is then placed inside a larger cylindrical electrode. A pair consisting of one of the wire electrodes and the cylindrical electrode is used to inject current and thereby stimulate the membrane, while the other wire electrode together with the cylindrical one are used to sense the transmembrane potential, which is measured by a high-input impedance amplifier with unity gain factor. At any moment, t, the difference between the transmembrane potential, $\varphi_{TM}(t)$, and the desired "clamp" voltage, $\varphi_{clamp}(t)$, set by a signal generator, is amplified by a factor $K(\gg 1)$, and the

6.2 Excitable Membranes

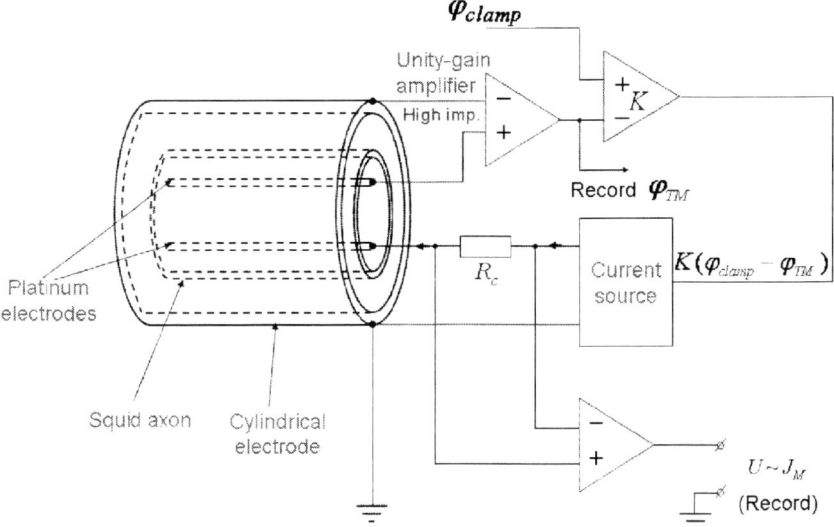

Fig. 6.7 Schematic of the voltage clamp technique.

output voltage, $K[\varphi_{clamp}(t) - \varphi_{TM}(t)]$, is fed into the current source. The source in turn injects current into the cell to reduce the potential difference to zero. Mathematically, the following relation is obeyed by the potentials:

$$K[\varphi_{clamp}(t) - \varphi_{TM}(t)] = \varphi_{TM}(t+\tau) \tag{6.23}$$

in which $\varphi_{TM}(t+\tau) \cong \varphi_{TM}(t)$ since τ is very small compared to the time scale of the action potential generation, so

$$\varphi_{TM} = \frac{K}{K+1}\varphi_{clamp} \cong \varphi_{clamp} \tag{6.24}$$

or

$$\varphi_{TM} \cong \varphi_{clamp}, \text{ for } K \gg 1 \tag{6.25}$$

The constancy of $\varphi_{TM}(t)$ is achieved because the applied current counterbalances the ionic current through the membrane. Therefore, by measuring the applied current (as a potential difference falling on the resistor R_c in Fig. 6.7), one essentially measures the transmembrane current (in fact, with opposite sign) after a stimulation event has occurred.

According to equation (6.25), the transmembrane potential is fixed at the desired value, which basically eliminates the capacitive current from equation (6.22), allowing one to measure the total ionic current flowing across the membrane as well as its separate components. Since the electrodes present very low resistances, the potential is also constant in the axial direction of the axon, i.e., $d\varphi_{TM}/dx = 0$.

With the aid of this technical arrangement it is relatively easy to measure the global ionic current, $J_M(t)$, in the membrane, by clamping the membrane at any desired voltage. Using specific pharmacological inhibitors of the K^+ and Na^+ channels, one can also measure either the Na^+ or the K^+ current (Hille, 2001). Note that, because the membrane is highly permeable to Cl^- ions, and because their internal concentration is very low, small variations in the Cl^- concentrations lead to rapid readjustment of the Nernst potential for chlorine ions, which is not observed at the time scale of the action potential. Cl^- is therefore not an important player in the generation and evolution of the action potential, and it ignored in the following discussion, being included in the leakage ions.

To measure the K^+ component of the transmembrane current, one either equilibrates the concentrations of Na^+ between the interior and exterior of the axon, or places the axon in an external solution containing Na^+ channel inhibitors, like *tetrodotoxin* (TTX), *saxitoxin* (STX), and *chlorpromazine*. In either case, the inward sodium current will be cancelled and, by stimulating the axon, one may record the temporal behavior of the K^+ component of the transmembrane current, $J_K(t)$, shown in Fig. 6.8.

Observation: Tetrodotoxin may be extracted from various marine animals, including a fish – the pufferfish, or, in Japanese, *fugu*. Although fugu itself presents immunity to it, tetrodotoxin acts like a strong poison on humans and has no known antidote. However, fugu is consumed as a delicacy in Japan, where the honor to prepare it is only conferred to master fugu chefs that learn the art of cooking fugu through years of diligent apprenticeship and after passing a rigorous test.

The Na^+ current, $J_{Na}(t)$, may be either determined by subtracting the potassium current from the total current, or measured directly after perfusing the axon with a solution containing K^+ channel inhibitors, such as *tetraethylammonium* (TEA), *charybdotoxin* (CTX), or *4-aminopyridine*, and applying a stimulus (Fig. 6.8).

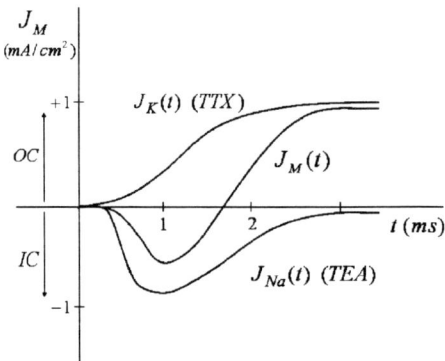

Fig. 6.8 The time course of the total ionic current, $J_M(t)$, potassium current, $J_K(t)$, in the presence of TTX inhibitor of K^+ channels, and sodium current, $J_{Na}(t)$, in the presence of TEA inhibitor of K^+ channels, following excitation by a step current function of infinite duration. Other symbols: IC – inward cell current; OC – outward cell current.

6.2 Excitable Membranes

Observation: If one blocks both Na$^+$ and K$^+$ channels, the total current should vanish. However, in practice very tiny currents can be recorded, due to minute structural modifications of the Na$^+$ and K$^+$ channels, caused by axon depolarization. These currents are known as *gating currents* (Armstrong and Bezanilla, 1973; Schneider and Chandler, 1973), because they are related to certain charged particles that open and close the channels by undergoing first-order transitions. To those interested in the details of these minute currents (of the order of a few μA cm^{-2}), and indeed in all aspects of ionic currents, we recommend an excellent book by Hille (2001).

As already mentioned at the beginning of this section, the temporal changes in the membrane ionic currents is due to changes in electrical conductances (or permeabilities) of the membrane for each type of ion. This conclusion follows immediately from analysis of the results of the voltage clamp experiments discussed above: since $J_k = \varphi_{TM} g_k$ ($k =$ Na, K), and the transmembrane potential is constant (and equal to φ_{clamp}), the conductances should change with time if it were for the currents to vary (as they did). With this observation, we can return to the description of the action potential (i.e., in which $\varphi_{TM}(t)$ is not fixed), and try to understand it in terms of permeabilities and currents.

Observation: Note that, in so doing, we appeal to knowledge acquired subsequent to the work done by Hodgkin and Huxley, in which the concept of ionic channels was just a hypothesis. This simple fact attests to the importance of the Hodgkin and Huxley theory, which will be presented in the next section.

After a short latency period, the depolarization induced by the exciting supraliminal current step (Fig. 6.9) leads to opening of Na$^+$ selective channels, which results in much higher permeability of the membrane to Na$^+$ compared to the resting state (by a factor of ∼500), and, equivalently to a higher membrane conductance, g_{Na^+}. Therefore, massive amounts of sodium ions will diffuse into the axon, due to the existing sodium concentration gradient across the membrane. The Na$^+$ influx generates an inward sodium current, which decreases in time (in absolute value), since an inactivation mechanism of the sodium channels now comes into play (with some delay compared to the activation mechanism).

The potassium conductance, g_{K^+}, also increases in time but at a lower rate than g_{Na^+}, due to the slower behavior of the K$^+$ selective channels. Therefore, potassium current flows through the membrane towards the outer medium. The potassium conductance also relaxes in time, due to the sensitivity of the potassium ion channels to the transmembrane potential, which changes in time. In the case of the squid giant axon, this relaxation is almost ten times slower compared to the relaxation of the sodium conductance.

In conclusion, in spite of its fundamental simplicity, the voltage clamp method has been instrumental in the studies of the nerve impulse, which led Huxley and Hodgkin to formulate their famous *theory of excitation* for the neuronal axon. This theory will be summarized in the next section, while more recent refinements of the voltage clamp method, which led to the patch-clamp technique that allowed Neher and Sakmann (Sakmann and Neher, 1995) to study the behavior of the individual ionic channels, will be described in chapter 7.

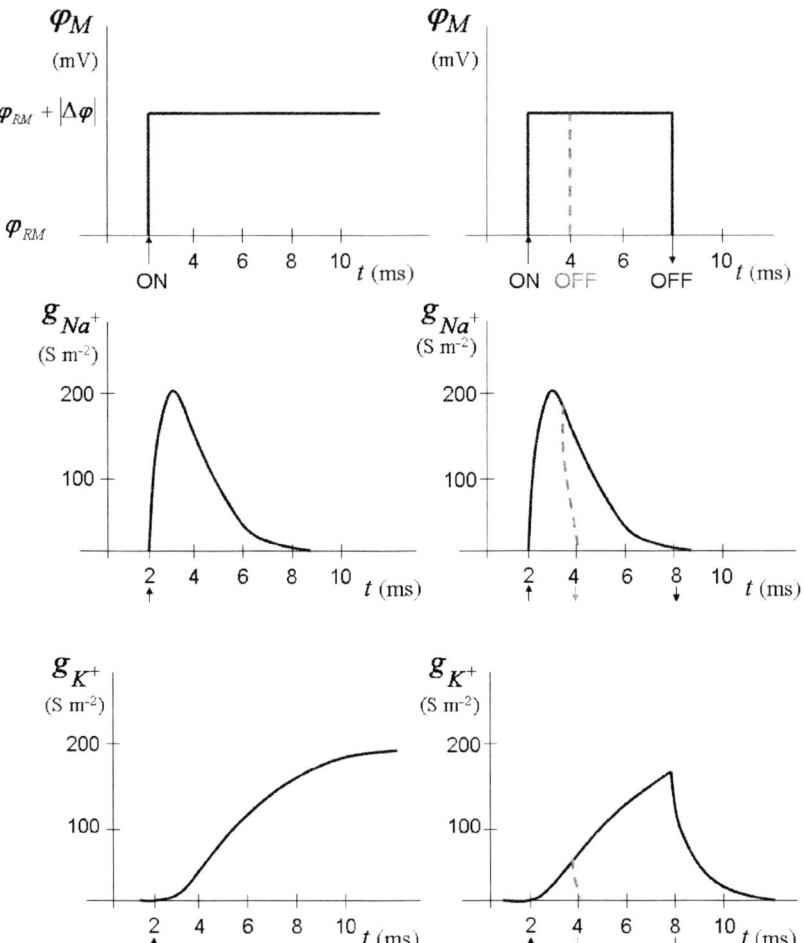

Fig. 6.9 Time course of unmyelinated axon sodium (middle panels) and potassium (lower panels) conductances in response to the action of supraliminal infinite-step (top-left) and finite-step stimulus (top-right) stimuli. Significance of the symbols: g_{Na^+} – Na$^+$ specific conductance; g_{K^+} – K$^+$ specific conductance; $\varphi_{TM}(t)$ = transmembrane potential difference; φ_{RM} – resting transmembrane potential difference; $|\Delta\varphi|$ – membrane depolarization potential. In the case of squid giant axon at $t = 8.5\,°C$, $\varphi_{RM} = -65$ mV, $|\Delta\varphi| = 56$ mV (Hille, 2001).

6.2.3 The Hodgkin-Huxley Model of Action Potential

Based on the above observations regarding the temporal variation of the transmembrane currents of sodium and potassium ions, Hodgkin and Huxley (Hille, 2001; Clay, 2005) postulated that membrane ionic conductances depend on the instantaneous values of the transmembrane potential and time in specific ways.

6.2 Excitable Membranes

6.2.3.1 Formulation of the Theory

The potassium conductance was assumed to be regulated by four independent structures (gates) that lead to a dependency of the potassium conductance on time of the following form:

$$g_K = g_k^{max}[n(\varphi_{TM}, t)]^4 \qquad (6.26)$$

where $0 \leq n \leq 1$ is the probability that a gate allows K^+ to diffuse across the membrane.

Quiz 2. In view of the above discussion, what would be the interpretation of $g_{K^+}^{max}$ from equation (6.26)?

If n represents the fraction of channel subunits activated at the time, t, then $1-n$ represents the fraction of inactivated subunits. Further, by modeling the transition from inactivated (closed: C) to activated (open: O) states as an unimolecular reaction,

$$C(1-n) \underset{\beta_n}{\overset{\alpha_n}{\rightleftarrows}} O(n) \qquad (6.27)$$

that goes from left to right and vice versa with different rate constants, $\alpha_n(\varphi_{TM})$ and $\beta_n(\varphi_{TM})$, one can write:

$$\frac{dn}{dt} = (1-n)\alpha_n(\varphi_{TM}) - n\beta_n(\varphi_{TM}) \qquad (6.28)$$

For $\alpha_n(\varphi_{TM}) = \beta_n(\varphi_{TM}) = constant$, this equation has the solution:

$$n(t) = n_\infty - (n_\infty - n_0)e^{-t/\tau_n} \qquad (6.29)$$

where $n_\infty = \alpha_n(\varphi_{TM})/[\alpha_n(\varphi_{TM}) + \beta_n(\varphi_{TM})]$, and $\tau_n = 1/[\alpha_n(\varphi_{TM}) + \beta_n(\varphi_{TM})]$.

The sodium conductance depends on the transmembrane potential and time according to:

$$g_{Na} = g_{Na}^{max}[m(\varphi_{TM}, t)]^3 \, h(\varphi_{TM}, t) \qquad (6.30)$$

where m is the probability ($0 \leq m \leq 1$) that three gates are activated (under membrane depolarization), and h is the probability ($0 \leq h \leq 1$) that one gate is inactivated under the same conditions. Their role is reversed during membrane hyperpolarization. The total probability that the four gates become simultaneously open and allow sodium ions to pass through is the product of the independent probabilities, m^3h.

Following the approach described above for potassium, to model the transition from inactivated (closed, C) to activated (open, O) states of the two different types of sodium gates one can write for m:

$$m(t) = m_\infty - (m_\infty - m_0)e^{-t/\tau_m} \qquad (6.31)$$

where $m_\infty = \alpha_m(\varphi_{TM})/[\alpha_m(\varphi_{TM}) + \beta_m(\varphi_{TM})]$, and $\tau_m = 1/[\alpha_m(\varphi_{TM}) + \beta_m(\varphi_{TM})]$, and for h:

$$h(t) = h_\infty - (h_\infty - h_0)e^{-t/\tau_h} \qquad (6.32)$$

where $h_\infty = \alpha_h(\varphi_{TM})/[\alpha_h(\varphi_{TM}) + \beta_h(\varphi_{TM})]$, and $\tau_h = 1/[\alpha_h(\varphi_{TM}) + \beta_h(\varphi_{TM})]$.

Observation: In the 1950s, when this model was elaborated, the nature of the membrane structures that gated the ionic passage was not known, nor was it clear that such structures even existed. In the years passing since then, sodium and potassium channels have been discovered and fully characterized, confirming the theory of Hodgkin and Huxley. It is now known that the potassium ion channels are composed of four protein subunits, which may be activated and inactivated largely independently by electrical field, thereby allowing or not allowing ions to pass. The parameter n can therefore be interpreted as the probability that a subunit is in an activated state, and n^4 gives the probability that the four subunits are simultaneously activated. It is also known that the sodium channel is composed of three identical protein subunits that are activated independently by membrane depolarization, and a fourth protein subunit that is activated under membrane hyperpolarization (therefore, inactivated by depolarization).

Taking equations (6.26) through (6.32) into account, equation (6.22) becomes:

$$J_M(t) = C_M \frac{d\varphi_{TM}}{dt} + (\varphi_{TM} - \Delta\phi_K) g_K^{\max} [n(\varphi_{TM}, t)]^4$$
$$+ (\varphi_{TM} - \Delta\phi_{Na}) g_{Na}^{\max} [m(\varphi_{TM}, t)]^3 h(\varphi_{TM}, t) + (\varphi_{TM} - \Delta\phi_L) g_L \qquad (6.33)$$

Hodgkin and Huxley fitted equation (6.33) to the experimental current vs. time curves for different values of the transmembrane potential (fixed by the voltage clamp setup) and obtained the following empirical expressions for the parameters α and β:

$$\alpha_n = \frac{0.1 - 0.01\varphi_{TM}}{e^{1-0.1\varphi_{TM}} - 1} \qquad (6.34a)$$

$$\beta_n = 0.125 e^{-\varphi_{TM}/80} \qquad (6.34b)$$

$$\alpha_m = \frac{2.5 - 0.1\varphi_{TM}}{e^{2.5-0.1\varphi_{TM}} - 1} \qquad (6.34c)$$

$$\beta_m = 4e^{-\varphi_{TM}/18} \qquad (6.34d)$$

$$\alpha_h = 0.07 e^{-\varphi_{TM}/20} \qquad (6.34e)$$

$$\beta_h = \frac{1}{e^{3-0.1\varphi_{TM}} + 1} \qquad (6.34f)$$

where φ_{TM} is measured in mV, while α_i and β_i ($i = n, m, h$) are measured in ms^{-1}.

Equations (6.33) and (6.34) represent the mathematical formulation of the Hodgkin-Huxley model (HHM) of action potential, which has dominated the theoretical neurobiophysics over four decades since its elaboration.

6.2.3.2 The Cable Model of the Axon

As it was already described, it has been observed that, unlike local potentials, the action potentials self-propagate along the axon without attenuation, even at large distances from the place of their generation. Therefore it is logical to assume that the transversal ionic currents can induce longitudinal currents leading to depolarization of the axonal membrane away from the place where the original stimulus acted, which is then crossed by ionic currents which, in their turn, will further provoke longitudinal currents, and so on. This leads to coupling between the spatial and temporal variation of the transmembrane potential, which may be evidenced by relating the currents and potentials for a small "slice" of an axon (Fig. 6.10).

Since the current must be conserved at each node of the circuit in the thin section (i.e., small δx) of the axon, one may write, successively:

$$I_e(x) + I_M = I_e(x + \delta x) \cong I_e(x) + \frac{\partial I_e}{\partial x} \delta x \tag{6.35}$$

and

$$I_i(x) = I_M + I_i(x + \delta x) \cong I_M + I_i(x) + \frac{\partial I_i}{\partial x} \delta x \tag{6.36}$$

where I_M represents the total transmembrane current (not the current density), as given by the HHM, $I_e(x)$ is the longitudinal current outside of the axon, and $I_i(x)$ is the longitudinal current inside the axon. These equations may be rearranged as:

$$i_e \equiv \frac{I_M}{\delta x} = \frac{\partial I_e}{\partial x} \tag{6.37a}$$

and

$$i_i \equiv \frac{I_M}{\delta x} = -\frac{\partial I_i}{\partial x} \tag{6.37b}$$

to give the current per unit length outside and, respectively inside the axon.

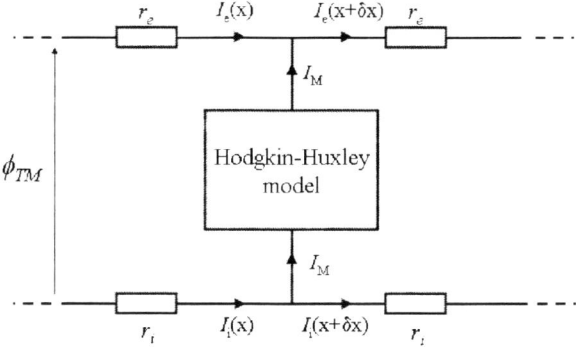

Fig. 6.10 Schematic representation of longitudinal and transversal currents in a portion of an axon of length δx and radius a. r_i and r_e represent the resistance per unit of length of the interior and exterior of the axon, respectively.

At this point, one may introduce the potentials on both sides of the axon, by using Ohm's law, i.e.,

$$I_e r_e = -\frac{\partial \varphi_e}{\partial x} \tag{6.38}$$

$$I_i r_i = -\frac{\partial \varphi_i}{\partial x} \tag{6.39}$$

Since, by definition,

$$\frac{\partial \varphi_{TM}}{\partial x} = \frac{\partial \varphi_i}{\partial x} - \frac{\partial \varphi_e}{\partial x} \tag{6.40}$$

with the help of equations (6.38) and (6.39), one may write the second order derivative of the transmembrane potential with respect to x, as:

$$\frac{\partial^2 \varphi_{TM}}{\partial x^2} = -r_i \frac{\partial I_i}{\partial x} + r_e \frac{\partial I_e}{\partial x} = (r_i + r_e) i_M$$

which gives immediately:

$$i_M = \frac{1}{(r_i + r_e)} \frac{\partial^2 \varphi_{TM}}{\partial x^2} \tag{6.41}$$

The last equation is the general cable equation for the axon.

Farther, by noticing that, in our formulation of the problem, the axon membrane forms a thin cylindrical shell of radius a, and length, δx, and therefore has an area $2\pi a \delta x$, the membrane current per unit length i_M, may be related to the transmembrane current density (i.e., current per unit area of membrane) in the HHM, through the following simple equation:

$$J_M = \frac{I_M}{S} = \frac{i_M}{2\pi a} \tag{6.42}$$

On the other hand, the inner resistance per unit length of axon is related to the resistivity of the internal medium through:

$$r_i = \frac{\rho_i}{\pi a^2} \tag{6.43}$$

while r_e may be considered negligible by comparison (i.e., $r_e \ll r_i$), due to the extensive space outside of the axon. Therefore, using this approximation from equations (6.41–6.43) it results:

$$J_M = \frac{1}{2\pi a (r_i + r_e)} \frac{\partial^2 \varphi_{TM}}{\partial x^2} \simeq \frac{a}{2\rho_i} \frac{\partial^2 \varphi_{TM}}{\partial x^2} \tag{6.44}$$

6.2 Excitable Membranes

Finally, equation (6.44) may be combined with equation (6.33) to obtain an equation,

$$\frac{a}{2\rho_i}\frac{\partial^2 \varphi_{TM}}{\partial x^2} = C_M \frac{\partial \varphi_{TM}}{\partial t} + (\varphi_{TM} - \Delta\phi_K) g_K^{\max} [n(\varphi_{TM},t)]^4$$
$$+ (\varphi_{TM} - \Delta\phi_{Na}) g_{Na}^{\max} [m(\varphi_{TM},t)]^3 h(\varphi_{TM},t) + (\varphi_{TM} - \Delta\phi_L) g_L \quad (6.45)$$

that relates the spatial to the temporal derivatives of the transmembrane potentials.

6.2.3.3 Interpretation of the Action Potential as a Depolarization Wave

The transverse and longitudinal currents flowing through and around an unmyelinated axonal membrane, as discussed above, may be thought of as forming local current loops that propagate "centrifugally" with a finite velocity, thus provoking membrane depolarization and leaving behind them the membrane in a *refractory state*. In this state, the membrane becomes temporarily nonresponsive on the affected portion, and forces the action potentials to propagate away from the point of initial stimulation. The local circuit currents propagating from point to point with a small velocity are known as *local Hermann currents* (Fig. 6.11) (Hermann, 1899, 1905a, b; Tasaki, 2006).

> **Observation:** In (more evolved) myelinated axons, the local Hermann currents are replaced by currents with larger steps, called *Stämpfli currents*. This *saltatory* propagation permits a significant increase in the velocity of the action potential propagating along the axon (Fig. 6.11) (Huxley and Stämpfli, 1949; Stämpfli, 1954; Bishop and Levick, 1956; Fitzhugh, 1962; Stämpfli, 1983; Terakawa and Hsu, 1991).

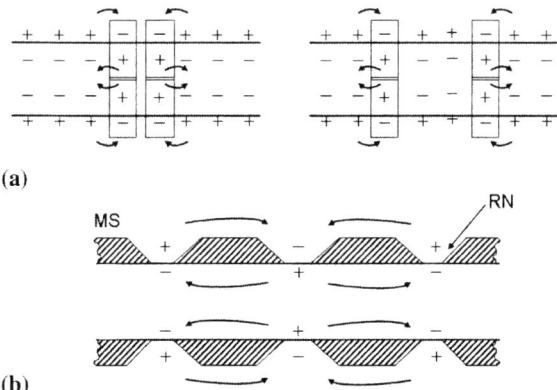

Fig. 6.11 Two snapshots of the action potential propagation through (**a**) local Hermann currents in an unmyelinated axon (propagation velocity = 0.5–2.5 m/s) and (**b**) saltatory Stämpfli currents between the Ranvier nodes (RN) of a myelinated axon (velocity = 10–100 m/s). MS is myelin sheath.

Quiz 3. Consider an axon of length l, which is simultaneously stimulated by supraliminal stimuli at its two extremities (that is, at $x = 0$ and $x = l$). (a) Will each nerve impulse succeed to reach the opposite extremity of the axon? (b) If not, how far away will each propagate along the axon?

In an alternative view, one can describe the propagation of action potential as propagation of a time-dependent membrane potential. In other words, one can consider the action potential as a *depolarization wave* propagating along the axon. In adopting such a point of view, one can make use of the *wave equation*:

$$\Delta \varphi_{TM} = \frac{1}{v^2} \frac{\partial^2 \varphi_{TM}}{\partial t^2} \tag{6.46}$$

where Δ is the Laplace operator, $\frac{\partial^2}{\partial x^2} + \frac{\partial^2}{\partial y^2} + \frac{\partial^2}{\partial z^2}$. Because the action potential propagates in one dimension only (along the axon length), the wave equation simplifies to:

$$\frac{d^2 \varphi_{TM}}{dx^2} = \frac{1}{v^2} \frac{\partial^2 \varphi_{TM}}{\partial t^2} \tag{6.47}$$

By eliminating the spatial variation of the transmembrane potential between the relations (6.45) and (6.47) one obtains the equation for the temporal variation of the potential at any point, x, along the axon:

$$\frac{a}{2\rho_i v^2} \frac{\partial^2 \varphi_{TM}}{\partial t^2} = C_M \frac{\partial \varphi_{TM}}{\partial t} + (\varphi_{TM} - \Delta\phi_K) g_K^{\max} [n(\varphi_{TM}, t)]^4$$
$$+ (\varphi_{TM} - \Delta\phi_{Na}) g_{Na}^{\max} [m(\varphi_{TM}, t)]^3 h(\varphi_{TM}, t) + (\varphi_{TM} - \Delta\phi_L) g_L \tag{6.48}$$

This nonlinear differential equation, which evidences the existing feedback between ionic currents and membrane potential, may be integrated numerically to give the value of the action potential at any moment, t. With it, Hodgkin and Huxley were able to accurately fit the temporal profile of the nerve impulse in the unmyelinated axon of squid. The best fit resulted into a value for the velocity, v, of $18.8\,\text{m s}^{-1}$, which was not far from the experimentally determined value of $21.2\,\text{m s}^{-1}$, and confirms beautifully the correctness of the model long before the ion gates hypothesized in the model were discovered experimentally.

The small discrepancy between the predicted and measured values of velocity, however, signaled a need for improvement of the original model. Several corrections have been implemented in the mean time, which included replacing the linear current-voltage relation equation (6.9′) with the Goldman-Hodgkin-Katz equation (6.17), and correcting an overestimate by a factor of two in the Na^+ current (Clay, 2005). Nevertheless, in all of its subsequent refinements, the main features of HHM were preserved. Thus, as Hille stated, "the success of HHM is a triumph of the classical biophysical method in answering a fundamental biological question" (Hille, 2001). For this outstanding scientific achievement, Alan Hodgkin and Andrew Huxley were awarded Nobel Prize in Physiology and Medicine in 1963.

References

Armstrong, C. M. and Bezanilla, F. (1973) Currents related to movement of the gating particles of the sodium channels, *Nature*, **242**: 459

Barzda, V., Greenhalgh, C., Aus der, A. J., Elmore, S., van Beek, J. and Squier, J. (2005) Visualization of mitochondria in cardiomyocytes by simultaneous harmonic generation and fluorescence microscopy, *Opt. Express*, **13**: 8263

Bear, M. F., Connors, W. B. and Paradiso, M. A. (2001) *Neuroscience: Exploring the Brain*, 2nd ed., Lippincott Williams & Wilkins, Baltimore, MD/Philadelphia

Bishop, P. O. and Levick, W. R. (1956) Published Online: 4 Feb 2005, Saltatory conduction in single isolated and non-isolated myelinated nerve fibres, *J. Cell. Comp. Physiol.*, **48**: 1

Boyd, R. W. (2003) *Nonlinear Optics*, 2nd ed., Academic, Boston, MA

Bullock, T. H. (1959) Neuron doctrine and electrophysiology, *Science*, **129**: 997

Clay, J. R. (2005) Axonal excitability revisited, *Progr. Biophys. Mol. Biol.*, **88**: 59

Cole, K. S. (1949) Dynamic electrical characteristics of squid axon membrane, *Arch. Sci. Physiol*, **3**: 253

Emaus, R. K., Grunwald, R. and Lemasters, J. J. (1986) Rhodamine 123 as a probe of transmembrane potential in isolated rat-liver mitochondria: spectral and metabolic properties, *Biochem. Biophys. Acta*, **850**: 436

Fitzhugh, R. (1962) Computation of Impulse Initiation and Saltatory Conduction in Myleninated Nerve Fiber, *Biophys. J.*, **2**: 11

Glaser, R. (2001) *Biophysics*, 5th ed., Springer-Verlag, Berlin/Heidelberg

Goldman, D. E. (1943) Potential, impedance and rectification in membranes, *J. Physiol.*, **27**: 37

Guyton, A. C. (1992): *Human Physiology and Mechanisms of Disease*, 5th ed., Saunders, Philadelphia.

Hibino, M., Itoh, H., Kinosita, K. Jr. (1993) Time courses of cell electroporation as revealed by submicrosecond imaging of transmembrane potential, *Biophys. J.*, **64**: 1789

Hermann L (1899) Zur Theorie der Erregungsleitung und der elektrischen Erregung, *Pflüger Arch. ges. Physiol.*, **75**: 574

Hermann, L. (1905a) Beitrage zur Physiologie und Physik des Nerven, *Arch. ges. Physiol.*, **109**: 95

Hermann, L. (1905b) *Lehrbuch der Phyisiologie*, 13th ed., August Hirchwald, Berlin

Hille, B. (2001) *Ion Channels of Excitable Membranes*, 3rd ed., Sinauer, Sunderland, MA

Hodgkin, A. L. and Katz, B. (1949) The effect of sodium ions on the electrical activity of the giant axon of the squid, *J. Physiol. (London)*, **108**: 37

Hodgkin, A. L. and Huxley, A. F. (1952) A quantitative description of membrane current and its application to conduction and excitation in nerve, *J. Physiol. (London)*, **117**: 500

Huxley, A. F, and Stämpfli, R. (1949) Evidence for saltatory conduction in peripheral myelinated nerve fibres, *J. Physiol.* **108**: 315

Kelly, J. P. (1991) The Neural Basis of Perception. In: *Principle of Neural Science*, 3rd ed., Kandel, R. K., Schwartz, J. H., Jessel, T. M. (Eds.), Appelton & Lange, Norwalk, CT

Kupfermann, I. (1991) Learning and Memory. In: *Principle of Neural Science*, 3rd ed., Kandel, R. K., Schwartz, J. H., Jessel, T. M. (Eds.), Appelton & Lange, Norwalk, CT

Koester, J. (1991) Membrane Potential. In: *Principle of Neural Science*, 3rd ed. Kandel, R. K., Schwartz, J. H., Jessel, T. M. (Eds.), Appelton & Lange, Norwalk, CT

Malmivuo, J. and Plonsey, R. (1995) *Bioelectromagnetism*, Oxford University Press, New York

Marmont, G. (1949) Studies on the axon membrane. I. A new method, *J. Cell. Comp. Physiol*, **34**: 351

Sacconi, L., D'Amico, M., Vanzi, F., Biagiotti, T., Antolini, R., Olivotto M., Pavone, F. S. (2005) Second-harmonic generation sensitivity to transmembrane potential in normal and tumor cells, *J. Biomed. Opt.* **10**: 024014

Sakmann, B. and Neher, E. (Eds.) (1995) *Single Channel Recording*, 2nd ed., Plenum, New York

Schneider, M. F. and Chandler, W. K. (1973), Voltage dependent charge movement of skeletal muscle: a possible step in excitation contraction coupling, *Nature*, **242**: 24

Stämpfli, R. (1954) Saltatory Conduction in Nerve, *Physiol. Rev.* **34**: 101

Stämpfli, R. (1983) The Ranvier node, past and future. A personal outlook after forty years of research, *Cell. Mol. Life Sci.*, **39**: 931

Tasaki, I. (2006) A note on the local current associated with the rising phase of a propagating impulse in nonmyelinated nerve fibers, *Bull. Math. Biol.*, **68**: 483

Terakawa, S. and Hsu, K. (1991) Ionic currents of the nodal membrane underlying the fastest saltatory conduction in myelinated giant nerve fibers of the shrimp *Penaeus japonicus*, *J. Neurobiol.*, **22**: 342

Vasilescu, V. and Margineanu, D. G. (1982) *Introduction to Neurobiophysics*, Abacus, Tunbridges Wells, Kent

Chapter 7
Structure and Function of Molecular Machines

7.1 Channels and Pores

We have seen in chapter 4 that passive diffusion of ions and small molecules across membranes occurs most efficiently when facilitated by pores, channels and carriers. Furthermore, in chapter 6 we discussed that certain gated structures had to exist for controlling the membrane permeability to ions during the generation and propagation of action potentials. These are now known as *ion channels*, and their presence and biological role has been convincingly revealed in electrophysiological studies even before high resolution structural studies became available. By contrast to channels and pores, ion pumps transport specific ions against concentration gradients across the membrane, and they do so by consuming energy produced in cellular processes.

The rate of channel-facilitated diffusion is 10^7–10^8 ions/s, more than 1,000 times higher than the rate of ion transport by ionic pumps. This rate is almost as high as that of free diffusion of ions in an aqueous solution. However, ion channels are not simple pipelines; instead, they possess a complex structure, which is sensitive to environmental changes. Channels present roughly two conformational states: *open* and *close*. Depending on the specific role of a channel and its location, transition between these two states (called *gating* or *activation*) may be triggered by various factors as summarized next.

(a) *Chemical activation* (ligand-gated) – by attachment of a specific ligand to the channel, as in the case of aquaporin (see below), G-protein-coupled channels (or receptors), and channels that detect the neurotransmitter acetylcholine
(b) *Electrical activation* (voltage-gated) – by a depolarizing electrical field, as for Na^+ and K^+ channels
(c) *Mechanical activation* (mechanically-gated) – by the action of an external force, torque, or pressure (e.g., channels in mechanosensory neurons)
(d) *Light-based activation* (light-gated) – induced by absorption of a photon, such as in the case of rhodopsin

We will discuss briefly the connection between structural and functional aspects of single channels in the next two sub-sections, while a summary of the current knowledge regarding the structure and function of ion pumps will be presented in section 7.1.2; particular emphasis will be put on Na, K^+-$ATPase$, which performs simultaneous countertransport of Na^+ and K^+ against their respective concentration gradients.

7.1.1 Electrical Behavior of Individual Ion Channels

Single ion channels had been investigated using electrical methods long before their structures were determined with high resolution. The main technique used in electrical studies of single channels – the *patch-clamp technique* – has evolved from the *voltage-clamp technique* (see chapter 6), primarily through refinements introduced in the micropipette method by Neher and Sakmann (Neher, 1991; Neher and Sakmann, 1976, 1992).

The patch clamp technique uses a voltage-clamp amplifier (similar to the one shown in Fig. 6.7) to fix the transmembrane voltage, and a glass pipette with a micrometer-sized tip for electrical current measurements. Alternatively, one may fix the current and measure the voltage. The smooth, round tip (diameter $\sim 1\,\mu m$) of the pipette is pressed against the cell membrane, while suction (i.e., "negative pressure") is applied to seal the contact with the cell. The interior of the pipette is filled with solutions containing electrolyte and other chemicals, to ensure electrical contact between the membrane patch and electronics, and provide means for controlling the behavior of the ion channels present in the patch of the membrane.

Reducing the *Johnson thermal noise* of the current source (i.e., the membrane patch) was the major hurdle that had to be overcome in the development of the patch-clamp technique (Neher, 1991). This noise is in addition to the noise introduced by the electronics and appears in the amplified signal in parallel to the signal from the membrane patch. To measure small currents generated at the passage of ions through single channels, large internal impedances of the source of current are necessary, which can be achieved with small signal sources: membrane patches. For that purpose, the contact between the microelectrode and the cell membrane has to be unusually good. The critical "ingredient" is the slight suction applied to the pipette to tightly seal the electrolyte in the interior of the pipette from the medium around the cell.

Quiz 1. The Johnson thermal noise can be expressed as the root-mean-square deviation (σ) of the current and depends on the absolute temperature (T), the measurement bandwidth (Δf) and the internal resistance of the signal source (R), as given by the following expression:

$$\sigma = \left(4k_B T \frac{\Delta f}{R}\right)^{1/2},$$

where k_B is Boltzmann's constant. Determine the internal resistance of the source necessary to measure a current of 1 pA at $\Delta f = 1\,\text{kHz}$ with 10% accuracy.

7.1 Channels and Pores

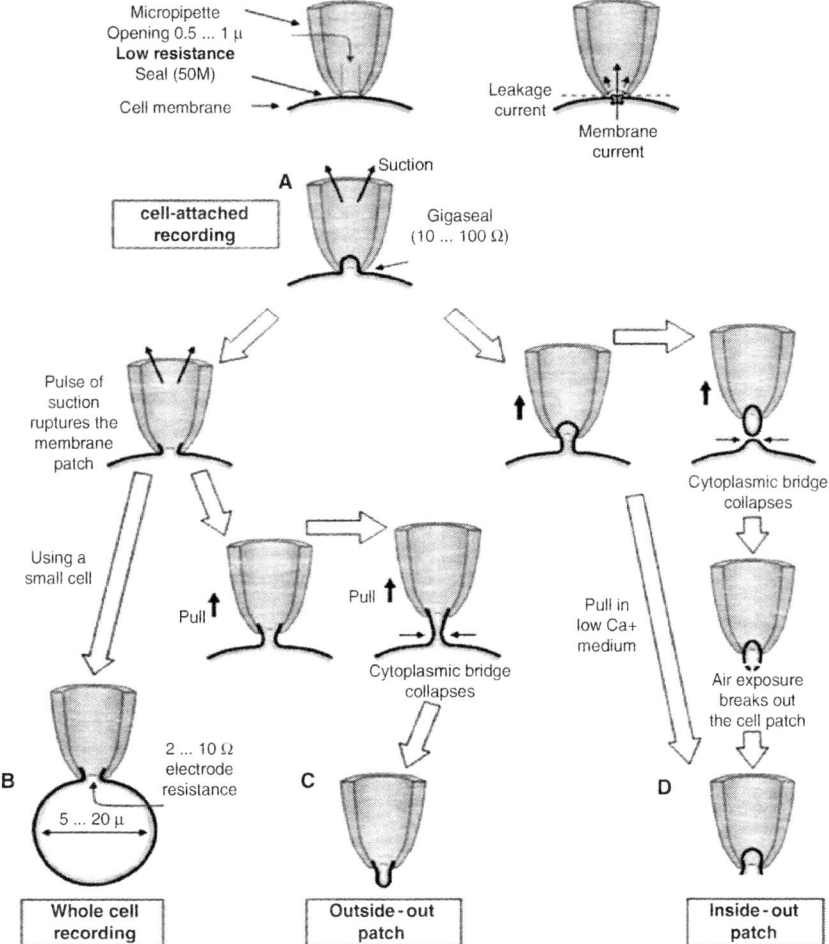

Fig. 7.1 Possible configurations of micropipette-membrane attachment. Figure reproduced from Malmivuo and Plonsey (1995) and Malmivuo and Plonsey (http://butler.cc.tut.fi/~malmivuo/bem/bembook/) by permission of Jaakko Malmivuo, Oxford University Press, Inc and Springer.

Depending on the purpose of the measurements, various configurations of the cell-pipette contact are used, as shown in Fig. 7.1 and briefly described below.

Cell-attached recording (top of Fig. 7.1). By applying slight suction to the end of the pipette opposite to the tip, the seal between the tip and the membrane is improved to the extent that single-channel signals may be distinguished from background noise. This configuration is good for measurements on single channels that happen to be in the area of the patch.

Whole-cell recording. Applying a short pulse of suction may rupture the membrane patch, and an electrical contact may be established with the interior of the cell. The method measures all the ion channels in the membrane and is similar to the microelectrode penetration technique illustrated in chapter 6 (Fig. 6.1), with the significant difference that it is less invasive to the cell (due to the smaller opening in the cell membrane). In addition, it opens a way for inserting various chemicals inside the cell, as necessary in experiments.

Inside-out patch. While in the whole-cell configuration (i.e., after the patch has been ruptured), the pipette may be pulled away from the membrane so that the cell and the patch separate from one another at their juncture and seal themselves from the external medium. The membrane patch trapped at the tip of the micropipette may be used to investigate the behavior of single channels under cytoplasmic fluid regulation.

Outside-out patch. Starting again from the cell-attached configuration, while the membrane is subjected to slight suction, and pulling the pipette away from the membrane, the patch detaches from the membrane and seals to form a small vesicle trapped at the tip of the micropipette. This configuration may be used to investigate the behavior of single channels activated by extracellular receptors.

In summary, three of the configurations described above could be used for detection of single channel currents under the effect of various endogenous or exogenous regulators. Figure 7.2 gives an example of single-channel signal obtained from patch-clamp measurements. The on-off behavior of the current suggests the channel transition between its "open" and "close" states.

The impact of the patch-clamp technique upon electrophysiological studies of ion channels has been tremendous, in particular by proving the existence of the gated structures hypothesized by Hodgkin and Huxley to explain the action potential. Detailed information on electrophysiological characterization of ionic channels is presented in a number of excellent books (see, e.g., Sakmann and Neher, 1995; Malmivuo and Plonsey, 1995; Hille, 2001). For their work on developing the patch-clamp technique on single ion channels behavior, Neher and Sakmann have been awarded the Nobel Prize in Physiology or Medicine in 1991 (Neher, 1991; Sakmann, 1991).

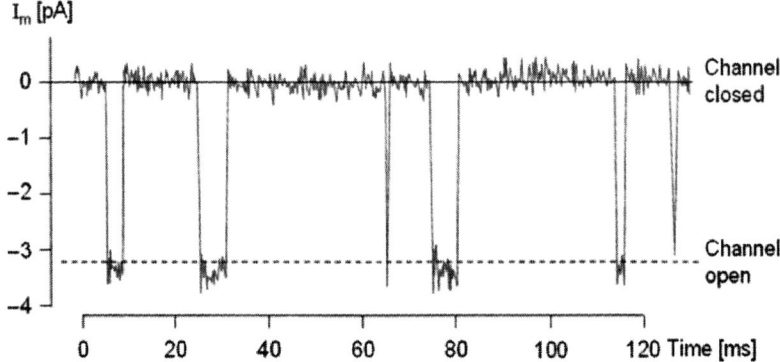

Fig. 7.2 Example of channels signal from patch-clamp measurements of a single ion channel at the neuromuscular endplate of frog muscle fiber. Figure reproduced from Malmivuo and Plonsey (1995) and Malmivuo and Plonsey (http://butler.cc.tut.fi/~malmivuo/bem/bembook/) by permission of Jaakko Malmivuo.

7.1.2 Structural Characterization by X-Ray Crystallography

In this section, we shall present the basic notions and concepts of X-ray crystallography (Rhodes, 1993; Drenth, 1994), with particular emphasis on the way in which they are applied to protein structure determinations.

The idea of using crystals to achieve X-ray diffraction has been credited to Max von Laue, who, together with two technicians, subjected a sphalerite [i.e., (Zn, Fe^{2+})S] crystal to X-rays and recorded a pattern of well-defined spots caused by diffraction. The method has evolved rapidly, notably through the work of the Braggs, father and son, who were able to determine the structure of many crystalline substances (and received the Nobel Prize in 1915). The first protein structure determination was reported by John Kendrew and co-workers (Kendrew et al., 1958) for myoglobin. For his work, Kendrew received Nobel Prize in Chemistry in 1962 (along with his colleague, Max Perutz).

A real breakthrough in the interpretation of X-ray diffraction on crystalline materials was conceiving the diffraction as a *selective reflection* of X-rays on the atoms pertaining to the *reticular planes* of the crystals (Fig. 7.3). According to this view, the constructive interference of the diffracted rays is obtained only if the *Bragg-Bragg formula* is obeyed:

$$2d_{hkl} \sin \alpha_n = n\lambda; \; n = 1, 2, \ldots \quad (7.1)$$

where λ is the wavelength of the selectively reflected X-rays and α_n are the incidence/reflection angles.

In order for X-ray diffraction to occur, the protein of interest needs to be crystallized, most often as a dimeric or larger unit (Drenth, 1994). The X-ray photons are scattered by the electronic clouds of the atoms comprising the macromolecules (Fig. 7.4). Since the macromolecules are located at regular positions within the crystal, the scattered photons interfere to form distinct patterns at the plane of detection.

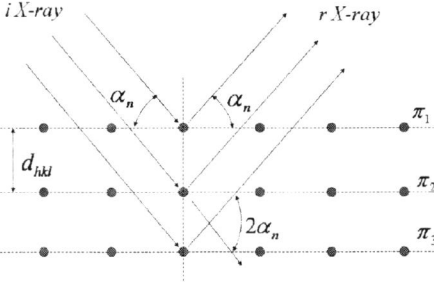

Fig. 7.3 X-ray selective reflection on the crystal atoms (filled circles) pertaining to the reticular plane family $\pi_i (i = 1, 2, \ldots)$ described by the Miller indices hkl. Notations: $i\,X\text{-}ray$ – incident X-rays; $r\,X\text{-}ray$ – reflected X-rays; d_{hkl} – distance between two adjacent reticular planes of the family (hkl).

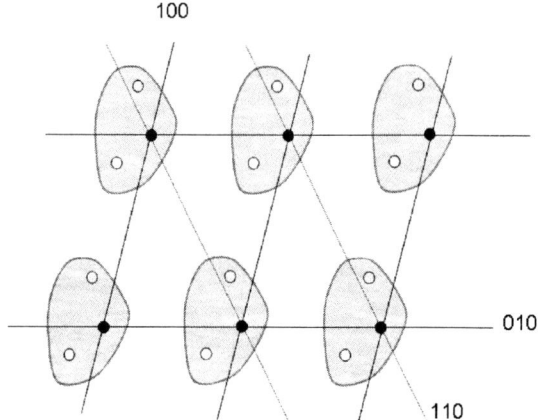

Fig. 7.4 Schematic representation of a small fraction of a protein crystal. The protein is shown as a large contour, in which three atoms are highlighted. As one can easily see, the atoms represented by same color (or same gray level) form a lattice in which they are lying on a family of atomicly smooth mirror planes indicated by their Miller index. The same argument holds for all atoms in the protein and also for the center of mass of each molecule (represented in black). This can be generalized: Each family of planes crossing the lattice points connect equivalent molecules and are, therefore, mirror planes.

X-ray scattering is stronger for atoms with larger atomic number, Z, due to the higher electron density of those atoms. In the case of proteins, however, the constitutive elements are usually elements with low atomic numbers, ^{1}H, ^{6}C, ^{8}O, and ^{16}S, and only rarely ^{26}Fe, ^{27}Co, and ^{30}Zn (which also appear in small quantities in prosthetic groups), so that the proteins diffract X-rays only slightly.

One can enrich proteins with heavy metal ions, and compare the diffraction patterns of the native protein with the one with heavy atoms added. Most protein crystals used in X-ray crystallography contain a known number of heavy metal ions (e.g., ^{78}Pt^{2+}, ^{80}Hg^{2+}) attached in known positions to the protein chain (such as, for instance, to the –SH groups of Cysteine). The process of inserting heavy ions into protein chains must conserve the crystal original form, with the exception that certain atoms are replaced; that means that the original and modified proteins are isomorphous. In practice, multiple isomorphic derivatives of a macromolecule need to be studied, in order to overcome the ambiguity related to an absence of the phase information (see below).

Because a crystal presents spatial periodicity, its centers of diffraction (i.e., the electronic clouds of its atoms) may be described as being situated in different reticular planes characterized by the *Miller indices h, k*, and *l*. As a result of scattering on regularly distributed centers, and of the interference (Fig. 7.3), a mathematical transformation is effectively performed of a 3-D structure (not yet known) into an experimentally measurable 2-D diffraction image. More precisely, diffraction performs a *Fourier transform* of a periodic arrangement in the real 3-D space (i.e., the 3-D structure) to render an image in the *inverse* Fourier space. Therefore, the problem of resolving the 3D structure is equivalent to performing an *inverse Fourier*

7.1 Channels and Pores

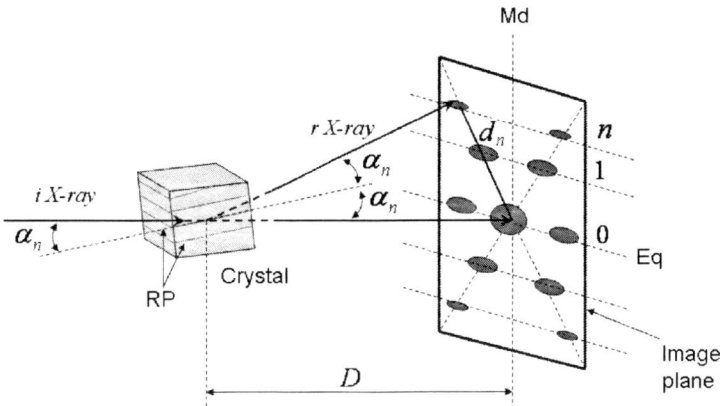

Fig. 7.5 Schematic depiction of the X-ray diffraction (or selective reflection) by a crystal and the diffraction image in the case of a fibrillar macromolecule. Notations: *i X-ray* – incident X-rays; *r X-ray* – reflected X-rays; RP – equivalent reticular planes inside the crystal; D – distance between crystal and image plane; Eq – equator of the image; Md – meridian of the image; n – order of diffraction; d_n – distance between the central diffraction spot (originating from the undiffracted X rays) and the nth order diffraction spot; α_n – incidence/reflection angle.

transform from the experimentally-recorded diffraction image back into the real 3-D space in which the unknown structure is embedded.

From Fig. 7.5, one may observe that the diffraction/reflection angle entering equation (7.1) obeys the relation:

$$\alpha_n = \frac{1}{2}\arctan(d_n/D) \tag{7.2}$$

A crystal structure may be obtained formally by repetitive arrangement of an elementary structure along three directions in space, Ox, Oy and Oz (which are not necessarily perpendicular to one another). This repetitive unit is called the *unit cell* of the crystal and is characterized by three sides $(\vec{a}, \vec{b}, \vec{c})$ and three angles (α, β, γ), as shown in Fig. 7.6.

The inverse Fourier transform of the diffraction image obtained from proteins described by, e.g., an orthorhombic system (with $a \neq b \neq c, \alpha = \beta = \gamma = 90°$, see Fig. 7.6) gives the electron density as:

$$\rho(x,y,z) = \frac{1}{V}\sum_h\sum_k\sum_l F(hkl)e^{-2\pi i(hx/a+ky/b+lz/c)} \tag{7.3}$$

where V is the volume of the unit cell, $F(hkl)$ are structure factors associated with the reticular planes (hkl), and $i = (-1)^{1/2}$. The summation in (7.3) is made over all the reticular planes (in principle, h, k, and l vary from $-\infty$ to $+\infty$), each family of parallel reticular planes contributing to a diffraction peak in the diffraction image. The higher the order of diffraction spots that can be measured the better the resolution of the recovered electron density. Therefore, reticular planes with higher Miller indices are important only because they improve resolution.

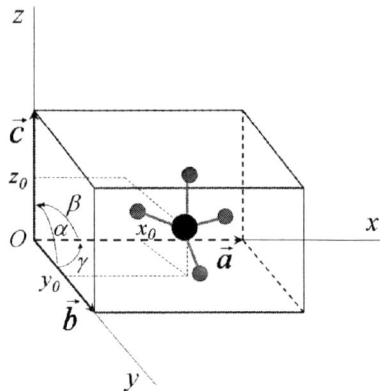

Fig. 7.6 Schematic representation of a unit cell associated with the crystal of a hypothetical molecule composed of five atoms, four of which are identical. The central atom (large filled circle) has the coordinates $x_0 = a/2$, $y_0 = b/2$, $z_0 = c/2$.

The structure factors, $F(hkl)$, are complex quantities which may be expressed as:

$$F(hkl) = |F(hkl)| e^{i\alpha(hkl)} \tag{7.4}$$

where $|F(hkl)|$ represent the magnitudes of $F(hkl)$ and $\alpha(hkl)$ are their phase. The magnitudes may be directly related to experimentally measurable intensities, $I(hkl)$, of the diffraction spots, according to:

$$|F(hkl)| = \sqrt{I(hkl)/K} \tag{7.5}$$

where K is a known experimental factor. By contrast, the phases are much more difficult to recover experimentally, although they too are required for carrying out the triple summation in equation (7.3). Phase recovery constitutes one of the greatest challenges in 3D protein structure determination from X-ray diffraction. In practice, there are methods to recover the structure factor phases, which include *multiple isomorphous replacement* of certain atoms by heavy metal ions, all of which are quite laborious.

If the primary structure of the protein is known and once phases are determined, at least partially, determination of the spatial arrangements of atoms in the protein (i.e., the tertiary and quaternary structure) may proceed. First, the known magnitudes and the approximate phases of the structure factors are introduced into expression (7.3) to compute the approximate electron density. From this, one can deduce the approximate spatial coordinates $[(x_i, y_i, z_i), i = 1, 2, \ldots, N)]$ of the N atoms in the protein. At this point, one can obtain better approximations of the phases, $\alpha_1^c(hkl)$, from the approximate spatial coordinates and by using a direct Fourier transform. The new phases are re-introduced into expression (7.3) together with the experimentally determined magnitudes of the structure factors, from which improved estimates of the electronic densities, $\rho_1(x_i^1, y_i^1, z_i^1)$, are obtained, which lead to a new set of phases, $\alpha_2^c(hkl)$, and so on. The process of structure refinement continues until a so-called *reliability factor*,

$$R_i = \frac{\left|\sum_h \sum_k \sum_l |F^{ex}(hkl)| - |F_i^c(hkl)|\right|}{\sum_h \sum_k \sum_l |F^{ex}(hkl)|} \qquad (7.6)$$

which is determined at each refinement step, i, from computed (superscript "c"), and measured (superscript "ex") $|F(hkl)|$ values, decreases below an arbitrarily set value of 0.25; this value is usually regarded as a good practical limit. The structure of the macromolecule thus obtained may be represented graphically, by using computer programs, in various ways – as, protein backbone, ribbon representation, etc. – as shown in section 2.2.1.4.

Maintaining the X-ray diffraction's dominant role in structural molecular studies has been made possible by steady increase in sensitivity of diffractometric techniques and by notable advances in obtaining high quality protein crystals. However, exhausting the pool of crystallizable proteins is a growing challenge in X-ray crystallography which needs to be overcome if it were to at least maintain, if not to increase, the current pace of protein structure determinations. In chapter 2, we have mentioned a possible way for circumventing this problem.

7.2 X-Ray Investigations of Channels and Pores

Much of the knowledge of channels structure and function accumulated over the past few decades originated from electrophysiological and other biophysical studies – including patch-clamp – which were often coupled with artificially induced genetic mutations in the channels that affected their function. X-ray diffraction studies of channels and pores have been hampered by difficulties in obtaining crystals of such proteins, which, as discussed in section 2.2, constitutes the major problem with X-rays studies of most membrane proteins. This situation has persisted until relatively recently, when the success of Roderick MacKinnon and his colleagues in crystallizing channel proteins has led to a series of breakthroughs in the study of high-resolution channel structures (Doyle et al., 1998; Dutzler et al., 2002; Jiang et al., 2003).

As shown in Fig. 7.7, several classes of channels form *oligomeric complexes* of identical monomeric units. While voltage-dependent potassium channels and ligand-gated cation or anion channels form a single pore at the center of the oligomeric structure, the monomers of the voltage-gated chlorine channels and the ligand-gated aquaporin possess their own individual pores (Jensch, 2002). In this chapter, we discuss the general structural and functional features of the potassium and chlorine channels. For ligand-gated channels, such as the *aquaporin*, the reader is referred to the literature in the field (see, e.g.: Törnroth-Horsefield et al., 2006).

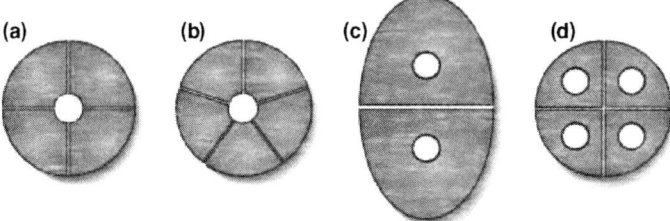

Fig. 7.7 Schematic representation of different types of oligomeric assemblies of protein subunits, or monomers: (**a**) tetrameric potassium channels and (**b**) ligand-gated ion channels (such as the gamma-aminobutyric acid-gated, or GABA-gated channels) forming a single central pore; (**c**) chlorine channels; and (**d**) water channels (aquaporin) forming two or more pores. Figure reprinted by permission from Macmillan Publishers Ltd: Jentsch (2002).

Figure 7.8 shows the tetrameric structure as seen from two perpendicular directions of a voltage-gated K^+ channel determined from X-ray crystallography (Doyle et al., 1998). The transmembrane helices form a hydrophilic pore, padded generously with water molecules, for the passage if ions through the otherwise hydrophobic core of the membrane.

The permeability of the channel is controlled by a fast and highly efficient voltage-sensitive gate that switches the channel between an *open* and a closed *state*. From studies of several K^+ channels, a common feature has emerged that the voltage sensor, formed by charged helices whose charges, may move across the membrane following changes in the transmembrane potential (Jiang et al., 2003). This motion probably induces conformational changes in the channel, which may thus become open or close.

According to Yellen (2002), the amazing combination of high selectivity and high permeability of potassium channels arises from several specific features of the channel architecture, as illustrated in Fig. 7.9 and described next.

- The permeation pathway contains water molecules, which help stabilize the ion in the pore – see label (1) in Fig. 7.9.
- The electrical dipoles inherent in every α-helix are oriented such that the negative end is oriented towards the center of the pore (2). This is assumed to introduce a preferential stabilization of the cations, thereby preventing the anions from entering the channel.
- As seen in chapter 1, positive ions dissolved in water are surrounded by the oxygen atoms of water molecules forming a hydration shell around the ion. The potassium channel contains a selectivity filter that mimics the structure of the hydration shell of the potassium ion in water (3), but not of other cations, such as Na^+. As seen in Fig. 7.9, the potassium ions in the channel form a chain in which individual ions are separated by a distance of about 7 Å (4), with hydration water interspersed along the chain.

In contrast to the potassium channels, chlorine channels are dimeric complexes, each monomer forming a barrel-like structure around its own pore (see Fig. 7.10).

7.2 X-Ray Investigations of Channels and Pores 183

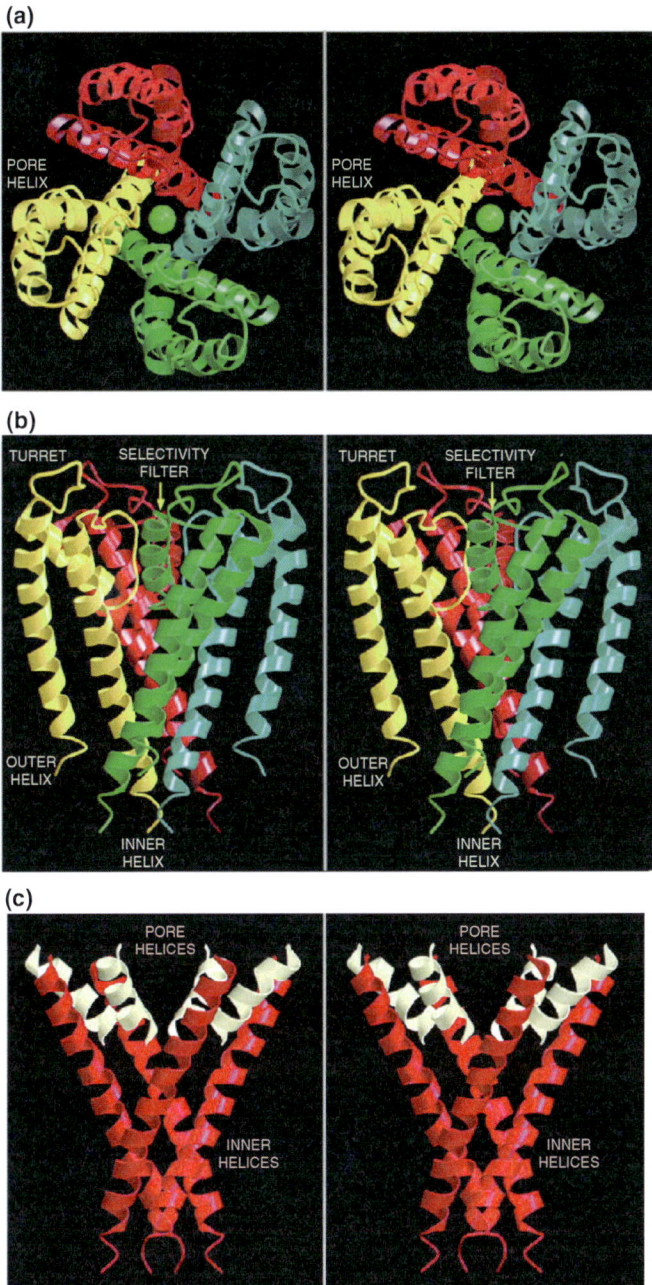

Fig. 7.8 Ribbon representation of a voltage-gated potassium channel (KcsA). (**a**) Stereo view illustrating the three-dimensional fold of the tetramer viewed from the extracellular side. Each color indicates a different subunit. (**b**) Stereo view from a direction contained in the membrane plane. (**c**) Stereo view of the inner helices only. Notice the top (white) helices creating a selectivity filter. From Doyle et al. (1998). Reproduced with permission from AAAS.

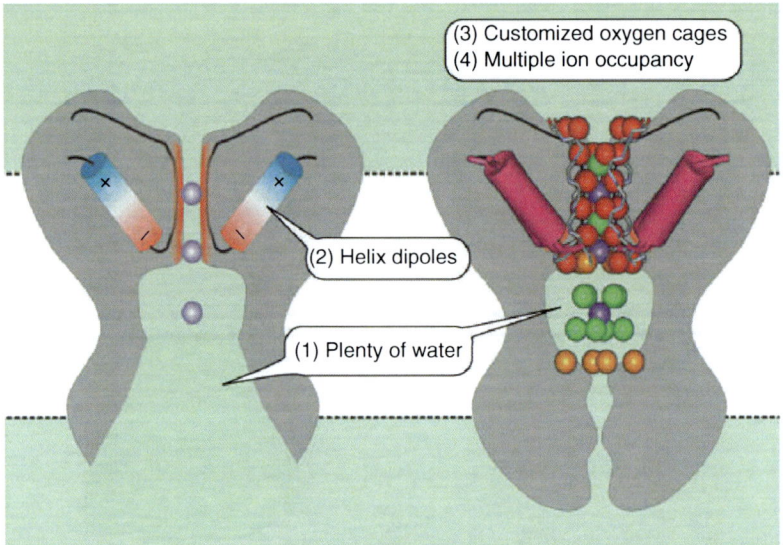

Fig. 7.9 Architectural features of K+ channels leading to high selectivity and permeability. Potassium ions are represented by purple spheres, while the water appears in green. Cartoon reprinted by permission from Macmillan Publishers Ltd: Yellen (2002).

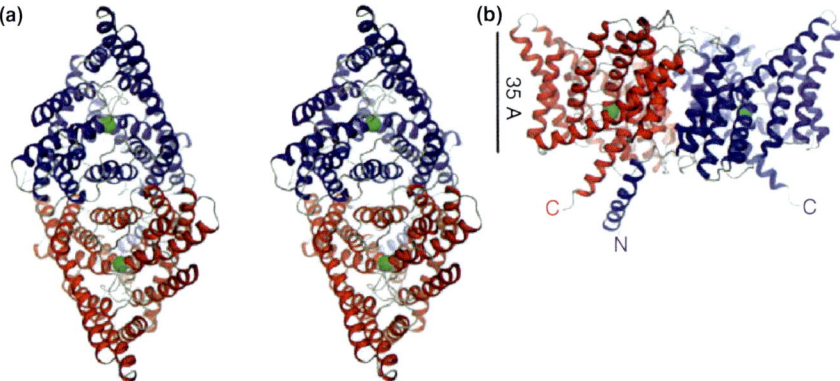

Fig. 7.10 (**a**) Stereo view of a ribbon representation of a chlorine-channel from the extracellular side. (**b**) View of the same channel from a direction within the membrane plane, with the extracellular side above the structure. The green spheres represent the selectivity filter of the Cl^- channel. Figure reprinted by permission from Macmillan Publishers Ltd: Dutzler et al. (2002).

Other differences between Cl^- and K^+ channels include the absence of an aqueous cavity form the chlorine channels and, most notably, a negatively charged side chain of a glutamate amino acid. The latter is presumed to act as a gate that swings in and out of the diffusing pathway (Dutzler et al., 2002).

Potassium and chlorine channels, each actually constitutes large families of channels that are present in all plants and animal kingdoms. The slight differences observed from channel to channel within a family of channels probably reflect the variety of specific roles ion channels play in different organisms as well as in different cell types of the same organism. On the other hand, the more notable differences between the two families of channels await further clarification.

Observation: In recognition of his important contributions to the structural and mechanistic characterization of ion channels, Roderick MacKinnon was awarded the Nobel Prize for Chemistry in 2003 (MacKinnon, 2003).

7.3 Ion Pumps

7.3.1 The Na-K-ATPase

On discussing the main ideas of active transport in chapter 4, we have postponed the discussion on the relation between the structure of the ion pump and its function. To be precise, there are several types of ion pumps, one of the most important being the *Na,K-ATPase*, already discussed in chapter 4. Na,K-ATPase is an integral membrane protein which couples the active transport of Na and K ions to ATP hydrolysis, and thereby acts as an enzyme for splitting the ATP into ADP.

The Na,K-ATPase is a *molecular machine* that pumps three Na ions out and two K ions into the cell per cycle, performed in about 10 ms (Alberts et al., 2002). Because, the concentration ratio of the transported ions is $[Na^+]/[K^+] = 3/2$, the pump has an *electrogenic* character (i.e., it contributes to the electrical polarization of the cell membrane) due to the net positive charge transferred outside the cell.

The global reaction describing the Na,K-ATPase function is:

$$3Na^+_{in} + 2K^+_{out} + ATP_{in} \rightarrow 3Na^+_{out} + 2K^+_{in} + ADP_{in} + P_{i,in} + \Delta G_{in} \quad (7.7)$$

where, ADP is adenosine diphosphate, P_i inorganic phosphate, and ΔG_{in} is the Gibbs free energy involved in the reaction, which is of the order of $\sim -50\,\text{kJ/mol}$ ($\sim -12\,\text{kcal/mol}$).

Observation: Because at each cycle the ionic pump translocates simultaneously Na^+ out of the cell and K^+ into the cytosol, the pump is called an *antiport system* or, simply, an *antiporter*. If the two different ions are transported in the same direction, the transport system is called a *symport system* or *symporter*. If only one particle is transported in a single direction, the system is called a *uniporter* (e.g., carriers).

There are several variants of the Na,K-ATPase, all of which are composed of two types of protein subunits: α and β subunits (Kaplan, 2002; Hebert et al., 2003). The α-subunit is composed of ten transmebrane domains (helices) with a total molecular mass of approximately 110 kDa. A part of the α-subunit sticks out of the membrane into the cytoplasm and contains the *catalytic active site* of this pump, which is the site of ATP attachment and hydrolysis. The β-subunit has a single transmembrane domain with a molecular mass of 55 kDa. A short amino acid chain is exposed to the cytosol, while a larger part of the subunit forms the glycosilated extracellular portion. The β-subunit plays an essential role in delivery and insertion of the α-subunit in the membrane. In addition to the α- and β-subunits, some Na,K-ATPase pumps have also a small γ-subunit with a molecular mass of about 7.3 kDa, which is believed to play a role in regulation of the enzymatic activity of the pump.

Apart from the crystal structure of some small fragments (posted on the Protein Data Bank, http://www.rcsb.org/pdb), a complete crystal structure of the Na, K-ATPase has not been obtained yet, due to the difficulty of obtaining 3D crystals. However, cryo-electron microscopy of two dimensional crystals in membranes (Herbert et al., 2003; Rice et al., 2001), together with other biophysical methods, furnished a rather complete picture of its structure. In addition, determination by Toyoshima et al. (2000) of the structure of a related ion pump, the Ca-ATPase, has allowed for useful comparisons to be made. Figure 7.11 shows outlines of the α- and β-subunits of the Na,K-ATPase.

In principle, the operation of the Na$^+$-K$^+$-ATPase is based on its ability to undergo *cyclic conformational changes* driven by the energy released from ATP hydrolysis. In the terminology used to describe enzymatic reactions, the series of successive conformational states of the pump are described by two main states: one with high

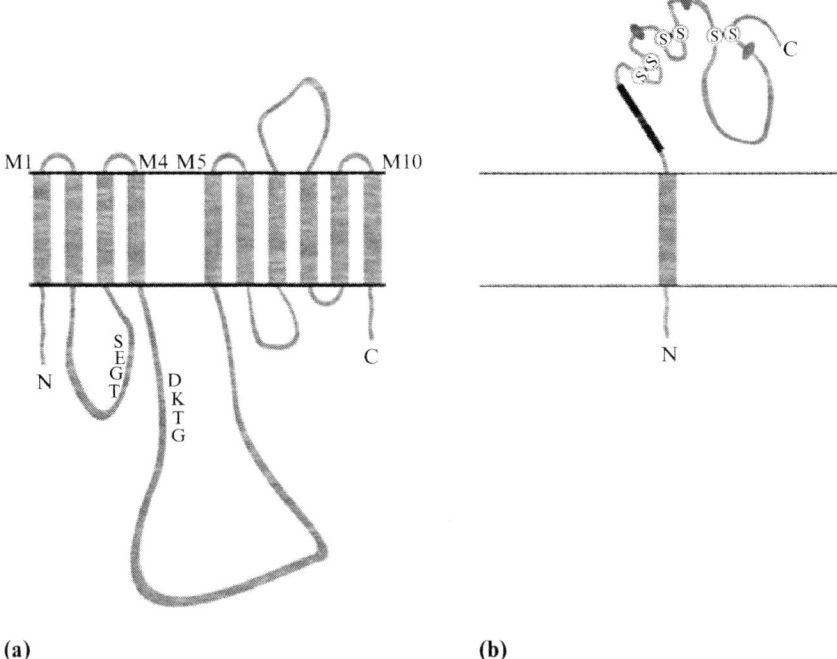

Fig. 7.11 Outline of (**a**) the α- and (**b**) β-subunit of the Na$^+$,K$^+$-ATPase. The heavy lines in the β-subunit indicate sequence involved in associations with the M7M8 loop of the α-subunit. Figures reproduced from Kaplan (2002), with permission from Annual Reviews.

7.3 Ion Pumps

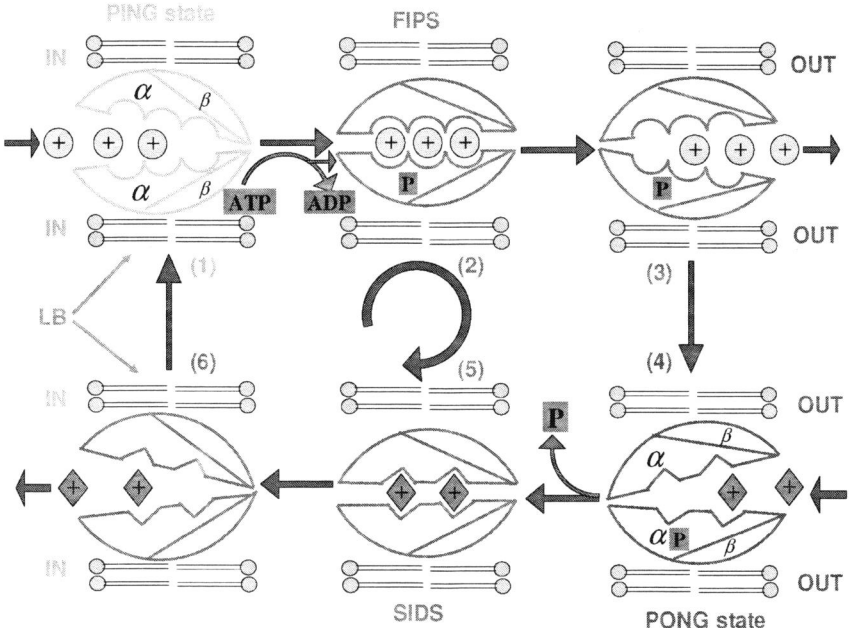

Fig. 7.12 The ping-pong mechanism accounting for the Na,K-ATPase functioning. Significance of the symbols: IN, OUT – extracellular, respectively intracellular space; FIPS – first intermediate phosphorylated state of the pump; SIDS – second intermediate dephosphorylated state. LB – lipid bilayers. Na ions are represented by circles and K ions by diamonds.

affinity for Na^+, which is called "*ping*," and another one with high affinity for K^+, called "*pong*." This "*ping-pong*" mechanism is illustrated in Fig. 7.12.

According to the ping-pong mechanism, the Na,K-ATPase is *phosphorilated* on its cytoplasmic side (α-monomer). This brings the pump into the *ping* state which is characterized by low affinity for K ions and very high affinity for Na ions. In this state, the pump binds three Na ions and undergoes a conformational change into a *first intermediate phosphorilated state* (FIPS), which leads to internalization of the Na ions. This is followed by the transition to the *pong state*, which is characterized by very low affinity for Na^+, the three captive Na ions being now released to the extracellular side of the membrane (where their concentration is already high). But, in the pong state, Na, K-ATPase presents a very high affinity for K^+ on the external side of the pump, which will now bind two K ions. This state is followed by the dephosphorylation of the α-subunit, which brings the pump into a *second intermediate dephosphorilated state* (SIDS), when the K ions are internalized within the membrane. In the "final" ping state, which presents low affinity for K^+, the two captive K ions are released inside the cell, where their concentration is already higher than outside. Thus, one full cycle of Na, K-ATPase active pumping of ions is complete, the pump being ready to enter the next cycle.

7.3.2 Other Ionic Pumps

In addition to the Na,K-ATPase, other ion pumps, such as H,K-ATPase, Ca-ATPase, K-ATPase, H-ATPase, and even light-activated proton pumps, may be found in the plasma membranes and organelle membranes of different types of cells (Stryer, 1988; Alberts et al., 2002).

H,K-ATPase, found in the cell membranes of the gastric mucosa (which plays a role in food digestion), maintains a very low pH of the gastric juice (Stryer, 1988), which is required for optimal functioning of different enzymes. The general equation describing the kinetics of reaction for this antiporter is:

$$nH^+_{in} + nK^+_{out} + ATP_{in} \rightarrow nH^+_{out} + nK^+_{in} + ADP_{in} + P_{i,in} + \Delta G_{in} \qquad (7.8)$$

where n is an integer.

Ca-ATPase actively maintains a Ca^{2+} concentration difference of four orders of magnitude between the interior and the exterior of the cell (Alberts et al., 2002). The very low concentration of Ca ions in the cytosol (0.1 μM) is required for cellular signal transmission from an extracellular medium to the interior of the cell, by passive Ca^{2+} diffusion following specific chemical or physical triggers. The general equation describing the role of this *uniporter* is:

$$2Ca^{2+}_{in} + ATP_{in} \rightarrow 2Ca^{2+}_{out} + ADP_{in} + P_{i,in} + \Delta G_{in} \qquad (7.9)$$

A particularly interesting ion pump is the H-ATPase, found in the membrane of some bacteria (e.g., *Halobacterium halobium* and *Escherichia coli*). This pump uses energy either from photons absorbed by a membrane pigment (the *bacteriorhodopsin*), in the case of *Halobacterium halobium* (Stryer, 1988), or from ATP hydrolysis, in the case of *Escherichia coli*, to transport H ions outside the cell against their concentration gradient. In their turn, proton concentration gradients across the membrane constitute a source of energy that may be used either to perform *chemical work* (i.e., to synthesize ATP in the dark in *Halobacterium halobium* cells), or to perform *mechanical work* (in the case of flagellated bacteria, such as *Escherichia coli* and *Salmonella thyphimurium*). In the latter case, the same ionic pump that expelled H^+ outside the cell may operate in reverse (Berg et al., 2002).

Observation: In the case of *Halobacterium halobium*, when the H-ATPase is fuelled by ATP or light, it transports H^+ outside the cell, creating an important proton concentration gradient. While in the dark, the bacterium expends the energy stored in this proton concentration gradient to synthesize ATP (chemical energy) needed for the cell metabolism. Therefore, the motor will operate in reverse, the H-ATPase, becoming an ATP-synthase. We are witnessing here an unexpected coupling between two processes with different *tensorial* orders: a *vectorial* process that transports molecules along a given direction (the passive diffusion) and a *scalar* process (the chemical reaction). This coupling could not take place in a homogeneous medium according to the *Curie Principle* (Prigogine, 1947), which in essence states that a scalar cause cannot produce a vectorial effect. However, the membrane and the cytosol are highly anisotropic structures, thereby favoring the coupling.

7.4 Light Absorption in Photosynthesis

In the previous chapters and also above in this chapter, we have discussed about conversion and use of energy in biological processes. But how did this energy come about in the first place? Organisms are thermodynamic systems that function far from equilibrium and which build and maintain their internal organization by consuming some sort of order from their environment, which Schrödinger called *negentropy* (Schrödinger, 1992). This is obtained from energy reservoirs in the universe – the stars (the Sun, in our case) – and is converted by *photosynthetic* organisms on the Earth (such as green plants and bacteria) and stored as free energy (Meszena and Westerhoff, 1999). The stored free energy is used in biological processes, for example to pump ions across the cell membrane against their concentration gradients, as discussed above. This section attempts to provide a thermodynamic description of the process of light absorption by special molecular machines involved in the processes of *photosynthesis* in green plants and bacteria.

7.4.1 Brief Overview of the Mechanism of Photosynthesis

Photosynthesis is a process of converting the light energy into bio-chemical energy, which consumes CO_2 and water and generates oxygen. As already mentioned in chapter 2, photosynthesis occurs in the *thylakoid* membranes of the specialized cell organelles – the *chloroplasts*. The main light-absorbing molecules in photosynthetic systems in bacteria and higher plants are *chlorophylls* and *carotenoids*. Chlorophyll molecules are bound into a protein complex, called the *light-harvesting system*, and arranged in space to form an *antenna* (Fig. 7.13) that efficiently absorbs light from the Sun and guide it to the two *reaction centers* where the conversion into free energy occurs (Gobets and van Grondelle, 2001; van Amerongen and van Grondelle, 2001; Atkins and de Paula, 2002; Berg et al., 2002).

The process of light-harvesting starts with absorption of photons by a chlorophyll molecule from the light-harvesting complexes, which is brought into an excited state, from where it decays within a few picoseconds (through nonradiative processes) to a state with lifetime of the order of nanoseconds (Nordlund and Knox, 1981). During its excited lifetime, there is a high probability that the chlorophyll looses its energy through resonance energy transfer to an unexcited nearby chlorophyll that has appropriate orientation relative to the first one (see chapter 8, for a brief description of the mechanism of resonance energy transfer, in a different context). The excitation can thus hop from chlorophyll to chlorophyll until it reaches a special chlorophyll called the *primary electron donor* of the reaction center, or chlorophyll P. The excited chlorophyll P transfers its energy to a chain of electron-transfer reactions that lead ultimately to storage of light energy into chemical energy, e.g., ATP and NADPH synthesis (Atkins and de Paula, 2002).

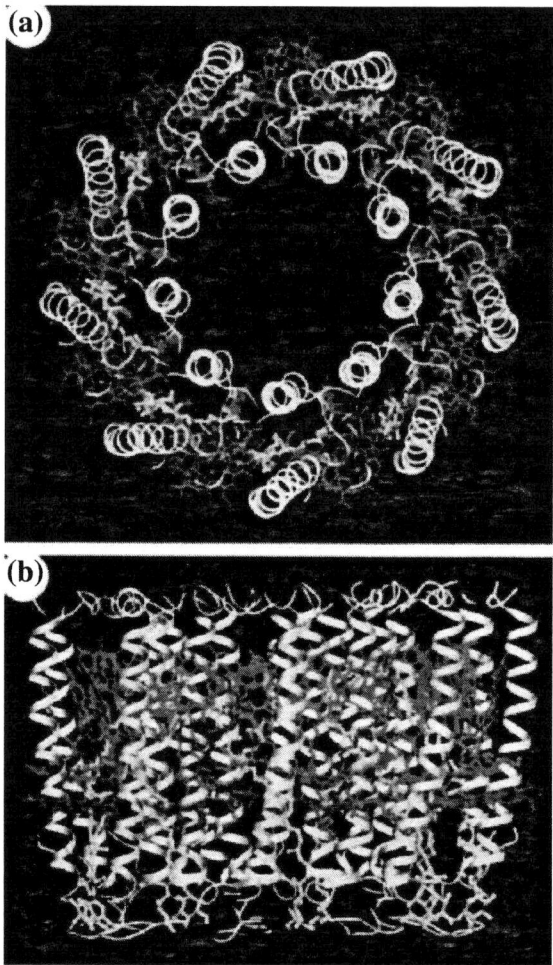

Fig. 7.13 Crystal structure of a light-harvesting complex from bacteria. (**a**) View of the complex from the cytoplasmic side of the membrane. (**b**) View from a direction perpendicular to the symmetry axis. Two types of chlorophyll a, B800 and B850, are shown in green and red color, respectively. The two chlorophylls absorb at 800 nm and 850 nm, respectively. The two chlorophylls are bound non-covalently to two hydrophobic apoproteins (shown in white) alongside with carotenoids (in yellow). The whole complex is an oligomer of these components. Figure reprinted from Macmillan Publishers Ltd: McDermot et al. (2002).

7.4.2 Thermodynamics of Light Absorption

In this section, we will introduce the thermodynamic aspects of the very first step in photosynthesis – the light energy absorption by chlorophyll. In doing so, we will follow closely the derivation presented by Lavergne and Joliot (2000).

7.4 Light Absorption in Photosynthesis

Consider the "reaction" of *chlorophyll* excitation-deexcitation at equilibrium when monochromatic photons of frequency, v, are absorbed and then are lost by chlorophyll:

$$Chl + hv \underset{k_d}{\overset{k_a}{\rightleftarrows}} Chl^* \tag{7.10}$$

where *Chl* stands for the chlorophyll in the ground state and *Chl** signifies the excited chlorophyll, k_a is the rate constant for light absorption and k_d is the rate constant for *Chl** de-excitation.

Since the process may be regarded as isotherm and isobar, populations of excited and unexcited chlorophylls may be described by chemical potentials, which we define here with respect to the number of molecules, N, as $\mu = \left(\frac{dG}{dN}\right)_{T,p}$. The two chemical potentials are:

$$\mu^{Chl} = \mu_0^{Chl} + k_B T \ln[Chl] \tag{7.11}$$

$$\mu^{Chl^*} = \mu_0^{Chl^*} + k_B T \ln[Chl^*] \tag{7.12}$$

where μ_0^{Chl} and $\mu_0^{Chl^*}$ are the standard chemical potentials of the two species and k_B is the Boltzmann constant, as usual.

If the system does not reach equilibrium, free energy may be extracted by converting *Chl** into *Chl*, which is proportional to the so-called *affinity*, defined by the difference of the chemical potentials described by equations (7.12) and (7.11), namely:

$$A = \mu_0^{Chl^*} - \mu_0^{Chl^*} + k_B T \ln \frac{[Chl^*]}{[Chl]} = \mu_0^{Chl^*} - \mu_0^{Chl^*} + k_B T \ln \frac{k_a}{k_d} \tag{7.13}$$

Note that the affinity is non-zero only if the system functions far from equilibrium, which should indeed be the case with the light-harvesting process in photosynthesis, where photon excitations are efficiently funneled towards a reaction center that converts energy in electrochemical form.

Assuming negligible changes in volume and entropy between the two states of the chlorophyll, then $dG = hv + pdV - TdS \cong hv$ (where h is Planck's constant), hence:

$$A = hv + k_B T \ln \frac{[Chl^*]}{[Chl]} = hv + k_B T \ln \frac{k_a}{k_d} \tag{7.14}$$

The concentration ratio in this equation may be determined from the Boltzmann formula:

$$\frac{[Chl^*]}{[Chl]} = \exp\left(-\frac{hv}{k_B T_{bb}}\right) \tag{7.15}$$

where the absorber is considered to be a black-body at a temperature, T_{bb} (Lavergne and Joliot, 2000; Meszena and Westerhoff, 1999). By combining the last two equations, we obtain immediately:

$$A = hv\left(1 - \frac{T}{T_{bb}}\right) \tag{7.16}$$

where the expression in the bracket is the yield of some Carnot machine that produces work from heat transfer between a "hot" source at temperature T_{bb} (a black body that absorbs energy from Sun) and a cold source (the external medium) at temperature T.

Quiz 2. Estimate the Carnot yield of the chlorophyll under light illumination with wavelength of 663 nm at the ambient temperature of $27\,°C$, assuming that the lifetime of the excited state is $1/k_a \cong 1\,\mathrm{ns}$ (Nordlund and Knox, 1981) and that $k_d = 0.1\,\mathrm{s}^{-1}$. Estimate also T_{bb}.

References

Alberts, B. A., Lewis, J., Raff, M., Roberts, K. and Walter, P. (2002) *Molecular Biology of the Cell*, 4th ed., Garland Science/Taylor & Francis, New York
Atkins, P. and de Paula, J. (2002) *Atkins' Physical Chemistry*, 7th ed., Oxford University Press, Oxford
Berg, J. M., Tymoczo, J. L. and Stryer, L. (2002) *Biochemistry*, 5th ed., W. H. Freeman, New York
Doyle, D. A., Morais Cabral, J., Pfuetzner, R. A., Kuo, A., Gulbis, J. M., Cohen, S. L., Chait, B. T. and MacKinnon, R. (1998) The structure of the potassium channel: molecular basis of K^+ conduction and selectivity, *Science*, **280** (5360): 69
Drenth, J. (1994) *Principles of X-Ray Crystallography*, Springer, New York
Dutzler, R., Campbell, E. B., Cadene, M., Chait, B. T. and MacKinnon, R. (2002) X-ray structure of a ClC chloride channel at 3.0 A reveals the molecular basis of anion selectivity, *Nature*, **415**: 287
Gobets, B. and van Grondelle, R. (2001) Energy transfer and trapping in photosystem I, *Biochim. Biophys. Acta*, **1507**: 80
Herbert, H., Purhonen, P., Thomsen, K., Vorum, H. and Maunsbach, A. B. (2003), Renal Na,K-ATPase structure from cryo-electron microscopy of two-dimensional crystals, *Ann. NY. Acad. Sci.*, **986**: 9
Hille, B. (2001) *Ion Channels of Excitable Membranes*, 3rd ed., Sinauer, Sunderland, MA
Jentsch, T. J. (2002) Chloride channels are different, *Nature*, **415**: 276
Jiang, Y., Lee, A., Chen, J., Cadene, M., Chait, B. T. and MacKinnon, R. (2003) X-ray structure of a voltage-dependent K^+ channel, *Nature*, **423**: 33
Kaplan, J. H. (2002) Biochemistry of Na,K-ATPase, *Annu. Rev. Biochem.*, **71**: 511
Kendrew, J. C., Bodo, G., Dintzis, H. M., Parrish, R. G., Wyckoff, H. and Phillips, D. C. (1958) A three-dimensional model of the myoglobin molecule obtained by X-ray analysis, *Nature*, **199**: 662
Lavergne, J. and Joliot, P. (2000) Thermodynamics of the excited states of photosynthesis, In: *BTOL-Bioenergetics*, chapter 2 (http://www.biophysics.org/btol/bioenerg.html)
MacKinnon, R. (2003) Potassium Channels and the Atomic Basis of Selective Ion Conduction, *Nobel Lecture*, http://nobelprize.org/nobel_prizes/
Malmivuo, J. and Plonsey, R. (1995) *Bioelectromagnetism*, Oxford University Press, New York
McDermott, G., Prince, S. M., Freer, A. A. Hawthornthwaite-Lawless, A. M., Papiz, M. Z., Cogdell, R. J. and Isaacs, N. W. (2002) Crystal structure of an integral membrane light-harvesting complex from photosynthetic bacteria, *Nature*, **374**: 517
Meszena, G. and Westerhoff, H. V. (1999) Non-equilibrium thermodynamics of light absorption, *J. Phys. A: Math. Gen.*, **32**: 301
Neher, E. (1991) Ion channels for communication between and within cells, *Nobel Lecture*, http://nobelprize.org/nobel_prizes/
Neher, E. and Sakmann, B. (1976) Single-channel currents recorded from membrane of denervated frog muscle fibers, *Nature*, **260**: 799

References

Neher, E. and Sakmann, B. (1992) The patch clamp technique, *Sci. Am.*, **266**: 28

Nordlund, T. M. and Knox, W.H. (1981) Lifetime of fluorescence from light-harvesting chlorophyll a/b proteins. Excitation intensity dependence, *Biophys. J.*, **36**: 193

Prigogine, I. (1947) *Etude Thermodynamique des Phénomènes Irréversibles*, Desoer, Liége

Rice, W. J., Young, H. S., Martin, D. W., Sachs, J. R., and Stokes, D. L. (2001) Structure of Na^+, K^+-ATPase at 11 Å resolution: comparison with Ca^{2+}-ATPase in E_1 and E_2 states, *Biophys. J.*, **80**: 2187

Rhodes, G. (1993) *Crystallography Made Crystal Clear. A Guide for Users of Macromolecular Models*, Academic, San Diego, CA

Sakmann, B. (1991) Elementary steps in synaptic transmission revealed by currents through single ion channels, *Nobel Lecture*, http://nobelprize.org/nobel_prizes/

Sakmann, B. and Neher, E. (Eds.) (1995) *Single Channel Recording*, 2nd ed., Plenum, New York

Schrödinger, E. (1992) *What Is Life? The Physical Aspect of the Living Cell with Mind and Matter & Autobiographical Sketches*, Cambridge University Press, Cambridge

Stryer, L. (1988) *Biochemistry*, 3rd ed., W. H. Freeman, New York

Törnroth-Horsefield, S., Wang, Y., Hedfalk, K., Johanson, U., Karlsson, M., Tajkhorshid, E., Neutze, R. and Kjellbom, P. (2006) Structural mechanism of plant aquaporin gating, *Nature*, **439**: 689

Toyoshima, C., Nakasako, M., Nomura, H. and Ogawa, H. (2000) Crystal structure of the calcium pump of sarcoplasmic reticulum at 2.6 Å resolution, *Nature*, **405**: 647

van Amerongen, H. and van Grondelle, R. (2001) Understanding the energy transfer function of LHCII, the major light-harvesting complex of green plants, *J. Phys. Chem. B*, **106**: 604

Yellen, G. (2002) The voltage-gated potassium channels and their relatives, *Nature*, **419**: 35

Chapter 8
Protein-Protein Interactions

8.1 Probing Protein Association In Vivo

Throughout this book, we have seen that proteins can interact with small ligands (e.g., hemoglobin-O_2, enzymes-substrate, etc.), and that monomers of same type bind and fold together to form a functional form of a protein. The archetypal example of such interactions is constituted by hemoglobin, which is composed of four monomers (see chapter 2) that function in concert to bind O_2 (see "cooperativity" in chapter 5). In chapter 7, we have also seen that ion channels are oligomeric complexes of identical monomers and that ion pumps are formed of two or more subunits. There is mounting evidence that many more proteins associate for shorter or longer periods of time with proteins of their own type or different types, in order to perform certain biological functions. In fact, it appears that most biological processes rely on and are regulated by dynamic interactions between proteins in the cell.

For example, protein-protein interactions play essential roles in signal transduction pathways (Gomperts et al., 2002). Members of the large class of membrane proteins called G-protein-coupled receptors (GPCRs) – which are involved in cellular responses to signaling molecules such as hormones, neurotransmitters, and other proteins that act as local mediators – function as homo- or hetero-oligomers (Milligan et al., 2003; Overton et al., 2005).

An active area of research has emerged in recent years, which deals with the development of methods for imaging the distribution and dynamics of proteins complexes in living cells, with the determination of protein complex stoichiometry, as well as with the study of the physical mechanisms involved in protein association and their relevance to the biological function.

> **Observation:** Molecular complexes formed as a result of association of protein monomers are called *homo-oligomers* (i.e., homo-dimers, homo-trimers, homo-tetramers, etc.), if the monomers are of the same type (e.g., self-association of ionic channels). The complexes are called *hetero-oligomers* if the monomers are of different type, as in the case of receptor-ligand complexes (e.g., interaction between a GPCR and the G-protein).

Early methods for detecting protein interactions relied on techniques such as immunoprecipitation, density gradient centrifugation or chromatography, which are based on the unique characteristics of complexes that allow their separation from other cellular components. However, as with any *in vitro* determinations, the results may not be entirely relevant to the *in vivo* situation. In addition, these methods do not identify the cellular location of interactions and may fail to detect interactions that form only inside cell compartments. A more recent technique, called protein fragment complementation, permits detection of protein interactions at their native locations in the cell (Rossi et al., 1997; Pelletier et al., 1998). In these methods, a fluorescent protein or enzyme is fragmented into two parts. Each fragment, which separately presents no fluorescence or enzymatic activity, is fused to a pair of potentially interacting proteins. Interaction of the two proteins of interest *in vivo* brings the two fragments together and allows for the reconstitution of the enzymatic activity or fluorescence; this is a tell-tale sign that protein-protein interaction has occurred. Although the interaction may occur at the correct physiological location, these methods yield no information about what proportion of protein is involved in the complex, or other stoichiometric information, which are of importance for biophysical studies.

With the advent of technologies for isolation, characterization, mutation and cloning of DNA from one type of cell to another, it has become possible to fuse naturally occurring fluorescent proteins to proteins of interest *in vivo*. Highlighting the proteins of interest with fluorescent tags allows one to study protein distributions *in living cells* based on detection of fluorescence (Zacharias et al., 2000; Lippincott-Schwartz and Patterson, 2003).

Observation: One of the most popular fluorescent proteins nowadays is the Green Fluorescent Protein (GFP) and its variants (including yellow, YFP, cyan, CFP, and blue, BFP, variants) (Zacharias et al., 2000; Lippincott-Schwartz and Patterson, 2003). The DNA of the original (or wild type) GFP has been isolated from a jellyfish called *Aequorea victoria*, mutated genetically, and cloned in various organisms. Several mutations have been introduced by many research laboratories around the world, which resulted in variants with reduced tendency for dimerization (which is a characteristic tendency of the wild type GFP) (Zacharias et al., 2002) improved thermostability (Cormack et al., 1996), and with various spectral as well as other physical properties (Zacharias et al., 2000; Zimmermann et al., 2002; Lippincott-Schwartz and Patterson, 2003).

Further, by taking advantage of the process of Fluorescence (or Förster) Resonance Energy Transfer (FRET) (Selvin, 1995; Clegg, 1996; Lakowicz, 2006) – a nonradiative transfer of energy from an optically excited fluorescent molecule (called "donor") to a non-excited one (called "acceptor") that resides within less than $\sim$1–5 nm – it is possible to detect interactions between proteins at their normal locations in living cells, and to actually image the distribution of protein complexes (Elangovan et al., 2002; Berney and Danuser, 2003; Zal and Gascoine, 2004; Raicu et al., 2005; Meyer et al., 2006). For instance, it is possible to apply the method to monitor the fraction of a protein population that participates in an interaction, and also to determine the average distance of separation between proteins within a protein complex. When used in conjunction with a microscope, FRET measurements can be done on single unicellular organisms such as yeast and bacteria or on small multicellular organisms such as *Drosophila melanogaster* or *Cenorhabditis elegans*. FRET studies still require tagged proteins, which could disturb somewhat the normal cellular function. However, it is generally accepted that the benefits currently outweigh this drawback.

In the following, we will first present a brief review of the phenomena of fluorescence and FRET, and then will show how they may be used for detection of protein-protein interactions, especially for the case of membrane proteins.

8.1.1 Elementary Theory of Fluorescence and FRET

Fluorescence (Lakowicz, 2006) is a process by which molecules excited through absorption of light (with a rate constant, Γ^{ex}) re-emits light with lower energy and, thereby, red-shifted (i.e., longer) wavelengths. The process is illustrated by the diagram shown in Fig. 8.1. The incident photon brings an electron from the fundamental singlet state, S_0, into the excited single state, S_1. This transition occurs with conservation of electron spin, as dictated by quantum mechanics. Part of the incident photon energy is lost through vibrational relaxation of the molecule, which is an extremely fast process occurring within $\sim 10^{-12}$ s. This process brings the photon to its lowest vibrational level in S_1, where it can reside for a few nanoseconds (or even microseconds for some substances, such as lanthanides). The remaining energy may be either re-emitted as a red-shifted photon (Stokes red-shift), with a rate coefficient, Γ^r, transferred to the triplet state, T_1, via a less probable transition (Intersystem crossing, ISC), which does not conserve spin, or transferred nonradiatively back to S_0 with a rate coefficient, Γ^{nr}. From T_1 the energy may be lost via another less probable transition to S_0. The light emitted as a result of the quantum transition from S_1 to S_0 is called *fluorescence*, while that due to transition from T_1

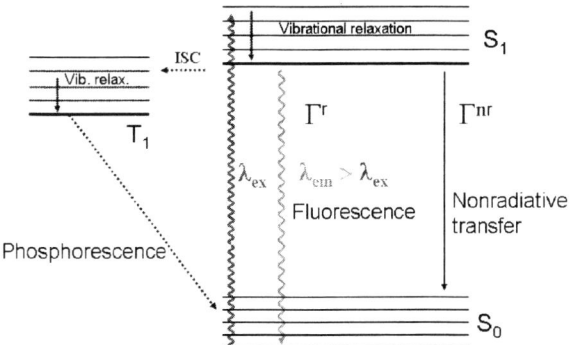

Fig. 8.1 Energy (or Jablonski) diagram for a fluorescent molecule. Significance of the symbols: S_0 – fundamental singlet state; S_1 – excited state; T_1 –triplet state; ISC – intersystem crossing; Γ^r – radiative rate of transfer from S_1 to S_0; Γ^{nr} – nonradiative transfer rate from S_1 to S_0. Wavy lines depict radiative transitions while the straight lines, the nonradiative ones.

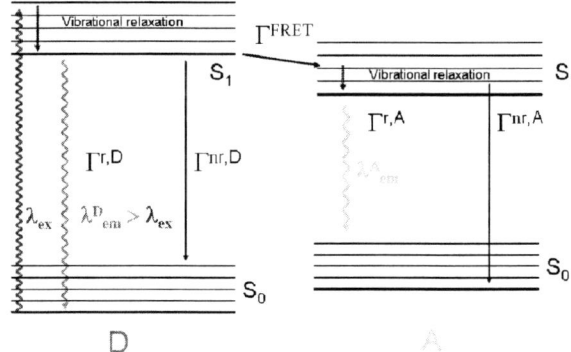

Fig. 8.2 The triplet state is not represented, for simplicity. Significance of the symbols: D – donor of energy; A – acceptor of energy; Γ^{FRET} – rate of Förster resonance transfer from excited donor to unexcited acceptor. The other symbols are defined in the legend to Fig. 8.1.

to S_0 is called *phosphorescence*. The latter type of transfer is very unlikely (since it does not conserve the spin) and its rate constant is therefore orders of magnitude lower that that of fluorescence.

If a second molecule (called "acceptor," A) capable of absorbing light at wavelengths at which the first one (called "donor," D) emits, lies close to the excited donor, the donor energy can be transferred to the acceptor through non-radiative dipole-dipole interaction called *Förster Resonance Energy Transfer* (FRET) (Selvin, 1995; Clegg, 1996; Lakowicz, 2006) (Fig. 8.2). This process brings the acceptor (A) into an excited state. If A is also a fluorescent molecule, it will emit light (with more red-shifted wavelengths compared to the donor emission) even if it had not been directly excited by light. In this case, FRET is also called *Fluorescence Resonance Energy Transfer*.

The phenomenon of transfer of excitation from an optically excited to an unexcited molecule has been known for about a century, but most theoretical attempts were unsuccessful until about the middle of the twentieth century. The first successful formulation of FRET is due to Förster, which has treated, in quantum mechanical terms, the interaction between transition diploes of a donor and an acceptor molecule. While Förster's theory has been summarized and discussed in several excellent reviews (see, e.g., Selvin, 1995; Clegg, 1996; Lakowicz, 2006), herein we will only employ its main result – the expression for Förster's rate of energy transfer [see equation (8.5)] – to formulate a simple kinetic theory of FRET in protein complexes.

To simplify the exposition, we first consider the simple case of a population of proteins forming homo-dimers (i.e., complexes containing only two identical monomers). We assume that appropriate fluorescent tags (serving as acceptors and donors of energy) are attached to each of the proteins of interest, and we aim to relate the rate constants defined above for D and A to other interesting physical properties, which may be determined experimentally.

To begin with, let us define the quantum yields of acceptors and donors as the rate of photon emission following excitation, namely:

8.1 Probing Protein Association In Vivo

$$Q^D = \frac{\Gamma^{r,D}}{\Gamma^{r,D} + \Gamma^{nr,D}}, \text{ and} \tag{8.1a}$$

$$Q^A = \frac{\Gamma^{r,A}}{\Gamma^{r,A} + \Gamma^{nr,A}} \tag{8.1b}$$

where the radiative, Γ^r, and nonradiative, Γ^{nr}, rate constants for de-excitation of donors and acceptors are defined in Fig. 8.2. These may be related to the lifetimes of the excited, S_1, states of the donor and acceptor through the equations:

$$\tau_D = \left(\Gamma^{r,D} + \Gamma^{nr,D}\right)^{-1} \tag{8.2a}$$

$$\tau_A = \left(\Gamma^{r,A} + \Gamma^{nr,A}\right)^{-1} \tag{8.2b}$$

As seen in Fig. 8.2, FRET effectively opens an additional pathway for de-excitation of the donor (in addition to radiative and nonradiative de-excitation), which needs to be incorporated into the quantum yield of the donor. This gives:

$$Q^{DA} = \frac{\Gamma^{r,D}}{\Gamma^{r,D} + \Gamma^{nr,D} + \Gamma^{FRET}} \tag{8.3}$$

where the new fluorescence lifetime of the donor becomes:

$$\tau_{DA} = \left(\Gamma^{r,D} + \Gamma^{nr,D} + \Gamma^{FRET}\right)^{-1} \tag{8.4}$$

The notation "τ_{DA}" reads: "the fluorescence lifetime of D in the presence of A (or FRET)."

Förster's theory gives the rate constant of the energy transfer for the case of a dipolar interaction, namely

$$\Gamma^{FRET} = (\Gamma^{r,D} + \Gamma^{nr,D})(R_0/r)^6 \tag{8.5}$$

where r is the D-A separation, and R_0 is the so-called *Förster distance*, which depends on the degree of overlap between donor's emission spectrum and acceptor's excitation spectrum, as well as on the relative orientation of the transition dipoles, the quantum yield of the donors and the refractive index of the medium (Selvin, 1995; Clegg, 1996; Lakowicz, 2006).

Observation: The interaction needs not be dipolar, and other mechanisms of transfer have also been investigated (Dexter, 1953).

One may define an efficiency of energy transfer (or FRET efficiency) as the proportion of photons dissipated by the excited donor through FRET, as:

$$E = \frac{\Gamma^{FRET}}{\Gamma^{r,D} + \Gamma^{nr,D} + \Gamma^{FRET}} \tag{8.6}$$

The extra term in the sum of rate constants in the denominator (Γ^{FRET}) decreases the lifetime of the donor (i.e., $\tau_{DA} < \tau_D$). Using equations (8.2a) and (8.4), equation (8.6) may be rearranged as:

$$E = 1 - \frac{\tau_{DA}}{\tau_D} \qquad (8.7)$$

which establishes the connection between the FRET efficiency and the fluorescence lifetimes of the donor in the presence and absence of acceptor. This equation provides a convenient means for measuring the FRET efficiency from *fluorescence lifetime measurements* (see, e.g., Elangovan et al., 2002; Bacskai et al., 2003; Lakowicz, 2006 and references therein).

Alternatively, equation (8.6) may be rearranged, by using equation (8.5), as:

$$E = \frac{R_0^6}{R_0^6 + r^6} \qquad (8.8)$$

which connects the FRET efficiency to the distance between donor and acceptor in the complex. The immediate implication of the relations (8.7) and (8.8) is that the distance between D and A in a complex may be calculated from lifetime measurements, if R_0 is known.

Observation: Knowledge of R_0 requires determination of the relative orientation between the transition dipoles of D and A, which is not usually an easy task (Clegg, 1996).

By combining equation (8.1a) with (8.3) and (8.6), one obtains the following relation between the quantum yields of the donor in the presence and absence of acceptors:

$$Q^{DA} = Q^D (1 - E) \qquad (8.9)$$

which indicates that the donor emission is reduced due to FRET by a factor $1 - E$. This reduction, known as *donor quenching*, can be used to quantify the interaction between D and A using measurements of donor fluorescence intensity in the presence and absence of acceptor. Obviously, the quantum yield of the acceptor remains unchanged, since FRET only introduces a new pathway for de-excitation of the donor and not of the acceptor.

Quiz 1. Derive relation (8.9).

We also note that the excitation rate of acceptors increases due to FRET, while the excitation rate of the donors does not. The former situation is described by the relation (which follows from Fig. 8.2):

$$\Gamma^{ex,AD} = \Gamma^{ex,A} + \Gamma^{ex,D} E \qquad (8.10)$$

where $\Gamma^{ex,AD}$ is the excitation rate of acceptors in the presence of donors, $\Gamma^{ex,A}$ is the excitation rate of acceptors alone, and $\Gamma^{ex,D}$ is the excitation rate of the donors. The increased acceptor excitation rate expressed by equation (8.10), called *acceptor sensitized emission*, can be used to detect FRET from measurements of acceptor emission intensity.

8.1 Probing Protein Association In Vivo

Observation: The excitation rate constant of donors and acceptors depends on the intensity of the incident light, $I_0(\lambda_{ex})$, the absorption cross-sections, $\varepsilon(\lambda_{ex})$, at the excitation wavelength, λ_{ex}, according to the following relation (Song et al., 1995):

$$\Gamma^{ex,D} = I_0(\lambda_{ex})/(hcN_A)\varepsilon^D(\lambda_{ex}) \quad (8.11a)$$

$$\Gamma^{ex,A} = I_0(\lambda_{ex})/(hcN_A)\varepsilon^A(\lambda_{ex}) \quad (8.11b)$$

where h is Planck's constant, c is the speed of light in vacuum, and N_A is Avogadro's number.

In the next section, we will introduce the elementary theory of fluorescence intensity, which permits characterization of FRET in populations of proteins appropriately tagged, and gives the possibility to determine the stoichiometry of formation of protein complexes from individual monomers.

8.1.2 FRET-Based Determinations of Interaction Stoichiometry

8.1.2.1 Theory

Let us consider two populations of fluorescent molecules (mixed together in a solution or a living cell) excited by light with an arbitrary wavelength, λ_{ex} (Raicu, 2007). Initially, all molecules of the two populations are separated from one another, and no energy transfer takes place between them (see Fig. 8.3). When observed at an arbitrary emission wavelength, λ_{em}, the measured fluorescence intensity, $F^m(\lambda_{ex})$, of the sample is given by the sum of the fluorescence intensities of the two populations, i.e.,

$$F^m(\lambda_{ex}) = F^D(\lambda_{ex}) + F^A(\lambda_{ex}) \quad (8.12a)$$

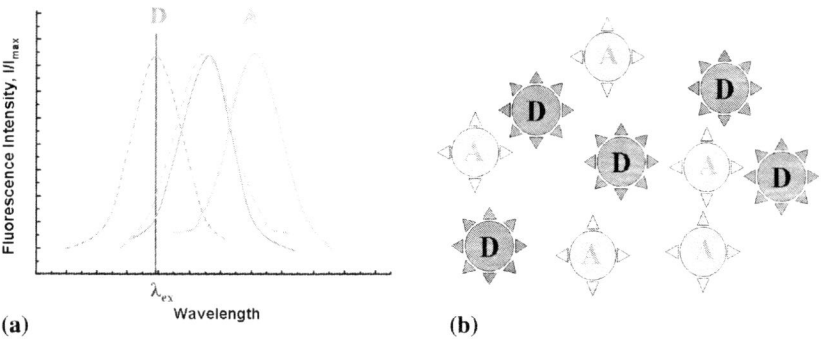

Fig. 8.3 (**a**) Excitation (dashed lines) and emission (solid lines) spectra of a donor-acceptor pair for FRET studies. An arbitrarily chosen excitation wavelength is illustrated by a vertical straight line. (**b**) Emission spectra in (**a**) and equation (8.12) predict that the fluorescence emission of a population of donors excited at an arbitrary wavelength λ_{ex} is stronger than the emission of acceptors, as represented by the number of "light rays" (triangles) leaving the molecules (circles). However, note that there is emission from A at almost any wavelength within emission spectrum of D (which is called *spectral cross-talk*).

where F with various subscripts are the integrated emission intensities over all the available emission wavelengths for A or D molecules. By using a method of decomposition it is possible to separate the measured spectrum into its two components $F^D(\lambda_{ex})$ and $F^A(\lambda_{ex})$ (see, e.g., Zimmermann et al., 2002; Neher and Neher, 2004; Raicu et al., 2005). Then, the theory in the preceding section allows one to write simple relationships for the total fluorescence emission from donors,

$$F^D(\lambda_{ex}) = \Gamma^{ex,D}[D]_T Q^D \tag{8.12b}$$

and acceptors,

$$F^A(\lambda_{ex}) = \Gamma^{ex,A}[A]_T Q^A \tag{8.12c}$$

by using the total concentrations of donors, $[D]_T$, and acceptors, $[A]_T$.

Let us assume next that some of the molecules are moved to the close proximity of the A molecules, as illustrated in Fig. 8.4, so that there is potential for energy transfer, with efficiency E, from donors (D) to acceptors (A). Note that, the two populations of D and A molecules can form not only DA dimers, which are FRET productive, but also AA and DD dimers, which do not show FRET.

Using again relations (8.9) and (8.10) introduced in the previous section, one can write the expressions for emission intensities of such mixtures as:

$$F^{DA}(\lambda_{ex}) = \Gamma^{ex,D}\left\{[D]Q^D + [D]_D Q^D + [D]_A Q^{DA}\right\} = \Gamma^{ex,D}[D]_T Q^D - \Gamma^{ex,D}[D]_A Q^D E \tag{8.13a}$$

$$F^{AD}(\lambda_{ex}) = \Gamma^{ex,A}\left\{[A]Q^A + [A]_A Q^A\right\} + \Gamma^{ex,AD}[A]_D Q^A = \Gamma^{ex,A}[A]_T Q^A + \Gamma^{ex,D}[A]_D Q^A E \tag{8.13b}$$

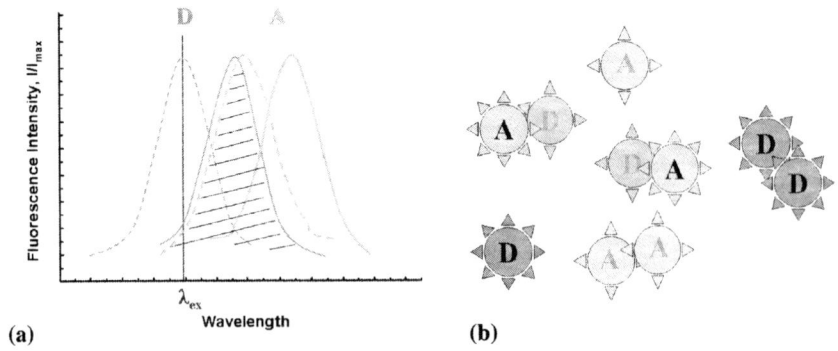

Fig. 8.4 (a) When some donor molecules are brought closer to acceptor molecules, as shown in (b), a nonradiative transfer of excitation energy from D to A (or FRET) may occur. The hatched area denotes the overlap between donor emission and acceptor excitation spectra, which indicates that FRET may occur. (b) FRET makes the acceptors brighter and the donors dimmer [compare emission intensities between interacting and noninteracting molecules, as represented by the number of "rays" (triangles) emanating from molecules (circles)].

where $[D]_A$ is the concentration of D molecules that form complexes with A molecules, $[A]_D$ is the concentration of A molecules that form complexes with D molecules, $[D]_D$ is the concentration of D in D-only complexes, $[A]_A$ is the concentration of A in A-only complexes, while $[D]_T = [D] + [D]_D + [D]_A$ and $[A]_T = [A] + [A]_A + [A]_D$ are the total concentrations of D and A molecules. By using the notations defined in (8.12) and introducing the following new ones,

$$F^D(FRET) = \Gamma^{ex,D}[D]_A Q^D E \qquad (8.14a)$$

$$F^A(FRET) = \Gamma^{ex,D}[A]_D Q^A E \qquad (8.14b)$$

equation (8.13) may be rewritten as:

$$F^{DA}(\lambda_{ex}) = F^D(\lambda_{ex}) - F^D(FRET) \qquad (8.15a)$$

$$F^{AD}(\lambda_{ex}) = F^A(\lambda_{ex}) + F^A(FRET) \qquad (8.15b)$$

which indicate that the fluorescence of the donors population should decrease due to FRET, while the fluorescence of the acceptors population should increase, similarly to pure DA pairs (see previous section).

Two *apparent FRET efficiencies* may be now obtained from the integrated fluorescence emission of the donors,

$$E_{app}^{Dq} \equiv \frac{F^D(FRET)}{F^D(\lambda_{ex})} = 1 - \frac{F^{DA}(\lambda_{ex})}{F^D(\lambda_{ex})} \qquad (8.16a)$$

and of the acceptors,

$$E_{app}^{Ase} \equiv \frac{F^A(FRET)}{F^A(\lambda_{ex})} = \frac{F^{AD}(\lambda_{ex})}{F^A(\lambda_{ex})} - 1 \qquad (8.16b)$$

where the superscripts "Dq" and "Ase" stand for "donor quenching" and "acceptor sensitized emission."

Equations (8.16) may rewritten, by making use of equations (8.11), (8.12), (8.14), and (8.15), as:

$$E_{app}^{Dq} = \alpha_D E \qquad (8.17a)$$

$$E_{app}^{Ase} = \alpha_A \frac{\varepsilon^D}{\varepsilon^A} E \qquad (8.17b)$$

where $\alpha_D = [D]_A/[D]_T$ is the fraction of donors forming complexes with the acceptors and $\alpha_A = [A]_D/[A]_T$ is the fraction of acceptors in complexes with donors. These equations now show that the apparent FRET efficiency represents the overall efficiency of mixtures of DD, DA and AA of which only a particular subset (DA) is actually involved in energy transfer.

Observation: In its most general form, which applies to oligomers of arbitrary size and geometry, the kinetic model leads to relations between the apparent FRET efficiency and the concentrations of oligomers that are significantly more complex than described above.

In particular, for oligomers comprising more than two monomeric units that are situated at equal distances from one another, the apparent FRET efficiencies are given by the more general equations (Raicu, 2007):

$$E_{app}^{Dq} = \frac{\mu_{oligo} \sum_{k=1}^{n-1} \frac{k(n-k)E}{1+(n-k-1)E} \binom{n}{k} P_D^k P_A^{n-k}}{[D] + \mu_{oligo} n P_D} \quad (8.18a)$$

$$E_{app}^{Ase} = \frac{\mu_{oligo} \sum_{k=1}^{n-1} \frac{k(n-k)E}{1+(n-k-1)E} \binom{n}{k} P_D^k P_A^{n-k}}{[A] + \mu_{oligo} n P_A} \frac{\varepsilon^D}{\varepsilon^A} \quad (8.18b)$$

where μ_{oligo} is the concentration of oligomers, n is the total number of monomers in an oligomer, k is the number of donors, while $P_D = \{[D]_D + [D]_A\}/\{[D]_D + [D]_A + [A]_A + [A]_D\}$ and $P_A = \{[A]_A + [A]_D\}/\{[D]_D + [D]_A + [A]_A + [A]_D\}$ are the fractions of donor and acceptor concentrations in oligomers, respectively and $\binom{n}{k} = \frac{n!}{k!(n-k)!}$.

The parameters $F^{DA}(\lambda_{ex})$ and $F^{AD}(\lambda_{ex})$ in equation (8.16) may be measured directly in FRET experiments. $F^D(\lambda_{ex})$ is usually determined by measuring $F^{DA}(\lambda_{ex})$ after inactivating the acceptor through photobleaching, which is accomplished by prolonged exposure to intense laser light of appropriate wavelength (Kubitscheck et al., 1993). Finally, $F^A(\lambda_{ex})$ is measured by exciting the acceptor at a wavelength $\lambda \neq \lambda_{ex}$ at which the donor is not excited, and rescaling the intensity based on the known excitation spectrum of the acceptor (Lakowicz, 2006; Merzlyakov et al., 2007; Raicu et al., 2007).

Depending on the purpose of the study, one may either proceed further with the theory, in order to connect the apparent efficiencies to the concentration of interacting and non-interacting proteins, or just determine the decrease in donor fluorescence (or increase in acceptor fluorescence) due to FRET. The first method is more quantitative and allows one to determine, e.g., the fraction of interacting proteins out of a population of monomers. In that case, one typically determines the apparent FRET efficiencies from equation (8.16) and then uses them to obtain the values of α_D and α_A from equation (8.17) and a theory that connects the two parameters to the probability for D's and A' to form DD, DA and AA complexes (Raicu et al., 2005; Meyer et al., 2006; Raicu et al., 2008). On the other hand, the second method presents the advantage of simplicity and is appropriate for mapping out the distribution of protein complexes in living cells, although it is more prone to errors, including those caused by random encounters between donors and acceptor (due to molecular crowding in the cell) which may have no particular biological relevance.

8.1.2.2 Experimental Methods

Fluorescence from molecules in solutions or in unicellular organisms may be determined using time resolved or spectrally resolved spectrofluorometers and microscopes. The latter category includes wide-field fluorescence microscopes, as well as laser-scanning microscopes, such as confocal and two-photon microscopes.

8.1 Probing Protein Association In Vivo

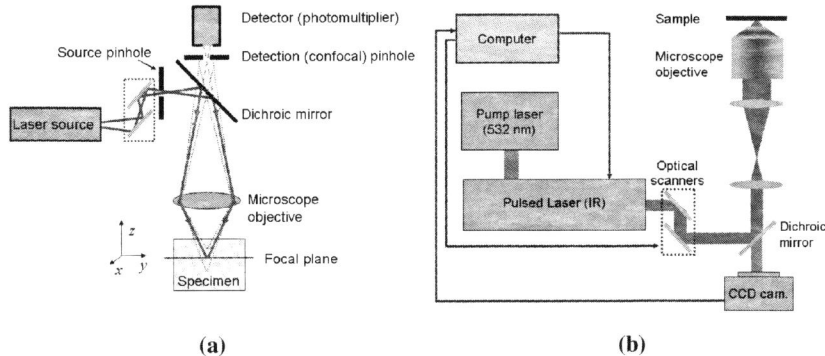

Fig. 8.5 (a) Schematic representation of the confocal microscope. The light rays that originate from beneath and above the focal point of the objective diverge too much and are filtered out by the confocal pinhole (in front of the detector). Unlike those rays, light originating from the focal point passes through the pinhole and is detected. (b) Schematic representation of the two-photon microscope (TPM). TPM achieves image-sectioning capability through the fact that excitation only occurs at the focal point, where the excitation light intensity is high enough for overcoming the extremely low probability of the two-photon absorption to occur.

Introduction of a *confocal* microscope Minski (1961) has tremendously benefited the field of fluorescence imaging by making it possible to obtain quasi-2D images (i.e., images of sections that are thin compared to the other two spatial dimensions) of 3D samples without physically sectioning the sample. From stacks of 2D images, one can reconstruct 3D images, which may not be normally obtained with wide-field microscopes.

In the confocal microscope design (Fig. 8.5a), the sample is raster-scanned through the focal point of a high-numerical-aperture microscope objective and the back-propagating part of the fluorescence is separated from the excitation light and sent onto a sensitive detector. The image sectioning capability is achieved by placing the detector of fluorescence at a position that is confocal to the focal point of the microscope objective. The fluorescence intensity is converted into electrical signal and the image of the sample is subsequently reconstructed as an $x-y$ map of fluorescence intensity vs. position of the excitation beam. This design allows for detection of the fluorescence originating from the confocal point and excludes the fluorescence from beneath or above the objective focal plane, thereby providing images of thin sections of the sample.

Another breakthrough in the development of imaging techniques has been the introduction of the two-photon laser scanning microscope by the group of W.W. Webb in early nineteen nineties (Denk et al., 1990; Zipfel et al., 2003). This microscope (Fig. 8.5b) exploits the ability of fluorescent molecules to absorb at once two photons (Masters and So, 2004) having together the necessary energy to induce an electronic excitation of the molecule, equivalent to the excitation of a single, more energetic photon (Fig. 8.6). The two photons do not have to present equal energies (or wavelengths); their wavelengths only have to obey the relationship: $1/\lambda_0' = 1/\lambda_1 + 1/\lambda_2$. In addition, this effect may occur in principle for arbitrary number of excitation photons.

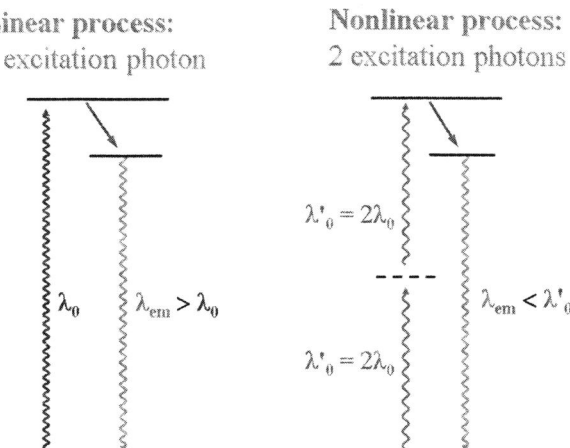

Fig. 8.6 Comparison between single-photon and two-photon excitation of fluorescence. Note the relationship between excitation and emission wavelengths in the two cases, and also that the intermediate level (represented by a dashed line) in the two-photon excitation does not have a correspondent in reality (i.e., it is a *virtual* level).

The conditions for multiphoton process to occur are created experimentally by tightly focusing the beam of a pulsed (sub-picosecond) laser through a high numerical aperture objective and thus confining the photons spatially and temporally. Because the intensity of the light is only high enough for the two-photon excitation process to occur into a small volume around the focal point, image-sectioning is achieved without a confocal pinhole.

Some of the advantages of multiphoton imaging compared to single photon imaging are: (i) the number of photons arriving at the detector is larger than when using a detection pinhole because image sectioning occurs without a need to use confocal pinholes that attenuate the signal; (ii) any photodestruction (i.e., photobleaching) of the sample by the excitation beam is confined to the sample layer being imaged, leaving intact the rest of the sample for further study; (iii) biological materials are more transparent in longer wavelength and thus images can be obtained from deeper layers of biological tissues; and (iv) using two near-IR photons to excite molecules that present single-photon excitation maxima in the UV range reduces the damage to the rest of the sample that would be otherwise induced excitation with by UV light.

Currently, one of the most suitable lasers for two photon microscope (TPM) is the modelocked Ti:Sapphire laser introduced by Sibbet et al. in 1990, which provides near pulsed light with typical pulse durations of the order of 10–100 femtoseconds ($10^{-14} - 10^{-13}$ s) and extremely high peak powers (of the order of many kilowatts). With its tunable near IR light (700–1,000 nm) and/or a very broad bandwidth (inherent in the short pulses) of 20–100 nm, the modelocked Ti:Sapphire laser is particularly suitable for exciting fluorescent dyes with excitation maxima in the near UV to the green part of the spectrum, as for example variants of the green fluorescent protein. An added advantage of using a pulsed laser vs. continuous wave lasers

8.1 Probing Protein Association In Vivo

is that the former makes it possible to perform time-resolved measurements with sub-nanosecond resolution (Schönle et al., 2000; Bacskai et al., 2003).

8.1.2.3 Determination of the Distribution of Membrane Receptor Complexes in Living Cells

The theoretical scheme described in section 8.1.2.1. (or variants of it) has been applied recently (Raicu et al., 2005) to the study of homo-oligomerization (i.e., formation of homo-oligomers) of a model G-protein coupled receptor – the sterile 2 α factor receptor protein Ste2 (Ste2p) (Overton et al., 2005) – in living yeast cells (*Saccharomyces cerevisiae*). Yeast cells were genetically engineered to co-express GFP-tagged Ste2p proteins as donors of energy and YFP-tagged Ste2p proteins as acceptors of energy, and then investigated using a confocal microscope. The seven transmembrane domains of the Ste2p proteins (Overton et al., 2005) allow the proteins to actually 'plow' through the phospholipid matrix of the membrane (i.e., to diffuse laterally within the membrane plane), while the fluorescent tags are attached to the cytoplasmic tails of the receptor (Fig. 8.7a). If donors and acceptor tags come close together and within the FRET range ($\sim$1–5 nm), either because of the Ste2p monomers association into stable oligomers, or due to random encounters between monomers, transfer of energy may take place.

The two apparent FRET efficiencies, E_{app}^{Dq} and E_{app}^{Ase}, are determined from equations (8.16) and then used to obtain the values of α_D and α_A from equations (8.17) (or some more general forms of them), via rather complex computation schemes that consider the statistics of protein association (Raicu et al., 2005; Raicu, 2007). (A thorough discussion on this method would be beyond the scope of this

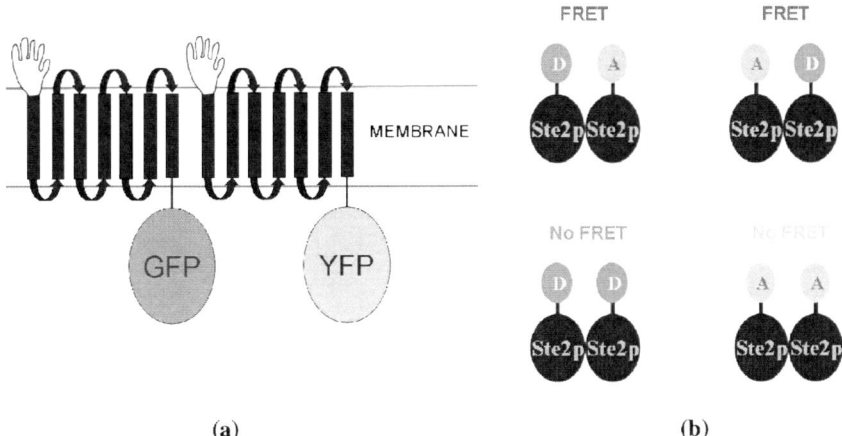

Fig. 8.7 (a) Ste2p proteins tagged with YFP and GFP forming a complex together may exchange energy through FRET. (b) Possible combinations between D- and A-tagged molecules in the case of equal concentrations of donors and acceptors in the membrane.

book; the interested reader is invited to consult the references.) Since donors and acceptors may combine to form not only DA pairs but also AA or DD pairs, the frequency of FRET occurrence is reduced due to the presence of non-productive DD and AA pairs (Fig. 8.7b). Hence, the need to determine both α_D and α_A for determination of all the interacting monomers (not just those in DA pairs).

In the study just described (Raicu et al., 2005) it has been found that almost 100% of the Ste2p receptors form stable *dimers*. Since the Ste2p complexes were formed in the absence of a ligand and since a majority of Ste2p seemed to form complexes with one another at any given time, it has been suggested that Ste2p proteins may be assembled as dimers soon after synthesis, although this has yet to be proved.

We next illustrate how a simpler version of the FRET theory described above – more precisely, equation (8.15a) – may be used to map out the locations at which proteins interact possibly to form dimeric complexes (R. Fung, D. Jansma and V. Raicu, unpublished, 2007). The cells were excited with the 458-nm line of an Ar-ion laser and their fluorescence was recorded with a confocal microscope. The wavelength has been chosen such that it excited optimally the GFP tags and minimally the YFP tags. Once acquired and corrected for certain errors, fluorescence images were separated by using a spectral deconvolution method (Raicu et al., 2005) into donor-only and acceptor-only signals for every pixel. The acceptor was then photobleached by prolonged exposure to intense 514-nm light, and the increase in donor fluorescence (due to destruction of FRET) was calculated relative to the donor fluorescence in the presence of FRET, i.e., before A photobleaching (see Fig. 8.8).

Quiz 2. The number of complexes of each type described above is given by the binomial distribution. Demonstrate that, if the total concentrations $[D]_T$ and $[A]_T$ are equal, then *and only then* the proportion of DD, DA and AA given in Fig. 8.5b is the correct one. (Note that AD and DA are the same, for the purpose of this discussion.)

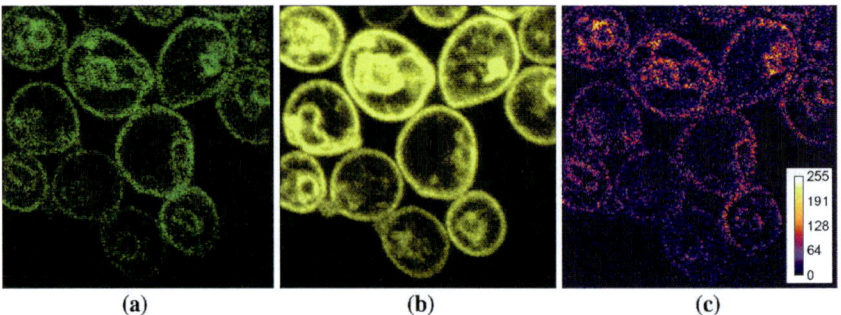

Fig. 8.8 Fluorescence images of yeast cells co-expressing GFP-tagged Ste2p proteins as donors of energy and YFP-tagged Ste2p proteins as acceptors of energy are acquired using a confocal microscope. The fluorescence intensities at each image pixel are separated by using a spectral deconvolution method into (**a**) donor-only (i.e., GFP) and (**b**) acceptor-only (i.e., YFP) signals. The process is repeated after the acceptor is photobleached and the increase in donor fluorescence is calculated to obtain (**c**) the distribution of FRET (i.e., protein complexes) (Data from Fung, Jansma and Raicu, unpublished).

According to the previous section, the differences in donor fluorescence intensities at each image pixel before and after acceptor photobleaching provide a measure of the extent of FRET, which, in their turn may be considered as a measure of the concentration of protein complexes (if random interactions can be neglected – see below).

The map of FRET distribution shown in Fig. 8.8c indicates that FRET occurred both on the plasma membrane and in internal membranes. Although there still exists some ambiguity as to the precise location of the complexes inside the cell (i.e., endoplasmic reticulum vs. transport vesicles), detection of FRET signal from internal membranes seems to support the above idea that Ste2p complexes are formed immediately after synthesis. However, this simple interpretation may not be entirely correct, since proteins drifting at random in the membrane plane may collide briefly to generate the so-called stochastic FRET (Wolber and Hudson, 1979). In fact, as already mentioned above, a recent investigation of the Ste2p system suggested that both long-lived functional complexes and random collisions may contribute to the observed distribution of FRET within the cell (R. Fung, D. Jansma and V. Raicu, unpublished, 2007).

8.2 Structural Studies of Protein–Protein Interactions

8.2.1 Principles of Nuclear Magnetic Resonance (NMR)

As we have already mentioned in section 2.4, both Nuclear Magnetic Resonance (NMR) and X-ray diffraction provide structural information about macromolecules. While the X-ray diffraction method exploits the spatial density distribution of the electrons, NMR probes the distances between atomic nuclei that possess a *magnetic momentum* (e.g., ^{1}H, ^{13}C, ^{15}N, ^{31}P, etc.) and are thereby able to interact with external magnetic fields.

> **Observation:** NMR relies on an important quantum mechanical property of nuclei, the spin, which is characterized by the *nuclear spin quantum number, I*. The spin quantum number of a nucleus is related to the spin of its protons and neutrons (which is $1/2$), and is given by the number of unpaired protons times $1/2$ and the number of unpaired neutrons times $1/2$. For instance, ^{1}H, which has one proton and no neutrons, has spin $1/2$ and ^{2}H has spin 1.

> **Quiz 3.** Compute the nuclear spins for the following isotopes: ^{12}C, ^{13}C, ^{14}N, and ^{15}N.

In the following, we will present a very simplified theoretical description of the NMR. In doing so, we will focus our attention on ^{1}H, which is of special interest in biology. The nuclear spin quantum number of ^{1}H is $1/2$ and the magnetic quantum number takes only two values, $m = I, -I$ or $m = \pm 1/2$. When subjected to a strong uniform magnetic field of intensity B_0 (of the order of 10 T, in NMR), the angular momentum component along the field axis (usually denoted by "z") has only two

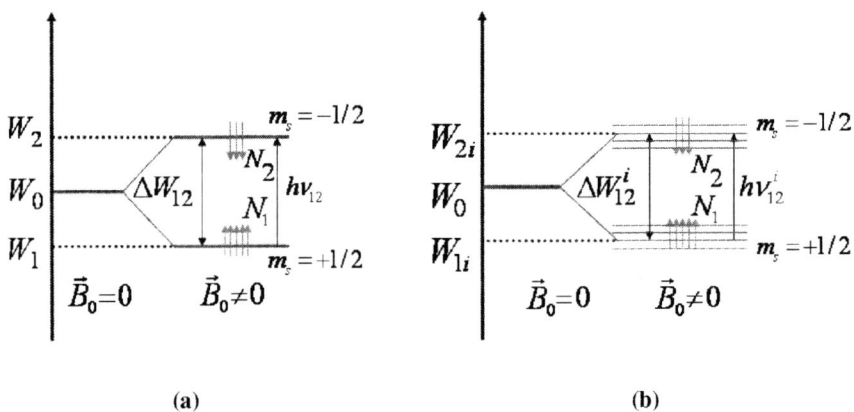

Fig. 8.9 Illustration of energy levels "splitting" for (**a**) a collection of identical protons (e.g., the bulk water protons) and (**b**) $N = N_1 + N_2$ protons of a (macro)molecule. For the sake of simplicity, only a few levels of energy (W_1, and W_2) are drawn. ν_{12} is the resonance frequency.

permitted directions, each corresponding to one m value. In that case, the energy levels of the spins split into two levels, as given by:

$$W_i = -\vec{\mu} \cdot \vec{B}_0 = -g\mu_N B_0 m_i \ (i = 1, 2) \tag{8.19}$$

where g is the *Landé factor* and μ_N is the *nuclear magneton* ($= e\hbar/(2m_p)$; m_p is the proton mass, e is the elementary charge, and $\hbar$ is Planck's constant divided by 2π). In other words the applied magnetic field splits the population of nuclei into two subpopulations, each corresponding to a different nuclear spin (Fig. 8.9).

Positively charged nuclei tend to align their spins in the direction of the field, such that $m = 1/2$, but some will be aligned antiparallel, such that $m = -1/2$. However, the population difference is not large (see below). In order to change from one orientation to another, the spins need to absorb an amount of energy equal to the difference between the energy of the two states, namely:

$$\Delta W_{12} = g\mu_N B_0 \tag{8.20}$$

The distribution of the nuclei among the two subpopulations follows the Boltzmann distribution,

$$N_2 = N_1 e^{-\Delta W_{12}/k_B T} \tag{8.21}$$

Another effect of B_0 on the "spinning" nucleus is, similarly to a gyroscope in classical mechanics, the precession of the magnetic momentum at a frequency, $\nu_L = geB_0/(2m_p)$, called Larmor frequency. If the physical system of precessing spins is subjected to a radiofrequency field (RF) with $B_1 \ll B_0$ ($\vec{B}_1 \perp \vec{B}_0$) and frequency $\nu_0 = \nu_L$, so that the resonance relation,

$$h\nu_0 = \Delta W_{12} = g\mu_N B_0, \tag{8.22}$$

holds true, the nuclei with lower energy, W_1, may be brought into the state with higher energy, W_2, due to the absorption of radiofrequency energy. The attenuation due to absorption of energy by the spins of the transmitted RF relative to its initial intensity may be measured by an RF detector (consisting of a coil around the sample); this constitutes the signal detected in NMR.

Quiz 4. The superconducting magnet of an NMR spectrometer generates a constant magnetic field $B_0 = 15\,T$. What radio-frequency field frequency is needed to access the resonance of ^1H? The Landé factor for the proton is $g = 5.56$.

Due to the fact that μ_N is very small (compared, for instance, to the Bohr magneton of the electron), the difference between the two energy levels, ΔW_{12}, is relatively small. Hence, equation (8.21) becomes $N_2 \cong N_1 (1 - \Delta W_{12}/kT)$, which implies that N_1 is only slightly larger than N_2 (by $\Delta W_{12}/kT$). For this reason, the NMR signal is very weak or, in other words, the signal-to-noise ratio (S/N) is very low, compared to other techniques. Therefore, large numbers of measurements, n, are performed and summed up so that $S/N = \sqrt{n}$ is substantially improved.

Observation: The Bohr magneton of the electron is about 1840 greater than the nuclear magneton, hence $N_1 \gg N_2$. Because of this, the technique based on electronic spin resonance (ESR) has a higher sensitivity than NMR spectroscopy (see below). Unfortunately, the applicability of ESR spectroscopy is limited to chemical species that posses unpaired electrons, while most stable molecular species have their electrons paired.

8.2.2 Chemical Shift and the NMR Spectroscopy

If all the protons (i.e., hydrogen ions) of a macromolecule behaved identically, nothing interesting would have been learned form NMR studies of macromolecules. Fortunately, although intrinsically indistinguishable, protons at different locations in a (macro)molecular structure behave differently. The reason for this is that, due to the specific electronic structure of an atom, the magnetic field sensed by each nucleus is different from the applied field, as expressed by the equation:

$$B^i_{eff} = B_0(1 - \sigma_i), (i = 1, 2, \ldots, N) \tag{8.23}$$

where σ_i (which is < 1) represents the so called shielding constant of the nucleus, i, and N is the number of nuclei of the macromolecule investigated. Therefore, equation (8.22) describing the single resonance condition has to be replaced by N resonance conditions (one for each equivalent class of protons in Fig. 8.9b):

$$h\nu^i_{12} = \Delta W^i_{12} = g\mu_B B^i_{eff} = g\mu_B B_0(1 - \sigma_i), (i = 1, 2, \ldots, N) \tag{8.24}$$

Observation: NMR is not a single-molecule type of method, and, when used for determination of macromolecular structure, S/N considerations require that the sample contains about 10^{18} macromolecules in their native state.

If a solution containing many copies of a macromolecule is exposed to polychromatic radiofrequency field, as in the case of the so called *impulse NMR*, then the frequencies satisfying relations (8.24) will be attenuated. In this way the detector of the transmitted RF field will record the absorption spectrum in the RF domain, which constitutes a fingerprint for each type of macromolecule. In practice, due to the fact that the shielding constants are very small, the frequencies, v_{12}^i, are very closely spaced and therefore difficult to resolve. NMR spectra are better represented as intensity vs. the so-called *chemical shift*, δ_i, which is defined by the relation:

$$\delta_i = 10^6 \times \frac{v_i - v_r}{v_r} (\text{ppm}) \qquad (8.25)$$

where v_r represents the resonance frequency of a reference molecule, usually, the tetra methyl silane [$(CH_3)_4Si$, or TMS], whose 12 equivalent protons are more shielded than most proteins one would encounter. A typical 1D NMR spectrum for a small protein ($\sim 12\,\text{kDa}$) is depicted in Fig. 8.10, which also gives some examples of proton chemical shifts (in *parts per million*, ppm). The chemical shifts are modified by the covalent bonds between atoms in the molecule as well as by nonspecific interactions occurring at the interface between molecules.

The complexity of 1-D NMR spectra is very high in the case of large proteins, an inconvenience that is overcome by N-dimensional (N = 2, 3,...) NMR spectra. Naturally, the protocol of obtaining a 2-D NMR spectrum is more complicated than that used for obtaining 1-D spectra. In essence, the system of nuclear spins in the sample is perturbed, in a series of X 1-D experiments (where X defines the number of points in the second dimension), by RF pulses separated by various time intervals. By applying a two-dimensional Fourier transform to the recorded signal (RF absorption), one obtains 2-D NMR spectra. Figure 8.11 shows the 2-D spectrum of the same protein as in Fig. 8.10.

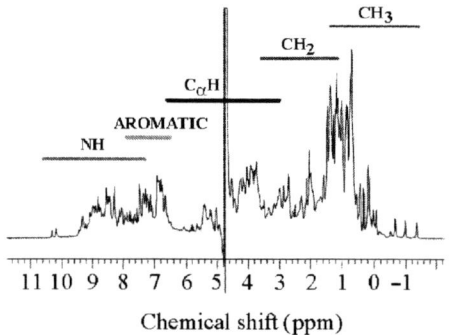

Fig. 8.10 The NMR spectrum of a small protein (Calcium Vector Protein). The range of chemical shifts of the protons involved in various chemical groups of the proteins is also represented (horizontal lines). The highest peak in the figure is due to bulk water protons. Note that positive chemical shifts are most often observed in NMR spectra, which are traditionally represented to the left of zero, while a few nuclei present negative chemical shifts, and are represented to the right of the reference (TMS, see text), which has "zero" chemical shift. Figure courtesy of C. T. Craescu, Institute Curie, Orsay, France.

8.2 Structural Studies of Protein–Protein Interactions

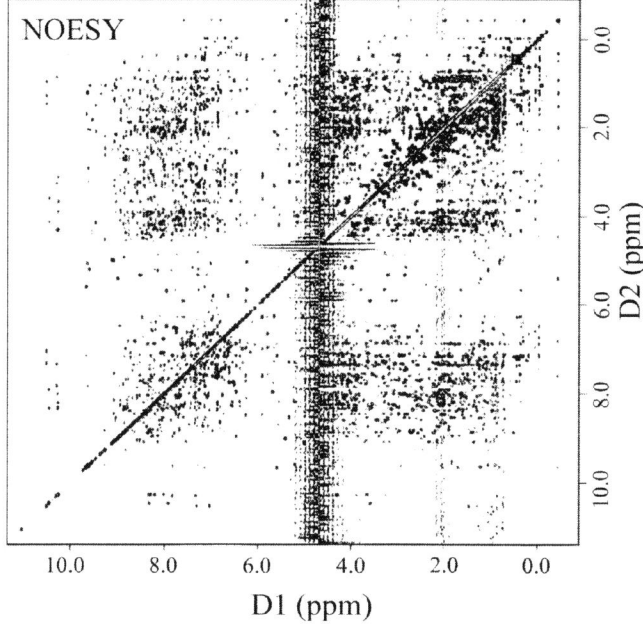

Fig. 8.11 A 2D NMR spectrum of the NOESY (Nuclear Overhausser Effect Spectroscopy) type of a small protein (Calcium Vector Protein) D_1 and D_2 stand for chemical shifts. Note that the 1D spectrum is recorded along the diagonal of the 2D spectrum. Equivalent information may be extracted form the peaks above and under the diagonal. The spectrum is vertically crossed by bulk water peaks. Figure courtesy of C. T. Craescu, Institute Curie, Orsay, France.

Two-dimensional NMR spectra present some interesting properties. Firstly, their diagonal gives 1-D spectra. Secondly, there is some redundancy in the information extracted from the 2-D spectrum, in that the peaks situated above the diagonal are equivalent to those under the diagonal.

In spite of the important improvement introduced by 2-D NMR, some peaks are not well resolved, and higher dimensional NMR spectra are required. For the sake of simplicity, we only discussed above the 1-D and 2-D NMR. For more information on NMR in general and also on 3-D NMR, the reader is referred to a methods book by Serdyuk et al. (2007) and also to the popular book on NMR by Cavanagh (Cavanah et al., 2007).

Figure 8.12 presents an example of an NMR-determined structure. Note that NMR methods give several conformers of the same protein, which all present relatively rigid structures at their cores as well as flexible peripheral segments. This multiplicity of conformers reflects either true motion of the peripheral segments, or the lack of experimental data to define this region of the structure. In X-ray crystal structures, such scenarios would correspond to the absence of electron density in those regions.

The ability of NMR to provide information on different conformational states of proteins may be used to monitor the dynamic behavior of proteins at multiple

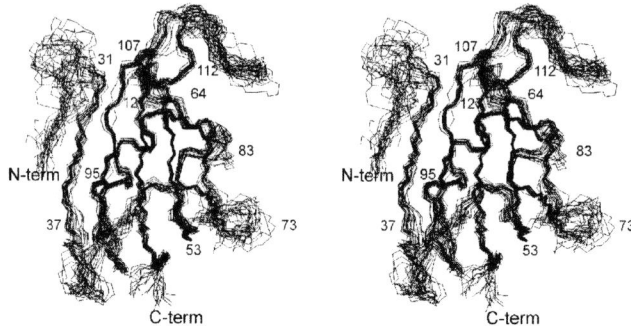

Fig. 8.12 A stereo view of set of conformers describing the backbone of the immunophilin-like domain of FKBP59. "C-term" and "N-term" stand for C-terminal and N-terminal of the polypeptide chain. Figure reprinted with permission from Craescu et al. (1996). Copyright 1996 American Chemical Society.

sites. This may provide information on the dynamics of protein-ligand interactions (Ishima and Torchia, 2000; Pellecchia et al., 2002; Takeuchi and Wagner, 2006). In the next section, we will review an NMR strategy for studying protein-protein interactions in solution, and even in the living cell.

8.2.3 NMR Studies of Protein-Protein Interactions

On discussing the protein-protein interactions in section 8.1, we have left aside the question of the site on a protein at which the physical interaction occurs. One of the most common biological approaches to determining the interaction site is by introducing site-directed mutations in the sequence of amino acids of the protein and then testing whether the interaction still occurs. However, unless the structure is determined for each mutant protein, the genetic method may not exclude the possibility that a mutation away from the binding site may induce the protein to fold differently compared to the original (i.e., *wild type*) protein, such that the binding interface is hidden inside the changed structure.

New developments in NMR spectroscopy led to the possibility of precisely determining the binding interface between a receptor protein and a ligand by mapping the chemical shift changes in interacting proteins. This relies on the fact that the chemical shifts can be affected by non-covalent interactions at the interface between protein and ligand, thereby indicating which protein residues are in close proximity to the ligand (Wüthrich, 2000; Pellecchia et al., 2002; Takeuchi and Wagner, 2006). One of the limitations of such methods is that chemical shift variations may be induced not only at the receptor-ligand interface but also away from the interface, due to binding-induced conformational changes.

An interesting method, called transferred cross-saturation (TCS), uses uniform ^{2}H- and ^{15}N-labeling of a ligand protein (protein I in Fig. 8.13), whose binding interface is to be identified, and an unlabeled receptor protein (protein II in Fig. 8.13).

8.2 Structural Studies of Protein–Protein Interactions

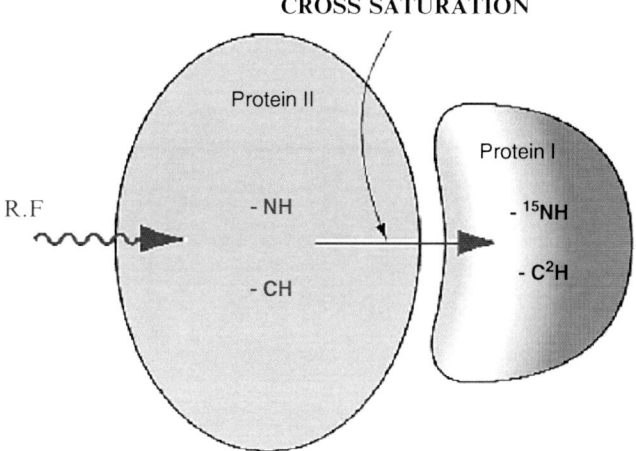

Fig. 8.13 Schematic representation of the transferred cross-saturation (TCS) method. The ligand protein (I) with a binding interface to be identified is labeled with rare isotopes of hydrogen (^{2}H) and nitrogen (^{15}N), while the receptor (protein II) is unlabeled. The measurements are only detecting 1H-15N 2D spectra (see text for details). Reprinted by permission from Macmillan Publishers Ltd: Takahashi et al. (2000).

Due to this labeling scheme, all hydrogen atom positions in the receptor are occupied with NMR-active ^{1}H atoms, while in the ligand only the amide hydrogen atoms are NMR-active. The RF field of the NMR spectrometer is tuned such that it excites nonselectively all the aliphatic protons (i.e., protons in non-aromatic and non-amide groups) in the receptor protein. The spins diffuse through the receptor, causing complete saturation of the nuclear resonances in the unlabeled receptor, and further to the binding interface of the doubly-labeled ligand (I).

> **Observation:** Spin diffusion is caused by the interaction between magnetic dipoles of different nuclei (called *nuclear Overhauser effect* or NOE), which depends on the inverse sixth power of the distance between nuclei. As mentioned in chapter 2, NOE is in fact one of the important effects used in determining protein structure by NMR, as it provides information on inter-proton distances.

Deuteration (i.e., ^{2}H-labeling) of the ligand protein is supplemented by deuterated water, the net effect being a drastic limitation in the saturation transfer to the ligand (since spin diffusion is only effective for closely spaced ^{1}H), which is confined to its residues at the binding interface. The interacting residues at the interface are then identified from the reduction of the intensities in ^{1}H-^{15}N correlation spectra as a result of saturation transfer to these spins. The experiment detects the ^{1}H-^{15}N groups, which are present in the ligand only.

The TCS method has been applied to the study of the ^{2}H, ^{15}N-labeled *B domain* of the *protein A* (FB) from *Staphylococcus aureus*, which binds specifically to the Fc portion of immunoglobulin G (IgG), and the binding residues have been determined (Fig. 8.14). The transient entity thus formed is called the FB-Fc complex.

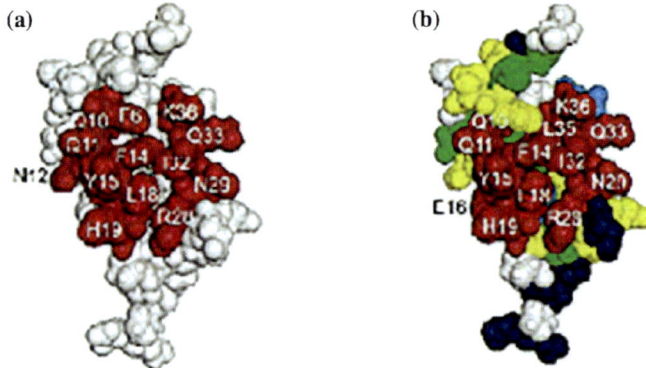

Fig. 8.14 Comparison of the binding sites of FB-Fc determined from (**a**) X-ray crystallography studies and (**b**) transferred cross-saturation experiments. Colors reflect different chemical shifts: red – $\delta < 0.5$ ppm; light blue – $\delta = (0.5–0.6)$ ppm; yellow – $\delta = (0.6–0.7)$ ppm; green – $\delta = (0.7–0.8)$ ppm; dark blue – $\delta > 0.8$ ppm. Reprinted by permission from Macmillan Publishers Ltd: Takahashi et al. (2000).

As seen from Fig. 8.14, the binding sites determined from TCS and X-ray crystallography agree remarkably well.

Other interesting developments of NMR include *in-cell NMR spectroscopy*, which has been used to study protein-protein interactions in living cells. In this method, the receptor protein with known NMR structure is labeled and, as the ligand binds to the receptor, the spectrum of the latter changes as a result of the interaction with the ligand at the interface between the two. Changes in the peak widths and chemical shifts allow identification with atomic resolution of the residues defining the binding interface (Serber and Dötsch, 2001; Burz et al., 2006).

Compared to X-ray crystallography, the in-cell NMR method presents the obvious advantage that it allows investigation of protein interactions at their location of normal function. Currently NMR lags behind X-ray crystallography with regard to the size of proteins that can be investigated. However, given the rapid progress in NMR studies over the past two decades, protein size may soon cease to pose difficulties. Particularly noteworthy in this regard are newly developed NMR techniques like TROSY and CRINEPT (Pervushin et al., 1997 Riek et al., 1999), which allow structural determinations of proteins as large as 100 kDa and open the possibility of looking at large protein complexes.

References

Bacskai, B. J., Skoch, J., Hickey, G. A. Allen, R. and Hyman, B. T. (2003) Fluorescence resonance energy transfer determinations using multiphoton fluorescence lifetime imaging microscopy to characterize amyloid-beta plaques, *Biomed. Opt.* **8**: 368

Berney, C. and Danuser, G. (2003) FRET or no FRET: a quantitative comparison, *Biophys. J.* **84**: 3992

References

Burz, D. S., Dutta, K., Cowburn, D. and Shekhtman, A. (2006) In-cell NMR for protein-protein interactions (STINT-NMR), *Nat. Protocols*, **1**: 146

Cavanah, J., Fairbrother, W. J., Palmer, A. G. III, Rance, M and Skelton, N. J (2007) *Protein NMR Spectroscopy: Principles and Practice*, 2nd ed., Academic, San Diego, CA

Clegg, R. M. (1996). Fluorescence resonance energy transfer. In: *Fluorescence Imaging Spectroscopy and Microscopy*. Wang, X. F. and Herman, B. (Eds.), Wiley, New York

Cormack, B. P., Valdivia, R. H. and Falkow, S. (1996) FACS-optimized mutants of the green fluorescent protein (GFP), *Gene* **173** (1 Spec No): 33

Craescu, T. C., Rouviere, N., Popescu, A. I., Cerpolini, E., Lebeau, M.-C., Beaulieu, E. E. and Mispelter, J. (1996) Three-dimensional structure of the immunophilin-like domain of FKBP59 in solution. *Biochemistry*, **35**: 11045

Denk, W., Strickler, J. H. and Webb, W. W. (1990), Two-photon laser scanning fluorescence microscopy, *Science*, **248**: 73

Dexter, D. L. (1953) A theory of sensitized luminescence in solids, *J. Chem. Phys.*, **21**: 836

Elangovan, M., Day, R. N. and Periasamy, A. (2002) Nanosecond fluorescence resonance energy transfer-fluorescence lifetime imaging microscopy to localize the protein interactions in a single living cell, *J. Microsc.*, **205** (Pt 1): 3

Gomperts, B. D., Kramer, I. M., and Tatham, P. E. R. (2002) *Signal Transduction*, Academic, New York

Ishima, R., Torchia, D. A. (2000) Protein dynamics from NMR, *Nat. Struct. Biol.*, **7**: 740

Kubitscheck, U., Schweitzer-Stenner, R. Arndt-Jovin, D. J. Jovin, T. M. and Pecht, I. (1993) Distribution of type I Fc-receptors on the surface of mast cells probed by fluorescence resonance energy transfer, *Biophys. J.*, **64**: 110

Lakowicz, J. R. (2006) *Principles of Fluorescence Spectroscopy*, Springer, New York

Lippincott-Schwartz, J. and Patterson, G. H. (2003) Development and use of fluorescent protein markers in living cells, *Science*, **300**: 87

Masters, B. R. and So, P. T. C. (2004) Antecedents of two-photon excitation laser scanning microscopy, *Microsc. Res. Tech.*, **63**: 3

Merzlyakov, M., Chen, L. and Hristova, K. (2007) Studies of receptor tyrosine kinase transmembrane domain interactions: The EmEx-FRET method, *J. Membr. Biol.*, **215**: 93

Meyer, B. H., Segura, J.-M., Martinez, K. L., Hovius, R., George, N., Johnsson, K. et al. (2006) FRET imaging reveals that functional neurokinin-1 receptors are monomeric and reside in membrane microdomains of live cells, *Proc. Natl. Acad. Sci. USA*, **103**: 2138

Milligan, G., Ramsay, D., Pascal, G. and Carrillo, J. J. (2003) GPCR dimerisation, *Life Sci.*, **74**: 181

Minski, M. (1961) *US Patent No. 3013467*

Neher, R. A. and Neher, E. (2004) Applying spectral fingerprinting to the analysis of FRET images, *Microsc. Res. Tech.*, **64**: 185

Overton, M. C., Chinault, S. L., and Blumer, K. J. (2005) Oligomerization of G-protein-coupled receptors: Lessons from the Yeast Saccharomyces cerevisiae, *Eukaryotic Cell*, **4**: 1963

Pelletier, J. N., Campbell-Valois, F. X., and Michnick, S. W. (1998) Oligomerization domain-directed reassembly of active dihydrofolate reductase from rationally designed fragments, *Proc. Natl. Acad. Sci. USA*, **95**: 12141

Pellecchia, M., Sem, D. S. and Wüthrich, K. (2002) NMR in drug discovery, *Nat. Rev.*, **1**: 211

Pervushin, K., Riek, R. Wider, G. and Wütrich, K. (1997) Attenuated $T2$ relaxation by mutual cancellation of dipole–dipole coupling and chemical shift anisotropy indicates an avenue to NMR structures of very large biological macromolecules in solution, *Proc. Natl. Acad. Sci. USA*, **94**: 12366

Raicu, V., Jansma, D. B., Miller, R. J., and Friesen, J. D. (2005) Protein interaction quantified in vivo by spectrally resolved fluorescence resonance energy transfer, *Biochem. J.*, **385** (Pt 1): 265

Raicu, V., Fung, R., Melnichuk, M., Chaturvedi, A., and Gillman, D. (2007) Combined spectrally-resolved multiphoton microscopy and transmission microscopy employing a high-sensitivity electron-multiplying CCD camera, *Proceedings of SPIE, Multiphoton Microscopy in the Biomedical Sciences VII*, Periasamy, A. and So P. T. C. (Eds.), 6442 (2007) 64420M-1

Raicu, V. (2007) Efficiency of resonance energy transfer in homo-oligomeric complexes of proteins, *J. Biol. Phys.* e-pub

Raicu, V., Chaturvedi, A., Stoneman, M., Fung, R., Saldin, D., Petrov, G., and Gillman, D. (2008) *Proceedings of SPIE, Multiphoton Microscopy in the Biomedical Sciences VIII*, Editors: A. Periasamy, P.T.C. So in press

Riek, R., Wider, G., Pervushin, K., and Wuthrich, K. (1999) Polarization Transfer by Cross-Correlated Relaxation in Solution NMR with Very Large Molecules, *Proc. Natl. Acad. Sci. USA*, **96**: 4918

Rossi, F., Charlton, C. A., and Blau, H. M. (1997) Monitoring protein-protein interactions in intact eukaryotic cells by beta-galactosidase complementation, *Proc. Natl. Acad. Sci. USA*, **94**: 8405

Schönle, A. Glatz, M. and Hell, S. W. (2000) Four-dimensional multiphoton microscopy with time-correlated single-photon counting, *Appl. Opt.*, **39**: 6306

Selvin, P. R. (1995) Fluorescence resonance energy transfer, *Methods Enzymol.*, **246**: 300

Serber, Z. and Dötsch, V. (2001) In-cell NMR spectroscopy, *Biochemistry*, **40**: 14317

Serdyuk, I. N., Zaccai, N. R., Zaccai, J. (2007) *Methods in Molecular Biophysics: Structure, Dynamics, Function*, Cambridge University Press, Cambridge/New York/Melbourne

Song, L., Hennink, E. J., Young, I. T. and Tanke, H. J. (1995) Photobleaching kinetics of fluorescein in quantitative fluorescence microscopy, *Biophys. J.*, **68**: 2588

Takahashi, H., Nakanishi, T., Kami, K., Arata, Y and Shimada, I. (2000) A novel NMR method for determining interfaces of large protein-protein complexes, *Nat. Struct. Biol.*, **7**: 220

Takeuchi, K. and Wagner, G. (2006) NMR studies of protein interactions, *Curr. Opin. Struct. Biol.*, **16**: 109

Wolber, P. K. and Hudson, B. S. (1979) An analytic solution to the Förster energy transfer problem in two dimensions, *Biophys. J.*, **28**: 197

Wüthrich, K. (2000) Protein recognition by NMR, *Nat. Struct. Biol.*, **7**: 188

Zacharias, D. A., Baird, G. S. and Tsien, R. Y. (2000) Recent advances in technology for measuring and manipulating cell signals, *Curr. Opin. Cell Biol.*, **10**: 416

Zacharias, D. A., Violin, J. D., Newton, A. C. and Tsien, R.Y. (2002) Partitioning of lipid-modified monomeric GFPs into membrane microdomains of live cells, *Science*, **296**: 913

Zal, T. and Gascoine, N. R. (2004) Photobleaching-corrected FRET efficiency imaging of live cells, *Biophys. J.*, **86**: 3923

Zimmermann, T., Rietdorf, J., Girod, A., Georget, V. and Pepperkok, R. (2002) Spectral imaging and linear un-mixing enables improved FRET efficiency with a novel GFP2-YFP FRET pair, *FEBS Lett.*, **531**: 245

Zipfel, W. R., Williams, R. M. and Webb, W. W. (2003) Nonlinear magic: Multiphoton microscopy in the biosciences, *Nat. Biotechnol*, **21**: 1369

Answers to Selected Quizzes

Chapter 1

Q.1. The differential operator ∇ ("nabla" or "del") is defined by: $\nabla = \frac{\partial}{\partial x}\vec{i} + \frac{\partial}{\partial y}\vec{j} + \frac{\partial}{\partial z}\vec{k}$. Since, $\vec{\xi} = x\hat{i} + y\hat{j} + z\hat{k}$, $\vec{p} = p_x\vec{i} + p_y\vec{j} + p_z\vec{k}$, $|\vec{\xi}| = \xi = \sqrt{x^2 + y^2 + z^2}$ and $\vec{p}\cdot\vec{\xi} = p_x x + p_y y + p_z z$, we can immediately obtain: $\nabla \xi = \frac{\partial \xi}{\partial x}\vec{i} + \frac{\partial \xi}{\partial y}\vec{j} + \frac{\partial \xi}{\partial z}\vec{k} = \frac{\vec{\xi}}{\xi}$. By using these relations and after some straightforward mathematical manipulations we obtain, successively, relations (1.12):

$$\nabla\left(\frac{1}{\xi}\right) = -\frac{\nabla \xi}{\xi^2} = -\frac{\vec{\xi}}{\xi^3},$$

$$\nabla\left(\frac{1}{\xi^3}\right) = -\frac{3\nabla \xi}{\xi^4} = -\frac{3\vec{\xi}}{\xi^5} \text{ and } \nabla(\vec{p}\cdot\vec{\xi}) = p_x\vec{i} + p_y\vec{j} + p_z\vec{k} = \vec{p}.$$

Q.2. Since the charge is equal to zero, $\vec{E} = -\nabla\phi = -\nabla\left(\frac{\vec{p}\cdot\vec{R}}{R^3}\right) = -[\nabla(\vec{p}\cdot\vec{R})/R^3 + (\vec{p}\cdot\vec{R})\nabla(1/R^3)]$. Using the relations from the answer to **Quiz 1**, one can easily obtain (1.20).

Q.3. At equilibrium, the energy of interaction reaches a minimum. This means that, from a mathematical point of view, the first derivative of the expression (1.27) must vanish.

Q.4. The high value of the surface tension coefficient of water is due to the strong dipolar interactions between water molecules in the liquid (including the surface layer). So, yes, a high value of the surface tension coefficient will prevent water molecules from leaving the interface (i.e., to pass from the liquid phase into the gaseous phase); this also explains the high value of the specific latent heat of water.

Q.5. At $3.98\,°C$, the density of water is $\rho = 1\,\text{kg dm}^{-3}$. The molecular mass of water is $\mu = 18\,\text{g/mol}$. Therefore, the molar concentration, C_M, defined as the number of mols per dm^3, is given by the formula: $C_M = \rho/\mu$. Computation gives: $C_M \cong 55.55\,\text{M}$.

Q.6. An adult human of $M = 85\,\text{kg}$ contains a mass of water, $m_w = 0.7 \times M = \approx 60\,\text{kg}$. Therefore, $m_{hw} \approx 10\,\text{g}$.

Q.7. The mass ratio of $^3_1H^{18}_8O^3_1H$ molecule and $^1_1H^{16}_8O^1_1H$ molecule is 4/3.

Q.8. Approximate $\mu_0^{\bar{n}} - \mu_0^1$ by a constant (e.g., $10k_BT$), and consider, for simplicity, that there are only two species in the system: free monomers and micelles with average concentration $\bar{n}$.

Chapter 2

Q.1. No, since the lateral chains are in different positions.

Q.2. Because there are 3.6 AA per turn and the pitch is 5.4 Å, it results $d = 5.4\,\text{Å} : 3.6 = 1.5\,\text{Å}$.

Q.3. Since the pitch of the DNA double helix is 34 Å, it results that there are 10 bases per turn. Therefore, $\delta = 360° \div 10 = 36°$.

Chapter 4

Q.2. $C^*_{MS} = N_T/N_A$, where N_T represents the total number of particles after dissociation and N_A, the Avogadro's number. Since $N_T = (N - N_D) + 2N_D$ and $N_D = \alpha N$, it follows that $N_T = (1 + \alpha)\,N$. Taking into account that $C_{MS} = N/N_A$, the required expression results immediately.

Q.3. According to (4.33b) and taking into account that the standard solution of physiological saline solution is a completely dissociated electrolyte ($\alpha = 1$), we use the formula $\pi = C^*_{MS}RT$ to obtain $\pi \approx 7.25\,\text{bar}$.

Chapter 5

Q.1. The corresponding global reaction (i.e., skipping the intermediates, ML) is:

$$\text{M} + 2\text{L} \underset{k_{dis}}{\overset{k_{as}}{\rightleftarrows}} \text{ML}_2, \tag{a}$$

where k_{as} and k_{dis} are the corresponding rate constants. The affinity constant, K_a, is given, according to LMA, by

$$K_a = \frac{[ML_2(\infty)]}{[M(\infty)][L(\infty)]} = \frac{[ML_2(\infty)]}{\{[M(0)] - [ML_2(\infty)]\}\{[L(0)] - 2[ML_2(\infty)]\}} \tag{b}$$

The particular values for K_a are obtained by plugging numerical values into equation (b): $6.2 \times 10^2\,M^{-1}$, $1 \times 10^4\,M^{-1}$, $4.5 \times 10^5\,M^{-1}$, $5 \times 10^7\,M^{-1}$, ∞. The increase in complex concentrations from left to right reflects the increasing affinity of ligand for the macromolecule. In the ideal case, when all the macromolecule is liganded, the affinity tends to infinity.

Analogously, the corresponding association constant for 1:1 stoichiometry can be computed from the expression:

$$K_a = \frac{[ML_2(\infty)]}{[M(\infty)][L(\infty)]} = \frac{[ML_2(\infty)]}{\{[M(0)] - [ML_2(\infty)]\}\{[L(0)] - [ML_2(\infty)]\}} \quad (c)$$

The particular values for K_a are obtained by introducing the numerical values into equation (b): $5.8 \times 10^2\,M^{-1}$, $6.7 \times 10^3\,M^{-1}$, $4 \times 10^4\,M^{-1}$, $1 \times 10^6\,M^{-1}$, ∞. Note that in this case the affinity constants are lower than in the previous case.

Q.2. The particular cases for $n = 1$ and $n = 2$ are expressed by the equations (5.11) and (5.18). We shall treat next the case for n = 3. One may start from the following three sequential reactions:

$$M + 3L \underset{k_{dis1}}{\overset{k_{as1}}{\rightleftarrows}} ML + 2L \underset{k_{dis2}}{\overset{k_{as2}}{\rightleftarrows}} ML_2 + L \underset{k_{dis3}}{\overset{k_{as3}}{\rightleftarrows}} ML_3 \quad (d)$$

and write successively:

$$K_{a1} = \frac{[ML][L]^2}{[M][L]^3} = \frac{[ML]}{[M][L]}. \quad (e)$$

This implies:

$$\frac{[ML]}{[M]} = K_{a1}[L], \quad (f1)$$

$$K_{a2} = \frac{[ML_2][L]}{[ML][L]^2} = \frac{[ML_2]}{[ML][L]}, \quad (f2)$$

$$K_{a3} = \frac{[ML_3]}{[ML_2][L]}, \quad (f3)$$

By definition, the fractional saturation, f, of ML_3 is given by:

$$f = \frac{[ML_3]}{[M] + [ML] + [ML_2] + [ML_3]} = \frac{[ML_3]/[M]}{1 + [ML]/[M] + [ML_2]/[M] + [ML_3]/[M]} \quad (g)$$

From (f1) and (f2) it results:

$$\frac{[ML_2]}{[M]} = K_{a1}K_{a2}[L]^2 \quad (h)$$

while combination of (f3) and (h) gives:

$$K_{a1}K_{a2}K_{a3} = \frac{[ML_3]}{[M][L]^3} \Rightarrow \frac{[ML_3]}{[M]} = K_{a1}K_{a2}K_{a3}[L]^3 \qquad (i)$$

Finally, by combining (f_1), (h) and (i) with (g) the following expression is obtained for f:

$$f = \frac{K_{a1}K_{a2}K_{a3}[L]^3}{1+K_{a1}[L]+K_{a1}K_{a2}[L]^2+K_{a1}K_{a2}K_{a3}[L]^3} \qquad (j)$$

One may therefore generalize (j), by induction, to obtain the expression (5.19).

Q.3. It binds and releases O_2 and CO_2 very rapidly.

Q.4. The surface area and volume of a sphere of radius r_0 are, respectively:

$$A_0 = 4\pi r_0^2, \quad V_0 = 4\pi r_0^3/3.$$

Magnifying the radius by the integer number, n, leads to the following volume of the magnified sphere:

$$V_n = 4\pi n^3 r_0^3/3,$$

which gives:

$$r_n = (3/4\pi)^{1/3} V_n^{1/3}.$$

This means that $D_f = 3$.

Q.5. Referring to Fig. 5.5 that describes Sierpinski's gasket, one can successively write:

$$a_0 = a_0 \qquad a_1 = 2^1 a_0 \qquad a_2 = 2^2 a_0 \qquad \ldots \qquad a_n = 2^n a_0 \qquad (k)$$
$$P_0 = 3a_0 \qquad P_1 = 2^1 P_0 \qquad P_2 = 2^2 P_0 \qquad \ldots \qquad P_n = 2^n P_0 \qquad (l)$$
$$A_0 = \frac{\sqrt{3}}{4} a_0 \qquad A_1 = 3^1 A_0 \qquad A_2 = 3^2 A_0 \qquad \ldots \qquad A_n = 3^n A_0 \qquad (m)$$

By applying logarithm to equations (k–m) one obtains, respectively;

$$\ln a_n = n \ln 2 + \ln a_0 \qquad (k')$$
$$\ln P_n = n \ln 2 + \ln P_0 \qquad (l')$$
$$\ln A_n = n \ln 3 + \ln A_0 \qquad (m')$$

From equations (k') and (l') it results

$$\ln \frac{a_n}{a_0} = \ln \frac{P_n}{P_0} \Rightarrow a_n = \frac{a_0}{P_0} P_n \Rightarrow a_n = \frac{1}{3} P_n \text{ (not interesting!)}, \qquad (n)$$

while from the equations (l') and (3') it results

$$\ln \frac{a_n}{a_0} = n \ln 2 \qquad (o1)$$

Answers to Selected Quizzes

and

$$\ln \frac{A_n}{A_0} = n \ln 3 \tag{o2}$$

From relations (o) one obtains:

$$\frac{a_n}{a_0} = \frac{A_n^{\ln 2/\ln 3}}{A_0^{\ln 2/\ln 3}} \Rightarrow a_n = \frac{a_0}{A_0^{\ln 2/\ln 3}} A_n^{\ln 2/\ln 3} \Rightarrow a_n = \frac{4^{\ln 2/\ln 3} 3^{-\ln 2/\ln 3} a_0}{a_0^{2\ln 2/\ln 3}} A_n^{\ln 2/\ln 3} \tag{p}$$

The final relation between a_n and A_n is:

$$a_n = \left(\frac{4}{3}\right)^{\ln 2/\ln 3} a_0^{1-2\ln 2/\ln 3} A_n^{\ln 2/\ln 3} \tag{r}$$

This gives $D_f = \ln 3/\ln 2 = 1.585$.

Q.6. Relation (5.41) can be written as:

$$p(r,t) = p_0 \, t^{-h} \exp\left(-\frac{r^{2+\theta}}{Dt(2+\theta)^2}\right). \tag{a}$$

The left-hand side part of equation (5.38), obtained by applying the rules of derivation to (a), is:

$$\frac{\partial p(r,t)}{\partial t} = p_0 \, t^{-(h+1)} e^{-X(r,t)} \left[-h + \frac{r^{2+\theta}}{Dt(2+\theta)^2}\right] \tag{b}$$

where $X(r,t) = \frac{r^{2+\theta}}{Dt(2+\theta)^2}$.

In order to obtain the right-hand side part of equation (5.38), first write the following partial derivative:

$$\frac{\partial p(r,t)}{\partial r} = -p_0 \, t^{-(h+1)} e^{-X(r,t)} \frac{r^{1+\theta}}{D(2+\theta)}. \tag{c}$$

Combination of equations (5.38), (5.39), and (c) gives:

$$\left[D(r) r^{D_f - 1} \frac{\partial p(r,t)}{\partial r}\right] = \left[D r^{-\theta} r^{D_f - 1} \frac{\partial p(r,t)}{\partial r}\right] = -r^{D_f} p_0 \, t^{-(h+1)} e^{-X(r,t)} \frac{1}{2+\theta} \tag{d}$$

Applying the partial derivative to both sides of (d), gives:

$$\frac{\partial}{\partial r}\left[D(r) r^{D_f - 1} \frac{\partial p(r,t)}{\partial r}\right] = -r^{D_f - 1} p_0 \, t^{-(h+1)} e^{-X(r,t)} \left[-h + \frac{r^{2+\theta}}{Dt(2+\theta)^2}\right] \tag{e}$$

The last step in obtaining the expression of equation (5.38) is to multiply equation (e) by $1/r^{D_f - 1}$.

Chapter 6

Q.1. The projection of the vectorial equation (6.12) onto the axis, Ox, gives:

$$j_k = -D_k \left\{ \frac{d[X_k]}{dx} + \frac{z_k F [X_k]}{RT} \frac{d\phi}{dx} \right\}. \tag{a}$$

Since $d\phi/dx \cong \varphi_{TM}/\delta_M$, it follows immediately from (a) that:

$$\frac{d[X_k]}{dx} = -\left\{ \frac{j_k}{D_k} + \frac{z_k F}{RT} [X_k] \frac{\varphi_{TM}}{\delta_M} \right\}$$

$$\Rightarrow dx = -\frac{d[X_k]}{\frac{j_k}{D_k} + \frac{z_k F}{RT}[X_k]\frac{\varphi_{TM}}{\delta_M}} = \frac{RT}{z_k F} \frac{\delta_M}{\varphi_{TM}} \frac{d\left\{\frac{z_k F}{RT}[X_k]\frac{\varphi_{TM}}{\delta_M}\right\}}{\frac{j_k}{D_k} + \frac{z_k F}{RT}[X_k]\frac{\varphi_{TM}}{\delta_M}} \tag{b}$$

Introducing j_k/D_k from (a) into (b), equation (6.13) follows immediately. One may write (b) as:

$$dx = A \frac{dB(X_k)}{B(X_k)}, \tag{c}$$

where $A = -\frac{RT}{z_k F}\frac{\delta_M}{\varphi_{TM}}$ and $B(X_k) = \frac{j_k}{D_k} + \frac{z_k F}{RT}[X_k]\frac{\varphi_{TM}}{\delta_M}$.

Integrating left-hand side of (c) from 0 to δ_M, and right-hand-side from $[X_k]_{me}$ to $[X_k]_{mi}$, one obtains:

$$\frac{B([X_k]_{mi})}{B([X_k]_{me})} = e^{\frac{\delta_M}{A}}, \tag{d}$$

which may be rewritten as:

$$\frac{j_k}{D_k} + \frac{z_k F}{RT}[X_k]_{mi}\frac{\varphi_{TM}}{\delta_M} = \left\{ \frac{j_k}{D_k} + \frac{z_k F}{RT}[X_k]_{me}\frac{\varphi_{TM}}{\delta_M} \right\} e^{-\frac{z_k F}{RT}\varphi_{TM}}. \tag{e}$$

By solving for j_k, one obtains relation (6.14).

Q.2. $g_{K^+}^{max}$ represents the potassium conductance when all channels are open (i.e., for $n = 1$).

Q.3. (a) If the length of the axon is such that the time it takes the nerve impulse to travel half the distance is shorter than the refractory period of the nerve impulse, neither of the two excitations will meet the other end of the axon, because each of them will leave behind a temporary refractory state that will prevent the further propagation of the other impulse. Otherwise, both nerve impulses will reach the other end. (b) In the first case described in the answer to (a), $x = l/2$.

Answers to Selected Quizzes

Chapter 7

Q.1. $\geq 2\,\text{G}\Omega$

Q.2. From equations (7.14) and (7.16), we obtain the Carnot yield:

$$\left(1 - \frac{T}{T_{bb}}\right) = 1 + \frac{k_B T}{h\nu} \ln \frac{k_a}{k_d} = 0.68.$$ From this relation,

we obtain: $T_{bb} = 944\,\text{K}$ (quite "hot!")

Chapter 8

Q.3. $I = 0$ for ^{12}C (with no unpaired nucleons), $I = 1/2$ for ^{13}C (one unpaired neutron), $I = 1$ for ^{14}N (one unpaired proton and one unpaired neutron), and $I = 1/2$ for ^{15}N (one unpaired proton).

Q.4. $\nu_0 = \frac{gB_0\mu_N}{h} = \frac{gB_0 e}{4\pi m_p} = 638\,\text{MHz}.$

Appendix

Chemical structures of the natural amino acids together with their names given both as three-letter and one-letter abbreviations. In the alternative stick-based representation, the colors represent different atoms, as follows: **black – C, blue – N, grey – H, red – O**, and **yellow – S**. Other notations: **HFB** – hydrophobic amino acids; **HFL** – hydrophilic amino acids; **ACD** – acidic amino acids; **BAS** – basic amino acids. Images, adapted from Wikipedia (http://en.wikipedia.org/wiki/X).

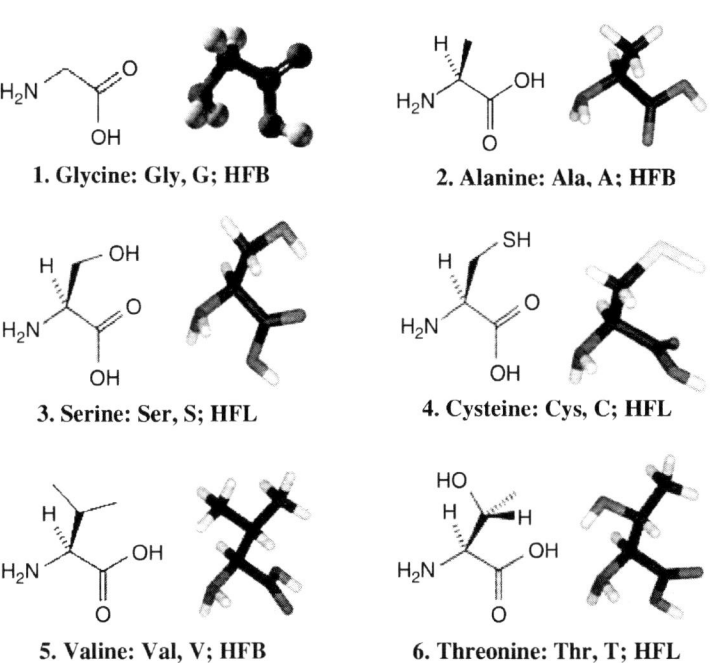

1. Glycine: Gly, G; HFB
2. Alanine: Ala, A; HFB
3. Serine: Ser, S; HFL
4. Cysteine: Cys, C; HFL
5. Valine: Val, V; HFB
6. Threonine: Thr, T; HFL

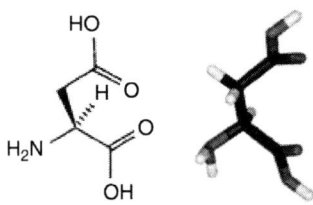
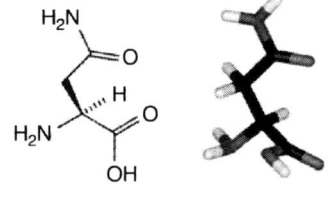

7. Aspartate: Asp, D; ACD **8. Asparagine: Asn, N; HFL**

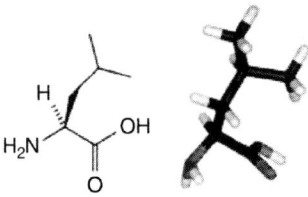
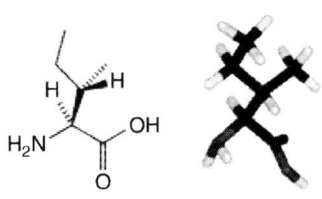

9. Leucine: Leu, L; HFB **10. Isoleucine: Ile, I; HFB**

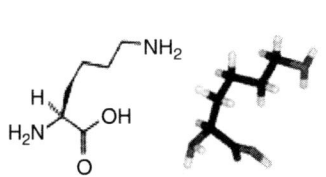
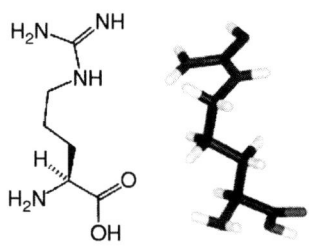

11. Lysine: Lys, K; BAS **12. Arginine: Arg, R; BAS**

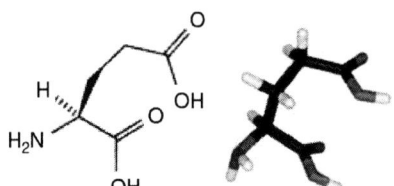
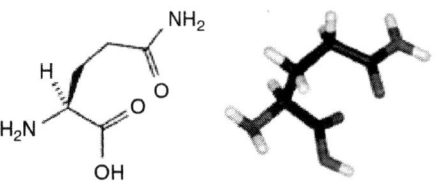

13. Glutamate: Glu, E; ACD **14. Glutamine: Gln, Q; HFL**

Appendix

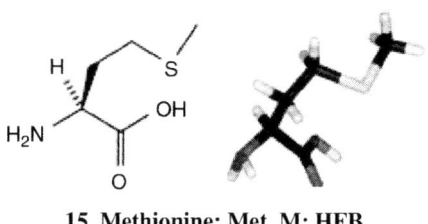

15. Methionine: Met, M; HFB

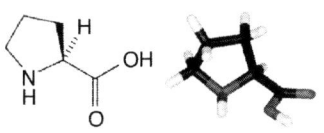

16. Proline: Pro, P; HFB

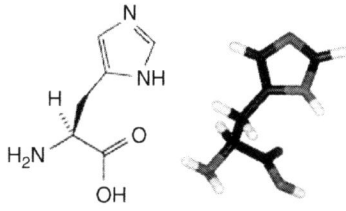

17. Histidine: His, H; BAS

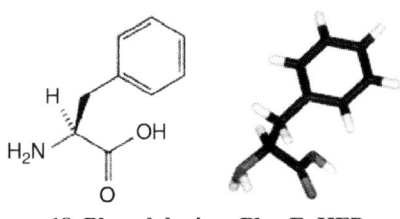

18. Phenylalanine: Phe, F; HFB

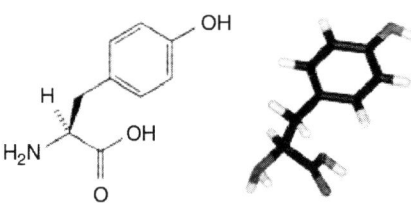

19. Tyrosine: Tyr, Y; HFL   **20. Tryptophan: Trp, W; HFB**

Index

A
Absolute temperature, 13, 31, 111, 112, 114, 174
Absolute value of energy, 15
Absorption angular frequency, 16
Absorption cross-sections, 201
Absorption spectrum, 212
Acceptor
  emission intensity, 200
  molecule, 198, 202
  of energy, 198
  photobleaching, 209
  sensitized emission, 290, 203
Acceptor's emission spectrum, 201
Acetylcholine, 173
Acetyl layer, 94
Acetyl region layer, 94
Acid-base systems, 29
Acidic medium, 48
Acidosis, 29
Action potential, 1, 4, 148, 155, 158–164, 166, 167, 169, 170, 173, 176
Activated states, 166
Activation, 128, 163, 173
Active
  centers, 55
  current densities, 152
  response, 157
  transport, 4, 62, 101, 103, 115
Activity coefficient, 31, 33
Adenine, 60, 62
Adenosine Monophosphate (AMP), 61, 62, 67
Adenosine Diphosphate (ADP), 61, 62, 125, 185
Adenosine Thriphosphate (ATP), 28, 41, 62, 66, 67, 76, 115, 120, 121, 125, 151, 185, 186, 188, 189

Adhesion, 15
Admittance, 89, 90, 93, 94, 97, 98
Adoptive mother, 43
Adsorbed anions, 78
Adsorption 15, 74, 78
Adult human being, 18, 21, 220
Aequorea victoria, 196
Affinity, 20, 55, 121, 125, 127–130, 187, 192, 221
Affinity constant, 125, 127, 129, 130, 220, 221
Agreggated monomers, 34
Aggregation, 31, 111, 139
  reversible, 15
Alcohols, 19, 30, 32, 95, 105
Alcohol dehydrogenase, 56
Algae, 40, 115, 148
Alkalosis, 29
All and nothing law, 159
Allosteric
  activation, 128
  binding, 128, 129
  inhibition, 128
Allostery
  negative, 128
  positive, 128
Alpha helix, 52, 53, 55, 182
Alpha-helix protein structure, 66
Alternating electric field, 89, 92
Amoebas, 40
Amino acid sequences, 58, 59, 68
Amino acids
  Alanine, 48
  Cysteine, 54, 178
  Glycine, 48
  Histidine, 48
  Phenylalanine, 48
  Prolyne, 48

Tryptophan, 48
Tyrosine, 48
Amino and carboxyl groups, 57
Anions, 20, 26, 29, 78, 81, 147, 150, 181, 182
Anisotropic membranes, 105
Anisotropic milieu, 103
Aminopyridine, 162
Amphiphiles, 31–37, 123
Amphiphilic
  aggregates, 32, 35, 37
  molecules, 31, 32, 34, 35, 37, 57
Anabolism, 46, 47
Anaphase, 44, 45
Anesthetic effect, 28
Anelectrotonus, 157
Angular frequency, 16, 89, 91
Angular velocity, 42
Anionic proteins, 147, 150
Anions, 20, 26, 29, 78, 81, 147, 150, 181, 182
Animal and vegetal kingdoms, 40
Anisotropic structures, 188
Anomalous behaviour, 20, 25
Anomalous diffusion, 142
Antenna, 4, 189
Antibodies, 17, 125
Antientropic association, 23
Antientropic effect, 79
Antientropic process, 123
Antigen, 17, 125
Antiparallel beta sheets, 53, 54
Antiparallel orientations, 15
Antiporter, 185, 188
Antiport system, 185
Apoprotein, 55, 190
Apoptosis, 45
Apparent FRET efficiencies, 203, 204, 207
Applied field, 88, 89, 92, 211
Applied magnetic field, 210
Approximate phases, 180
Approximate electron density, 180
Approximate spatial coordinates, 180
Aquaporin(s), 115, 173, 181, 182
Aquatic
  animals, 20
  fauna, 20
  plants, 20
Aqueous
  cavity, 184
  milieu, 8, 18, 84
  solution(s), 8, 31, 33, 112, 123, 150, 173
  solvent, 30
Architecture of the cell, 39–70

Artificial lipid bilayers, 88
  membrane bilayers, 88
  aspartate transcarbamylase, 56
Asexual division, 46
Asexual reproduction, 43, 44
Associated clusters, 25
Associated monomers, 34
Asymmetric distribution, 147
Asymptotic behaviour, 144
Atmospheric pressure, 111
Atomic number, 178
Atomic structure, 68
ATP hydrolysis, 28, 115, 185, 186, 188
ATP synthase, 188
Attenuation constant, 158
Attenuation of the transmitted RFF, 211
Attractive electrostatic interactions, 7
Attractive electrodynamic interactions, 7
(Auto)reproduction, 60
Average energy
  energy of interaction, 15
  probability, 141
  thermal energy, 13
Avogadro's number, 13, 83, 201, 220
Axonal membrane, 159, 167, 169
Axon(s), 28, 120, 148, 155–164, 167–170, 224

B

Backbone, 51–53, 55, 59, 63–65, 69, 181, 214
Background noise, 175
Bacteria, 40, 43, 77, 112, 115–117, 188–190, 196
Bacterial spores, 18
Bacteriorhodopsin, 188
Barrel-like structure, 182
Barrier maxima, 109
Basic medium, 48
Beta sheets, 52–55
  pleated sheets, 53
Beta strand, 52, 53, 59, 77, 116
Beta turn, 52–54
Bidimentional Fourier transform, 42
Bilayer
  lipid membranes, 92–94
  membranes, 37, 44
  slab, 93
  symmetry plane, 75
Binary acid-base systems, 29
Binding energy, 23
Binding sites, 55, 118, 125–129, 131, 141, 214, 216
Binomial distribution, 208
Biochemical/energetic transformations, 46

Index 233

Biochemical
  energy, 46
  reactions, 18, 46, 47, 106, 123, 145
  transformations, 46, 148
Biochemistry of the cell, 39
Biogenesis dogma, 43
Biological
  cells, 4, 30, 39, 59, 60, 95, 98, 106, 123, 145, 149
  fluids, 112, 114, 136
  functions, 7, 18, 26, 68, 195
  liquids, 29, 78, 114
  macromolecules, 8, 18, 47, 145
  matter, 2, 7
  particles, 79, 80, 84, 86, 87
  processes, 28, 30, 121, 122, 128, 189, 195
  systems, 4, 7–9, 16, 29, 47, 92, 103, 116, 123, 125
Biomolecular interactions, 123
Biophysics concepts and principles, 39
Biopolymers, 7, 48
Biosynthesis, 41, 46, 47, 60, 62, 125
Biosynthesis processes, 46, 47
Black-body temperature, 191, 192
Black lipid bilayers, 95
Blood
  capillaries, 115
  flow, 115
  plasma, 112, 114
Blue-green algae, 40
Body thermoregulation, 28
Bohr's magneton, 211
Boiling point, 19
Boltzmann
  constant, 13, 82, 104, 174, 191
  distribution, 82, 210
  factor, 13
Bond breaking, 8
Bond(s),
  coordinative, 23
  covalent, 8, 22, 23, 50, 54, 61, 69, 212
  hydrogen, 7, 22–28, 30, 52, 53, 56, 59, 63–65, 67, 75
Bound ligand(s), 126
Bound liganded macromolecules, 126
Bound water, 26
Bound water layer, 26
Bragg-Bragg formula, 177
Brain, 2, 4, 28, 35, 155
Breaking, 8, 24, 25, 67
Breathing movements, 8
Buffer, 29
Buffering systems, 29
Building blocks, 48, 60, 67, 74

Bulk
  aqueous solutions, 8
  electrolyte, 94
  liquid phase, 103
  of membrane hydrophylic layer, 107
  water, 18, 24
  water peaks, 213
  water protons, 210, 212
  water structure, 26

C
Ca-ATP-ase, 186, 188
Cable equation, 168
Cable model, 167
Cantor bar, 138
Cantorian fractals, 95, 137, 138
Cantorian interface, 138
Capacitance, 76
Capacitive current, 161
Capacity curve, 94
Capillary wall(s), 28, 115
Cardiac myocytes, 138
Carnot cyclic machine, 192
Carnot yield, 192, 225
Carotenoids, 189, 190
Carrier-mediated transport, 116, 140, 143
Carrier molecules, 119
Carriers, 47, 74, 76, 103, 117, 119, 120, 173, 185
Cartesian coordinate systems, 11
Cartesian reference frame, 104
Catabolic processes, 46, 47
Catabolic reactions, 46
Catabolism, 46, 47
Catalytic active site, 185
Cathelectrotonus, 157
Cations, 20, 26, 29, 78, 81, 150, 182
Cell
  -attached recording, 175
  differentiation and morphogenesis, 45
  division, 44
  membrane, 1, 4, 8, 21, 26, 32, 37, 39, 43, 73–98, 101, 105, 107, 112, 115, 116, 120, 152, 155, 174–176, 185, 188, 189
  morphology, 115
  organelles, 189
  suspensions, 1, 88
  wall, 43, 112
Cellular
  compartments, 103
  differentiation and morphogenesis, 43
  energetics, 41
  membrane formation, 19
  membrane capacitance, 98, 152, 157

membranes, 29, 30, 40, 103
metabolism, 18, 39, 46
mitosis, 66
movements, 42
osmotic pressure menace, 115
recognition, 47, 73, 190
responses, 195
self-reproduction, 43
signal transmission, 188
stabilization, 19
swelling, 112
volume, 112, 123
Cellulose, 43
Cenorhabditis elegans 197
Centripetal acceleration, 42
Chained serial processes, 103
Channel architecture, 182
Channel-facilitated diffusion, 117, 173
Channel inhibitors, 162
Channel proteins, 116, 181
Channels, 1, 4, 20, 26, 28, 42, 47, 56, 74, 76, 77, 103, 115–117, 120, 121, 152, 155, 159, 162, 163, 166, 173–192, 195, 224
Channel subunits, 165
Chaotic collisions, 103
Chaperonins, 56
Characteristic frequency, 91
Charge
   charge interactions, 9, 13
   cloud, 84
   density, 11, 81, 86–88
   distributions, 8, 11, 12, 14
   dipole interactions, 7, 10, 13
   fluctuations, 15, 16
   transport, 18
Charged molecules, 20
Charged phosphate group, 32
Charybdotoxin, 162
Chemical
   activation, 173
   bonds, 8
   composition, 74
   dependent channels, 155
   energy, 4, 188, 189
   equilibrium, 124
   neurotransmitters, 156
   potential(s), 30, 31, 33–37, 110, 111, 149, 191
   reactions, 32, 123, 125, 188
   shifts, 69, 211–214, 216
   signals, 73
   work, 188
Chlorine channels, 181, 182, 184
Chlorophylls, 43, 189–192
Chloroplasts, 43, 189
Chlorpromazine, 162
Cholesterol, 74
Chromatin, 40, 41, 45
Chromatography, 196
Chromosomes, 40, 44, 45, 57, 64, 66
Circulating fluids, 28
Clamp voltage, 160
Clathrates, 26, 28
Cloning, 43, 196
Close state, 176
Codon, 51
Coelenterates, 18
Cold source, 192
Colloidal chemistry, 85
Colloidal particles, 80
Colloid osmotic pressure, 114, 115
Compact double layer (CDL), 79, 80, 86
Compact structure, 53
Competitive binding, 127
Competitive interactions, 26
Complementary basis, 63, 65, 67, 68
   strands, 63, 67
   templates, 68
Complex quantities, 88, 180
   conductivity, 88
   permittivity, 88
Composition and organization of the cell, 39
Concentrated electrolyte, 115
Concentration difference, 106, 188
   gradient(s), 43, 103, 105, 116, 117, 120, 152, 154, 163, 173, 174, 188, 189
   of water, 110
   ratio, 21, 185, 191
   of amphiphiles, 34
Condensation, 20, 28, 45, 66
Conductance, 89–91, 93, 94, 152, 153, 155, 160, 163–165, 224
   dispersion curve, 94
Conductivity, 88, 89, 91, 92, 142, 151, 152
Confocal microscopy, 205, 207, 208
Conformational
   change(s), 117, 128, 182, 186, 187, 214
   fit, 26
   state(s), 117, 173, 186, 213
Conformers, 213, 214
Conjugated double bond, 10
Constitutive (unregulated) exocytosis, 102
Constructal theory, 139
Consumption of energy, 139
Content of water, 18
Continuity equation, 105, 141
Continuous distributions
   of electrical charges, 11

Cooperativity, 128–131, 195
Coulombian energy, 9
Coulumbian interaction energy, 8
Counterions, 85
Countertransport, 174
Coupled processes, 46
Covalent bond, 8, 22, 23, 50, 54, 61, 69, 212
Critical concentration, 34, 137
Critical micelle concentration (CMC), 32, 34
Cryo-electron microscopy, 186
Crystal hydrates, 26, 28
   structure, 1, 179, 186, 190, 213
Curie Principle, 188
Current density, 89, 144, 151–154, 167, 168
   per unit area, 151, 168
   per unit length, 167, 168
   source, 161, 174
Cyclic conformational changes, 117, 186
Cylindrical shell, 168
Cylindrical electrode, 160
Cyanobacteria, 40
Cysteine, 54, 178
Cytoskeleton, 40, 42, 80
Cytoplasm, 40, 137, 138, 185
Cytoplasmic tails of receptors, 207
Cytosine, 60
Cytosol, 7, 8, 10, 18, 25, 29, 40, 42, 67, 103, 105, 107, 112, 114, 15, 123, 185, 186, 188
Cytosol nucleotides, 67
Cytosolic particles, 103
Cytosolic pH, 42

D

Dalton rule, 114
Daughter cells, 44, 66
Debye
   (induction) interactions, 10, 13
   interactions, 14, 15
   dispersion function, 91, 92
   equation, 92, 96, 98
   length, 80, 84, 87, 88
Debye–Hückel parameter, 82–84
Decomposition, 202
Deconvolution method, 208
Degradation of fatty acids, 42
Degradation processes, 46
Degree of association, 24, 25
Dendrite, 156
Density, 10, 11, 20, 22, 25, 42, 81, 86–89, 103, 104, 111, 141, 151–154, 167, 168, 178–180, 196, 209, 213, 219
Dentine, 18
De-nucleated ovule, 43

Density gradient centrifugation, 196
Density of probability, 141
Deoxyribonucleic acid (DNA), 41, 47, 60-69, 87, 141, 156, 196
DNA
   binding proteins, 65
   binding surfaces, 65
   double helix, 64–66, 220
   ligase, 68
   nucleotides, 60
   polymerase-repair process, 66–68
   replication, 45, 68
   replication (multiplication), 66, 67
   strands, 63, 67
   structure and replication, 4, 39
Deoxyribose, 60-62, 65
Deoxyribonucleic-based (DNA), viruses, 39
Depolarization wave, 169, 170
Depletion regions, 137
Desalinated water, 111
Detergents, 32
Deterministic fractals, 139
Detoxifiation of the cell, 42
Deuterium, 21
De-excitation, 191, 199, 200
Dialysis, 125, 127
Dialysis sack, 126
Dielectric
   conductivity, 151
   constant, 8, 9, 21, 85, 95, 151
   dispersion, 92, 98
   increment, 91
   layers, 90, 92–94
   material, 89
   measurements, 76, 98
   model, 93, 94, 97
   multilayer, 90, 93
   particles, 92
   properties, 88, 95, 96
   relaxation, 90
   spectroscopy, 88, 92
   structure, 93
Differential equations, 82, 140, 170
Diffraction
   angle, 179
   image, 178, 179
   peak(s), 179
   spots, 179, 180
Diffusion
   anomalous, 142
   coefficient, 76, 104, 105, 142, 151, 153
   equation, 105, 140-142
   in homogeneous media, 103
   imited aggregation, 139

on fractal objects, 141
particles, 103, 140
processes, 103, 120, 141
Diffuse double layer (DDL), 79, 80, 84, 86
Diffusing anions and cations, 150
Diffusing pathway, 184
Diffusion coefficients, 76, 104, 105, 142, 151, 153
Dihedral angle, 13
Dilatation coefficient, 4, 123–125
Dimensionality, 4, 123–145
Dimension of the structure, 134
Dimeric
associations, 25
complexes, 182, 208
unit, 177
Dimer(s), 25, 56, 77, 116, 195, 198, 202, 208
Dinucleotide, 62, 63
Diploid cell, 44–46
Dipolar interaction, 11, 199, 219
Dipole
dipole interactions, 12, 69, 198
moments, 10, 11, 13–15, 20, 21, 26
Direct Fourier transform, 180
Direct osmosis, 110
Direction 5'-3', 62, 68
Disaccharides, 77
Disorganization of water molecules, 27
Dispersion
curves, 93, 94
interaction energy, 16
interactions, 10, 15
single-term expression, 16
Dissipative, 47
Dissociated molecules, 113
Dissociation, 33, 78, 113, 118, 124, 125, 143, 220
degree, 113
Distributed physical parameters, 11, 14
Distribution
Poisson-Boltzmann, 82
Donnan equilibrium, 150, 151
Donor
emission, 198, 200, 202
of energy, 198
fluorescence, 200, 204, 208, 209
quenching, 200, 203
Donor's emission spectrum, 199
Donor fluorescence intensity, 200, 209
Dormant phase of the cell, 45
Double helix model, 65
Double layer
capacitance, 87, 88, 93
structure, 79, 94

Drosophila melanogaster 40, 197
Dye-based method, 148
Dynamical viscosity coefficient, 19, 85, 104
Dynamic
equilibrium, 24
interactions, 195

E

Earth, 2, 18, 19, 25, 48, 60, 189
Effective colloid osmotic pressure, 15
Efficiency of molecular transport, 42
Egg phosphatidyl-choline, 94
Einstein's formula, 104, 142
Electronic spin resonance (ESR), 211
Electric
currents, 151
dipole, 9, 13
multipole, 10
octopole, 10
potential, 11, 95
quadrupole, 10
Electrical
activation, 173
behaviour, 4, 155, 174
capacitance, 76, 87, 88, 91, 94, 98
capacitances of membranes, 88
charge density, 11, 81
current density, 89, 154
dependent channels, 155
dipoles, 13, 21, 26, 92, 182
displacement, 89
dissociation, 78
double layer, 79, 80, 85–88, 93, 94
field, 13, 14, 79, 84, 87, 92, 94–96, 166, 173
field intensities, 14
interactions, 8
layers, 21
model, 151, 160
parameters, 88
polarisation, 185
potential, 9, 12, 81–86, 147, 149, 150, 152
potential gradients, 105, 153
properties, 37, 88–97, 152
setup, 155, 157
sources, 152
stimuli, 155
Electrochemical potential, 84, 149, 150
Electrodes, 88, 93, 148, 155, 160, 161, 175
Electrodynamic
interactions, 7
intermolecular interactions, 10
Electrogenesis, 151
Electrogenic character, 185
Electrokinetic mobility, 85

Index 237

potential (zeta potential), 85
radius, 84, 85
Electrolyte, 43, 65, 81, 82, 86–88, 93, 94, 98, 115, 174, 220
Electrolytic solutions, 29, 78, 83, 92–94, 113
Electromigration, 84, 87
Electronegativity character, 21
Electron
  acceptor, 23
  density, 10, 178–180, 213
  donor, 23, 189
  sharing, 8
  transfer reaction, 189
Electroneutrality of the medium, 83
Electronic
  absorption frequency, 16
  clouds, 177, 178
  density, 22, 180
  structure, 21, 22, 211
Electrophoresis, 85, 87
Electrophoretic mobility, 85
Electropositivity character, 21
Electrostatic
  energy, 9, 11, 12
  fields, 14
  interactions, 7, 8, 29, 78, 86
  landscape, 79
  repulsion, 65
Electrotonic passive response(s), 158
Electrovalency, 81, 83, 149, 151
Elementary blocks, 60
Elementary square, 133
Elementary structure, 179
Embryo, 18, 43, 45
Empirical expressions, 166
Endergonic, 46, 103, 125
Endergonic process, 46, 50, 62, 125
Endogenous or exogenous regulators, 176
Endoplasmic reticulum, 40-42, 156, 209
Endothermic, 27
Endothermic interactions, 27
Endocytosis, 73, 101, 102
Endothelial wall, 115
Energetic landscape, 17, 56, 59
Energetic plants of the cells, 41
Energy
  barriers, 58, 107, 109
  consuming process, 111, 115
  consumption, 103, 120
  levels, 210, 211
  levels splitting, 210
  liberated from ATP hydrolisis, 115
  of interactions, 9–15, 219
  reservoirs, 189

source, 66
transfer, 3, 189, 196, 198, 199, 201–203
Electrotonic, 155, 157, 158
Enthalpy, 27
Entropically driven, 27
Entropic effect, 23
Entropic thermal process, 23
Entropy, 23, 27, 28, 46, 47, 79, 103, 110, 116, 120, 124, 191
Entropy excess, 47
Entropy generating process, 110
Enzymatic activity, 186, 196
Enzymes, 17, 42, 47, 55, 66–68, 76, 123, 128, 185, 188, 195, 196
Equator of the image, 179
Equilibrium
  constant, 33, 124, 125, 128
  distance, 17
  shifted to the right, 9
  to the left, 9
Equivalent admittance, 90, 97
  capacitance, 91, 93
  class of protons, 211
  conductance, 93, 94, 153
  conductivity, 93, 94, 153
  electrical circuit, 93, 151, 152
  permittivity, 92
  protons, 212
Erytrocyte(s), 98
Erytrocyte ghosts, 98, 122
Erytrocyte membrane(s), 28, 107, 143, 149, 150, 155
Escherichia coli 77, 188
Euclidian geometry, 123, 131, 134, 135
Euclidian objects, 134
Eukaryotes, 39, 40
Eukariotic cells, 40, 66
Euler number, 84
Evaporation, 8, 20, 28
Evolutionary process, 25
Excitable animal cells, 147
Excitable cells, 4, 120, 148, 155, 159, 160
Excitants, 155
Excitation rate, 200, 201
Excited state(s), 189, 192, 197, 198
Exclusion principle, 17, 23
Exclusion volume, 137
Executive power, 47, 60
Exergonic, 46, 103, 125
Exergonic process, 46, 47, 116, 121, 125
Exocytosis, 73, 101, 102
Exothermic, 27
Experimental current, 166
Expression of genes, 141

Externalized particle, 102
External pressure, 111
Extracellular
  electrical potential, 147, 149
  milieus, 25, 40, 147, 151
  side, 74, 117, 143, 183, 184, 187
Extrinsic charging mechanism, 78

F

Facilitated diffusion, 116–120, 143–145, 173
Facilitated transport, 103, 116–120
Faraday's number, 83, 149
Fatty acids, 32, 41, 42, 105
Feedback, 160, 170
Fibrilar macromolecule, 179
Fick's first law, 104, 106, 107, 109, 141
Fick's second law, 105, 107, 109
Field
  alternating, 88, 93
  intensity, 153
First intermediate phosphorylated state, 187
Flagellated bacteria, 188
Flexible bilayers and vesicles, 37
Flippase, 76
Flip-flop movements, 76
Floatation, 20
Flocculation, 15
Flow, 19, 104, 108, 109, 115, 116, 133, 149, 152, 163
Fluid mosaic model, 75, 76, 93
Fluorescence
  intensity, 148, 200, 201, 205, 208, 209
  lifetime, 199, 200
  recovery after photobleaching (FRAP), 76
  resonance energy transfer (FRET), 196–204
  spectral shift, 148
Fluorescent
  molecules, 196–198, 201, 205
  protein(s), 196, 206
  tags, 196, 198, 207
Flux
  density, 103, 104, 153
  of mass, 104
  of substance, 107–109, 138, 142, 153
Foetus/ Fetus, 28, 45
Folded membranes, 7
Food digestion, 188
Forces of interactions, 7
Formation, 4, 7, 11, 19, 21, 26, 28, 30, 32, 33, 35, 37, 45, 52, 53, 69, 118, 123, 124, 139, 143, 201, 207
Forward and reverse rates of transfer, 108
Förster distance, 194

Förster resonance energy transfer (FRET), 196, 198–204, 207–209, 216
Förster's theory, 198, 199
Förster's transfer rate, 198
Fourier transform, 178–180, 212
Fourth dimension, 131
Fractal(s)
  concepts, 123
  deterministic, 139
  diffusion, 140-144
  dimension, 134–137, 139
  geometry, 13, 138–140, 145
  lattices, 140-145
  of biological interest, 136
  objects, 132, 141, 142
  stochastic, 139
  structures, 95, 134–136, 138, 140, 141
Fractional
  binding of ligand, 129
  calculus, 140
  diffusion equation, 140
  dimensions, 134, 141
  number of dimensions, 131
  power function, 144
  saturation, 126, 128–130, 221
  viscous-diffusion, 140
Freedom of motion, 18
Free diffusion
  energy, 27, 30–32, 36 46, 47, 58, 62, 103, 125, 185, 189, 191
  energy of transfer, 30–32
  ligand, 129
  monomers, 34, 129, 130, 220
  radicals, 42
Frequency, 16, 53, 59, 69, 76, 88, 89, 91–95, 98, 148, 191, 208, 210–212
FRET efficiency, 199, 200, 203, 204, 207
Fugu, 162
Functional
  complexes, 209
  form of a protein, 195
  proteins, 47
Fungi, 115, 116

G

GABA-gated channels, 182
Gain factor, 160
Gamete, 44–46
Gated structures, 173, 176
Gating, 173
Gating currents, 163
Gaseous phase, 25, 219
Gastric juice, 188
Gastric mucosa, 188

Gene autoreproduction, 66
Gene-expression machinery, 60
Genes, 47, 48, 66, 67, 141
Genetic information, 41, 43, 60, 139
  inheritance, 66
  material, 39–41, 44, 45, 68
  mutations, 181
Genitor(s), 43–45
Geometric and physical parameters, 11, 14
Geometrical factor, 90, 91, 95
GFP-tagged Ste2 proteins, 207, 208
Giant algae (*Nittela* and *Chara*), 148
Giant axon, 148, 157, 158, 163, 164
Gibbs free energy, 27, 30, 46, 58, 125, 185
Gibbs free energy variation, 125
Glandular cells, 73, 155
Glass pipette, 174
Global
  energetic landscape, 59
  entropy variation, 47
  ionic current, 162
  resistances, 152
Glucose, 41, 74, 105, 107, 113, 116–118, 140, 143–145
Glucose carrier, 117, 118, 143
Glucose facilitated diffusion, 143
Glucose transporter, 76, 77, 140
Glucosylcerebroside, 74
Glutamate amino acid, 184
Glycocalix, 74, 77
Glycolipids, 74
Glycophorin A, 76
Glycoproteins, 74, 77
Glycosylated extracellular portion, 186
Goldman-Hodgkin-Katz
  equation, 154, 170
  model, 153
Golgi complex, 40, 42
Gouy-Chapman theory, 79, 82
Gouy-Chapman-Stern theory, 79
G-protein-coupled receptors (GPCRs), 195, 207
Gradient, 11, 19, 43, 103–105, 116, 117, 120, 148, 152–154, 163, 173, 174, 188, 189, 196
Gradient of the conjugated quantity, 104
Green fluorescent protein (GFP), 196, 206–208
Green plants, 189
Grey matter, 18
Guanine, 60

H
Hair cells, 155, 173
Hairpin turns, 53
Halobacterium halobium 188
Haploid cell, 44, 46
Haploid gametes, 44, 45
Headgroup(s), 36, 78
Heat transfer, 192
Heavy metal ions, 178, 180
Heavy water, 21
Helicases, 67
Helices of higher order, 66
Helix axis, 53, 64–66
Helix of life, 60
Helmholtz layer, 79, 80
Helmholtz-Smoluchowski formula, 86
Hemagglutinin, 56
Hemoglobin, 1, 7, 27, 28, 53, 56, 57, 112, 130, 131, 195
Hemolysis, 112
Heme of myoglobin, 128
Henry formula, 86
Henry theory, 85
Hepatic plates, 136
Hepatic sinusoids, 136
Hepatocytes (liver cells), 42, 136–138
Hermann currents, 169
Heterogeneous media, 8
Heterobiopolymers, 60
Hetero-oligomers, 195
Heteropolymers, 48
Hexagonal arrangements, 24
High concentrations, 103
High dimensional manifold, 135
Higher energetic states, 66
Higher order DNA structure, 66
Hill coefficient, 130, 131
Hill equation, 130, 131
H,K-ATPase, 188
Hodgkin-Huxley
  model, 1, 163, 164, 166
  theory, 163
Holoproteins, 55
Homeothermic animals, 25, 28
Homo-dimers, 195, 198
Homogeneous
  electrical field, 95
  medium, 8, 103, 188
  particles, 95
Homo-oligomerisation, 145, 207
Homo-oligomers, 195, 207
Homo-tetramers, 195
Homo-trimers, 195
Hormone, 73, 101, 195
Hot source, 192

Hückel formula, 86
Human
  beings, 18
  body temperature, 25
  chromosomes, 64
  genome, 2, 48
Hybrid "supercells" 76
Hydrocarbon
  chains of lipids, 27
  phase, 34
  solvent, 30–32
  tails, 32, 35, 74
Hydration
  layer, 20, 35
  shell(s), 20, 26, 78, 182
Hydrogen
  binding, 24
  bonded water molecules, 28
  bonding, 22, 23, 25, 28
  bond network, 25, 26, 30
  bonds, 7, 2–26, 30, 52, 53, 56, 59, 63–65, 67, 75
  isotopes, 21
  nuclei, 23
Hydrolases, 42
Hydrolysis, 28, 62, 115, 185, 186, 188
Hydronium ion, 29, 78
Hydrophilic
  amino acids, 77, 227
  interactions, 76, 78
  pore, 182
  molecule, 20
  repulsion, 35
Hydrophobic
  amino acids, 77, 227
  apoproteins, 190
  attraction, 35
  core, 76, 88, 95, 182
  core volume, 76, 182
  effect, 28, 30
  interactions, 30, 54, 56, 57, 65, 75, 78
  layer, 76, 94, 95, 98, 107, 108
  molecules, 26–28, 30
  oil molecules, 30
  substances, 30
  tail(s), 32, 33, 37, 75, 147
Hydrophobicity, 30, 34, 117
Hydrostatic pressure, 111
Hydroxyl ions, 29
Hypertonic solution, 112
Hypertonic treatment, 112
Hypotonic solution, 113
Hypothesis, 163

I
Ice, 20, 24, 25
Ice crystals, 24
Ice hexagonal network, 25
Ice interstices, 25
Ice layer, 20
Ideal gas
  gas equation of state, 10
  (dilute) solutions, 111
Identical double strands, 68
Image pixel, 208, 209
Image plane, 179
Imaginary part of the admittance, 93
Imaginary sphere, 96
Immune defence, 73
Immunocompetent cells, 101
Immunophilin-like domain, 214
Immunoprecipitation, 196
Impedance amplifier, 160
Impedance spectroscopy, 88
Implanted microelectrodes, 157
Imponderability, 111
Impulse NMR, 212
Inactivated subunits, 165
Incident angles, 165
Incident light, 201
Independent binding model, 129
Independence principle, 152
Independent probabilities, 165
Independent structures (gates), 165
Independent sites, 127
Individual ionic channels, 163
Induced
  dipoles, 9, 10, 14
  electrical dipoles, 13
  multipoles, 10, 13
Induced fit model, 128
Induction interactions, 10, 13–15
Inert gases, 26–28, 30
Inhibitor, 120, 121, 128, 157, 162
Inner resistance per unit length, 168
Inorganic phosphate, 185
Internalized particle, 102
Initial amount of DNA, 68
Inside-out patch, 176
Instantaneous
  dipoles, 10, 14–16
  energy of interactions, 15
  induced dipoles, 10
  induced interaction energy, 14, 15
  induced multipoles, 10
  interactions, 12
  motion, 89
  multipoles, 10

Index 241

Integral proteins, 76, 78
Integrated emission intensities, 202
Interacting molecules, 8, 16, 17, 27
Inter-atomic distance, 17
Interface membrane, 103, 107
Interfacial Maxwell-Wagner polarization, 90, 92
Intermolecular
  attractive energy, 8, 10
  charge transfer, 28
  forces, 7
Internal and external media, 73, 107, 154
Internal face of the membrane, 77
Internal
  impedance, 174
  minima, 107
  resistance, 174
Interstitial liquid, 112, 114, 147
Intracellular
  electrical potentials, 147, 149
  milieus, 25, 40, 147, 151
Intramolecular interactions, 7
Interfaces of the particles, 7
Internal concentration, 162
Internal medium, 152, 168
Internal membranes, 40, 209
Interstitial monomers, 25
Intrinsic charging mechanism, 78
Inverse Fourier space, 178
Inverse Fourier transform, 179
Inverse osmosis, 110, 111
Inverted cone, 35, 37
Inverted micelles, 37
Inward cell current, 162
Inward sodium current, 162, 163
Ion-charged chemical group, 9
Ion channels, 1, 20, 28, 56, 116, 163, 166, 173–176, 181, 182, 185, 195
  expulsion from the cell, 115
Ionic
  asymmetry, 73, 151
  channel operation, 152
  channels, 28, 74, 103, 116, 117, 152, 155, 163, 176, 195
  conductivities, 155, 164
  current densities, 151
  disparity, 147
  distributions, 148, 151
  molecules, 20
  pumps, 76, 115, 121, 122, 147, 152, 173, 185, 188
  product of water, 29
  species, 81, 82, 148, 149, 152–154, 160
  strength, 83, 84

Ion-ion, 9
Ion mobility, 153
Ionization
  energy, 16
Ionized groups, 26
Ion pumps, 4, 47, 56, 74, 173, 174, 185, 186, 188, 195
Ion gates, 170
Isoelectric point/Isotonic solution, 48, 112
Isotonicity of the cytosol, 115

J
Jellyfish, 18, 196
Johnson thermal noise, 174

K
K^+ -channel, 76, 120, 162, 163, 173, 182, 184
Keesom energy of interactions, 13
Keesom interaction energy, 13
Keesom interactions, 10–13, 15
Kidney failure, 125
Kinetic energy, 107
Kramers-Krönig relationship, 91
Kreutzfeld-Jacob disease, 56

L
Lac
  proteins, 141
  repressor, 141
Lactose in yeast, 141
Laminar flow, 19
Landé factor, 210, 211
Laplacean operator, 81
Laplace equation, 83, 95
Laser light, 148, 204
Laser pulse(s), 148
Latent heat of
  freezing, 20
  melting, 20
  vaporization, 21
Law of mass action, 33, 123–129, 137, 140–145
Leading strand, 68
Leakage ions, 151, 152, 162
Lenard-Jones potential energy, 17
Legislative power, 60–67
Lifespan of a cell, 7
Lifetime of the donor, 199, 200
Liftime of the excited state, 192
Lifetime(s), 22–24, 189, 192, 199, 200
Life-cycle, 44, 45
Ligand-gated, 173, 181
Ligand-gated anion channels, 181
Ligand-gated cation channels, 181

Ligand-gated ion channels, 182
Ligand interactions, 26
Ligands, 27, 55, 126, 128, 131, 137, 150
Light
   absorbing molecules, 189
   absorption, 189–191
   -activated proton pump, 188
   -based activation, 173
   -gated, 173
   -harvesting process, 191
   -harvesting system, 189
   illumination, 192
Limiting conductivity, 91
   permittivity, 91
Linear current-voltage relation, 170
Linear scaling factor, 135, 136
Lipid
   anchored proteins, 76
   bilayer, 74, 75, 77, 80, 87, 88, 93–95, 98, 103, 107, 116, 120, 121, 148, 187
   insoluble polar charged molecules, 105
   insoluble uncharged molecules, 105
   matrix, 78, 103, 116
   membrane, 92–94
   molecules, 74, 76, 117
   monolayer, 75, 76, 93, 94
   soluble molecules, 105
   tails, 95
Lipoproteins, 74
Liquid water, 19–21, 24, 25, 28
Living cells, 120, 122, 151, 195, 196, 201, 204, 207, 214, 216
Living matter, 2, 3, 26, 28, 37, 39, 43, 46, 63, 74, 110
Living state, 25, 28
Living systems, 39
Living water, 18
Liver, 42, 136, 137, 139
Local circuit currents, 169
Local folding, 52
Local potential(s), 157–159, 167
London
   dispersion interactions, 10, 15, 16
   interactions, 16
   van der Waals interactions, 15
Longitudinal ionic current, 167, 169
Long range
   attraction, 7
   Coulombian interactions, 57
   non-specific physical interactions, 7
   specific physical interactions, 7
Low concentrations, 31, 81, 137, 188
Lower energetic state(s), 66
Low frequency, 91, 98

Lymphatic liquid, 28
Lysosomes, 40, 42, 43

M

Macroergic molecule, 411, 62
Macrophages, 42, 73, 101
Macromolecular complexes, 7, 26, 195
   crowding, 17
   solutions, 114
   species, 114
Macrotransport, 101, 102
Mad-cow disease, 56
Magnetic
   fields, 69, 209–211
   momentum, 209, 210
Magnification factor, 132, 135
Magnification step, 132, 134, 135
Major groove, 65
Male gamete, 46
Mammals, 18, 40
Mathematical transformation, 178
Matter and matrix, 18
Maxwell-Wagner interfacial polarization, 92
Maxwell-Wagner relaxation, 95
Mean free path, 103, 107
Mechanical
   activation, 173
   protection, 28
   channels, 155
   pressure, 115
   work, 188
   stress, 43, 73
Mechanically-gated, 173
Mechanosensory neurons, 173
Mediators, 195
Melting point, 19
   secondary point, 25
Membrane
   asymmetry, 77, 78, 120
   capacitance(s), 98, 152, 157, 160
   conductance(s), 152, 160, 163
   depolarization, 157, 159, 164–166, 169
   equivalent conductance, 153
   hyperpolarization, 157, 159, 165, 166
   lipids, 74, 121
   model, 75, 88, 93, 98
   patches, 174–176
   permeability, 19, 84, 107, 109, 160, 173
   plane, 76, 141, 183, 184, 207, 209
   potential, 3, 107, 147, 148, 151, 153, 155, 157, 159, 160, 165, 170
   protein receptors, 145
   proteins, 185
   receptors, 47

Index 243

specific resistance, 157
thickness, 95, 106, 153
3D structure, 74
Menger's triadic sponge, 136
Meridian of the image, 179
Messenger RNA, 41
Metabolic
 activity, 18
 currency, 41
 inhibitors, 121, 157
Metabolism, 18, 39, 46, 118, 125, 136, 141, 188
Metaphase, 45, 66
Micelle(s), 32–47, 74, 123, 220
Micelle formation, 32, 35
Microcrystal, 24
Microelectrode, 148, 157, 174, 176
Microelectrophoresis, 87
Micrometer-sized tip, 174
Micropipette method, 174
Micropipette techniques, 95
Microscopic elements, 133
Microscopic features, 132, 134
Microscopic particle, 95
Microtransport, 101, 103
Microvilli, 42
Miller indices, 177–179
Mimosa pudica 148
Minimum of potential, 107
Minimum potential, 36
Minor groove, 65
Misfolded state, 56
Misfolding, 39, 58
Mitochondria, 40-42
Mitotic division, 66
Mitosis, 40, 43–46, 66
Mobilities, 29, 84, 85, 87, 106, 145, 153
Model lipid bilayer, 94
Model synthetic membranes, 32
Molar concentration, 20, 29, 83, 104, 110, 112, 118, 121
Molar volume, 111, 114
Molecular collisions, 25
Molecular
 associations, 23, 24, 27, 123
 complexes, 101, 123, 195
 crowding, 204
 disorder, 47
 level, 2, 18
 machines, 4, 115, 173–192
 mass, 74, 112, 114, 185, 186, 219
 mechanism of defence, 68
 motor(s), 67
 organization, 92

self-association, 30-35
transport, 42
Mole fraction, 111
Monochromatic photons, 191
Monomeric units, 181, 204
Monomers, 25, 33–37, 56, 60, 129, 130, 181, 182, 187, 195, 198
Morphogenesis, 43, 45, 46, 73
Morphological and functional unit, 39
Mother and medium, 18
Mother cell, 44, 46
Multi-cellular organisms, 2, 46, 47, 197
Multi-compartmented cells, 39
Multidimensional object, 131
Multifractals, 139
Multi-layer dielectric, 90, 93
Multimeric
 organization, 56
 structure, 56
Multiple isomorphic derivatives, 178
Multi-shelled particles, 97
Muscle cells, 42, 147, 155, 156
Muscle contraction, 3, 62
Muscle fiber, 176
Mutagenic agents, 68
Mutation and cloning of DNA, 196
Mutant, 51, 56, 214
Myelinated axons, 169
Myelin sheath, 156, 158
Myoglobin, 1, 27, 53, 55–57, 128, 177

N
NADPH, 1898
Na,K-ATPase, 174, 185–188
Native
 locations, 196
 macromolecule, 26
 state, 17, 5, 56, 58, 211
Natural membranes
 bilayers, 88
 fractals, 139
 genitor, 43
Naturally selected sequence, 59
Negatively charged proteins, 32, 147
Negentropy, 189
Nernst
 equation, 149
 formula, 153
 potential, 150, 155, 160, 162
Nerve impulse, 156, 160, 163, 170, 224
Nervous central system, 28
 impulse propagation, 28
 signals, 73
Net charge, 11, 15, 81, 147

Net flux, 124
Network of capillary vessels, 115
Neurofilaments, 42
Neuromuscular endplate, 176
Neuron(s), 73, 120, 147, 155, 156, 159, 173
Neurotransmitter(s), 101, 156, 173, 195
New double strands, 68
Nitrogenous base(s), 60, 63, 67
Nitrogen-containing bases, 60, 61, 64
NMR signal, 211
NMR spectrum, 212, 213
Node of the circuit, 167
Non-covalently-bonded atoms, 17
Nonequilibrium thermodynamics, 46, 47
Nonideal gases, 112
Nonlinear differential equation(s), 82, 170
Non-polar
   layers, 8
   molecules, 16, 20, 30
Non-radiative
   de-excitation, 199
   processes, 189
   transfer, 197, 202
   transfer of energy, 196
Normal locations, 196
Normal physiological conditions, 29, 149
Nuclear magnetic resonance (NMR), 56, 69, 70, 209, 211–216
Nuclear spin, 69, 209, 210, 212
Nucleic acids, 2, 7, 23, 26, 27, 29, 57, 125
Nucleotides, 27, 60-65, 67
Nucleus, 22, 39–43, 45, 156, 209–211

O

Okazaki fragments, 68
Ohm law, 89
Oligomeric assembly, 182
Oligomeric complexes, 181, 195
Oligomer(s), 203, 204, 207
Oligomerisation, 145, 195, 207
Oligopeptide, 50
One-dimensional diffusion, 104
On-off behaviour, 176
Ontogenic development, 18
Open state, 165
Optically excited fluorescent molecules, 196
Optimal area, 37
Optimum area, 36
Orbitals
   asymmetrically bilobate, 22
   hybrid, 23
   hybrid bilobate, 23
   spherical symmetric, 23
Order of diffraction, 179

Order parameter, 58, 59
Organellar membranes, 30
Organelle membranes, 188
Orientation interactions, 10
Origin, 8, 19, 67, 78
Origin of life, 18
Orthorhombic system, 179
Osmometer, 114
Osmosis, 110-115
Osmotic lysis, 43
   pressure, 43, 110-116
   pressure menace, 115
Outer monolayer, 77
Output voltage, 161
Outside-out patch, 176
Outward cell current, 162
Overall efficiency, 203
Ovule, 43–46

P

Parallel beta sheets, 53, 54
Parallel orientations, 15
Parallel-plate capacitor, 89–91
Parental DNA, 68
Partial molar volume, 111
   positive charge, 20
   pressures, 114
Particle concentration, 105
   shape, 95
Partition
   coefficient, 154
Passive diffusion, 4, 103, 106, 110, 116, 120, 121, 153, 173, 188
Passive ionic currents, 152
Passive transport, 101, 103, 149
Patch-clamp measurements, 176
Patch-clamp technique, 163, 174, 176
Pentapeptides, 50
Peptide bond, 28, 50, 51
Percentile concentration, 112, 114
Percolation, 95, 136, 137
Percolation cluster, 136, 137
Percolation-cluster-like structure, 145
Percolation network, 138
Permeabilities, 19, 20, 84, 106, 107, 109, 116, 151, 153–155, 160, 163, 173, 182, 184
Perimeter, 132–135
Peripheral blood circulation, 28
Peripheral proteins, 76, 77
Permanent dipoles, 10, 13–15, 92
Permeability
   coefficient, 106, 107, 109, 116
   of the erythrocyte membrane, 155
Permeation, 106, 116

Permeation pathway, 182
Permittivity, 8, 9, 14, 76, 78, 81, 88–92, 96, 97, 210
Peroxysomes, 40
Perpendicular orientations, 15
Perturbation theory
  second order, 16
pH, 8, 29, 32, 42, 48, 49, 78, 111–113, 124, 147
  acidic, 29
  basic, 29
  neutral, 29, 78
Phagocytosis, 73, 101
Pharmacological inhibitors, 162
Phase information, 178
Phase recovery, 180
Phosphate radical, 60, 61, 64, 65
Phosphatidiylcholine, 32, 94
Phosphatidiylserine, 32
Phosphodiester binding, 5'-3' 64
Phosphodiester bond, 61–63, 67, 68
Phosphoglycerides, 32
Phospholipid head groups, 78
Phospholipid matrix, 118, 147, 207
Phospholipids, 4, 21, 32, 33, 39, 40, 74, 75, 77, 107, 147
Phosphorescence, 198
Phosphoric acid, 60-62
Phosphoric acid radical, 63
Photobleaching, 76, 204, 206, 208, 209
Photon energy, 197
Photons, 148, 173, 177, 188, 189, 191, 197–199, 204–206
Photosynthesis, 3, 4, 28, 43, 189–191
Photosynthetic organisms, 189
Physical
  interactions, 3, 4, 7, 52, 105, 214
  properties of water, 19
  signals, 73
Physiological saline solution, 112, 113, 220
Physiological conditions, 29, 58, 149
Ping-pong mechanism, 187
Ping state, 187
Planar bilayers, 37, 76
Planar membrane, 88–98
Plasma membrane, 40, 73–76, 84, 88, 93, 95, 98, 101, 115–117, 136–138, 143, 151, 188, 209
  capacitance, 98
  colloid osmotic pressure, 114, 115
Plasmolysis, 112
Point charges, 9, 10, 80
Point charge interactions, 9, 10
Poisson-Boltzmann equation, 81–83
Poisson equation, 81, 82, 87
Polar
  groups, 26, 27, 74
  head, 31–33, 35–37, 75, 76, 88, 117, 121
  head layer, 94
  moiety, 31
  molecules, 230
  nonpolar mixtures, 30
  region, 88
  side, 74
  solvents, 30, 32
  substances, 20
  water molecules, 26, 30
Polarizabilities, 14, 16, 78
Polarisable particles, 10
Polar layers, 8, 107
Polychromatic RF field, 212
Poly-ionic mosaics, 8
Polynucleotide(s), 63
Polypeptide(s), 27, 50, 51, 59, 116, 214
Polysaccharides, 7, 26, 30, 43, 46, 57
Pong state, 187
Population, 79, 82, 191, 196, 198, 201–204, 210
Population of nuclei, 210
Pores, 18, 40, 74, 78, 103, 173–181
Position, 9, 32, 107, 141, 142, 205
Postsynaptic membrane, 155
Potassium
  channels, 77, 166, 181–183
  conductance, 163, 165, 224
  current, 162, 163
  ion channels, 163, 166
  selective channels, 163
  specific conductance, 164
Potential
  difference, 88, 90, 147–153, 161, 164
  energy barrier, 2–7, 109
  energy profiles, 107, 108
  theory, 97
Power function, 144
Power law, 17, 133
Preferential adsorption of anions, 78
Preferential interactions, 30
Preferential stabilization of the cations, 182
Primary electron donor, 189
Primary structure, 2, 50, 51, 55, 56, 58, 63, 64, 180
Primitive cells, 40
Principles of NMR, 209
Programmed synthesis, 68
Prokaryotes, 40
Prokaryotic cell, 39, 41, 66
Proliferation rate, 44, 45
Prometaphase, 44
Propagation of action potential, 170, 173

Prophase, 44, 45
Protein
  association, 4, 195–197
  backbone, 52, 53, 55, 181
  biosynthesis, 41, 60, 62
  chain, 51–54, 178
  chlorophyll complexes, 43
  complexes, 7, 77, 189, 195, 196, 198, 201, 204, 208, 209, 216
  complex stoichiometry, 195
  folding, 2, 57, 58
  folding kinetics, 58
  fragment complementation, 196
  lipid interactions, 30
  polysaccharide interactions, 30
  population, 196
  protein interactions, 4, 26, 30, 69, 195–216
  specific ligand interactions, 26
  structure determinations, 1, 177, 180, 181
  subunits, 166, 182, 185
  synthesis, 42
  synthesis, processing and sorting, 42
  water channels, 115
Protein Data Bank, 51, 56, 68, 186
Proteinates of K^+ 150
Proteins
  energetic landscape, 59
  local folding, 52
  primary structure, 50, 51, 55, 56, 180
  secondary structure, 1, 52, 53
  tertiary structure, 55
  quaternary structure, 56
Proton concentration, 29, 30
  gradient, 188
Protons, 18, 23, 28–30, 69, 115, 209–212, 215
Protozoa, 40
Pufferfish, 162
Pulse of suction, 175
Purine/Purinic bases, 60, 63
Pyrimidines, 60
Pyrimidinic bases, 60

Q

Quantity of DNA, 66
Quantity of substance, 104
Quantum
  mechanics, 4, 16, 47, 197
  mechanical repulsion, 1, 7, 57
  number, 209
  transition, 197
  yield, 199, 200
  yields of acceptors and donors, 198–200
Quaternary structure, 50, 56, 57, 180
Quasicrystalline structure, 25

R

Radiofrequency range, 76, 88, 95, 98
Ramachandran angles, 50
Radiofrequency field (RFF), 69, 210-212, 215
Random amino acid sequences, 59
Random collisions, 67, 123, 209
Random encounters, 204, 207
Random suspensions
  dilute, 95
  collisions, 67, 123, 209
  concentrated, 95
  of particles, 95
Random coil, 52, 53
Ranvier node, 156, 169
Rate constant(s), 20, 33, 108, 109, 118, 124, 143, 165, 191, 197–201, 220
Rates of change, 109
Rate of complex formation, 118, 124, 143
Rate of emission of photons, 198
Rate of transfer of molecules, 108
Rates of formation and dissociation, 33
Real gases, 10, 13, 112
Real solutions, 113, 114
Reaction center(s), 189, 191
Rectangular temporal profile, 155
Rectangular stimulus, 157
Receptor-ligand complexes, 195
Receptor mediated (regulated), exocytosis, 102
Receptor mediated secretion, 101
Receptor protein Ste2 207–209
Red blood cells, 1, 4, 40, 113, 117, 143
Red-shifted photon, 197
Red-shifted wavelengths, 198
Refinement factor, 181
Reflected X-ray, 177, 179
Reflection angle/Refractive index, 177, 179, 199
Refractory period, 159, 224
Refractory state, 169, 224
Regenerative process, 159
Regulated secretion, 101
Relative dielectric constant, 9, 21, 95
Relative permittivity, 9, 14, 110, 125
Relaxation time, 91
Replication, 4, 39, 45, 47, 66–68
Replication fork, 67
Repulsive component, 36
Repulsive electrostatic interactions, 7
Residues, 50–52, 58, 214–216, see also side chains
Resistance, 107, 152, 155, 157, 158, 161, 167, 168, 174, 175
Resistant wall, 43
Resistivity, 168

Index

Resting membrane potential, 148
Resonance condition(s), 211
Resonance energy transfer, 189, 196, 198
Resonance frequency, 210, 212
Resting
  state, 153–155, 159, 163
  potential(s), 155
Reticular planes, 177–179
Reverse direction, 108
Reverse osmosis, 111
Ribbon model, 55
Ribbon representation, 54, 181, 183, 184
Ribonucleic acid (RNA), 41, 61
Ribonucleic-based (RNA), viruses, 39
Ribose, 61
Ribosomal RNA, 41, 42
Ribosomes, 40, 42, 57
Right-handed double helix, 64
Rhodopsin, 56
Room temperature, 8, 13, 19, 21, 29, 112–114, 149
Rotational diffusion, 75
Rough endoplasmic reticulum, 42

S

Salmonela thyphimurium 188
Saltatory propagation, 169
Saturation, 31, 106, 120, 126, 128–130, 214
Saxitoxin (STX), 120, 162, 216, 221
Scalar process, 188
Scaling arguments, 141, 142
Scaling factor, 135, 136, 139
Scattered photons, 177
Sierpinski's carpet, 134–136
Sierpinski's gasket, 135, 136, 222
Scatchard plot, 127
Second harmonics, 148
Second law of thermodynamics, 27, 46, 47, 79
Second intermediate dephosphorylated state, 187
Secretion, 42, 73, 101, 116
Secretory cells, 42
Sedimentation constant, 42
Sedimentation rate, 42, 200, 201
Sedimentation speed, 42
Selected sequences, 59
Selective
  blocking, 28
  channels, 125, 163
  control, 73
  membrane, 110
  permeable membrane, 110, 125
  reflection of X-rays, 177, 179
  transfer of water, 115

Selectivity filter, 182–184
Self association, 20, 32, 34, 35, 57, 74, 123, 145, 195
Self associations of amphiphiles, 32, 123
Self-maintaining process, 159
Self-propagating perturbations, 160
Self-reproduction, 39, 43, 46
Self-similar aggregates, 95
Semiconservative, 67, 68
Sensitive fluorescent dyes, 148
Sequential binding scheme, 131
Sequential model, 129
Sequential reactions, 128, 221
Series of successive divisions, 43, 45, 66
Sex chromosome(s), 45
Sexual reproduction, 43, 44
Shear force, 19
Shell of water, 20
Shelled spheres, 97
SH groups, 178
Shielding constant, 211, 212
Short-lived pores, 103
Short-range
  attractive interactions, 15
  biophysical interactions, 7
  interactions, 13, 16, 17
Sialic acid, 78
Side chains, 48–50, 52–54, 60, 69, 184, *see also* residues
Signal generator, 160
Signaling molecules, 195
Signal-to-noise ratio (S/N), 211
Signal transduction, 195
Simple diffusion, 103–106, 110, 116, 119, 120, 141
  diffusion through membranes, 105, 106
  (passive) diffusion, 121
Singer-Nicolson model, 93
Single-channel signals, 175, 176
Single channels, 174–176
Single-shell model, 97
  particles, 97
Singlet state, 197
Single unicellular organisms, 196
Sinuous channels and cisternae, 42
Skeletal muscles, 18
Small micelles, 18
Sodium
  channels, 159, 163, 166
  current, 162, 163
  concentration gradient, 163
  conductance, 163, 165
  specific conductances, 164
Solid water, 24

Solute, 31, 105, 107, 108, 110-114, 119, 120
Solute concentration, 105, 110, 112, 114
 concentration difference, 105, 106
 molecules, 106–108, 113
 particle, 107
Solvent molecule, 7, 103, 110
Solvent passive diffusion, 110
Somatic cell, 43
Soma, 159
Somatic chromosomes, 44, 45
Spatial
 arrangements of atoms, 180
 charge distribution, 8
 density distribution, 209
 derivatives, 169
 dimensions, 81, 107, 131, 132, 154, 205
 folding, 68
 organization, 18
 packing, 7
 variation, 105, 170
Special cell morphology, 115
Species, 18, 21, 26, 31, 33, 40, 43, 46, 48, 76, 81–83, 114, 117, 118, 123, 124, 126, 136, 137, 147–149, 152–154, 160, 191, 211, 220
Specific
 capacitance, 94, 95
 channels, 26, 116, 121, 152
 chemical triggers, 188
 conductances, 94, 152, 164
 conductivity, 152
 electrical capacitance, 87, 88, 94, 98
 electrical conductance, 163
 heat, 19, 20
 molecular interactions, 16
 physical triggers, 188
 resistances, 152, 157
 substrate, 17, 123
Spectral cross-talk, 201
Spectral deconvolution method, 208
Spectrin, 77
Spectroscopy
 dielectric, 88, 92
 impedance, 88
Spectrum, 199, 201, 202, 204, 206, 212, 213, 216
Speed gradient, 19
Spermatozoid, 45, 46
Sphalerite, 177
Spherical bilayers, 37
Spherical particles, 80, 81, 85, 104
Spheroidal micelles, 37
Sphyngosine, 74
Squid giant axon, 148, 157, 158, 163, 164

Stacked dielectric
 layers, 90, 92
Stämpfli currents, 169
Standard chemical potential, 31, 111, 149, 191
State of aggregation, 31, 111
Static electrical properties, 88–97
Statistical size distribution, 35
Steady flow, 108, 109
Steady state, 118, 143, 149
Stereo view, 183, 184, 214
Steric contribution, 35
Steric fit, 26
Stern compact double layer
 layer, 84, 87
 electrical potential, 86
 plane, 86
 potential, 80, 83
 space, 83
Stick representation, 55
Stimuli
 amplitude, 159
Stochastic fractals, 139
Stochastic FRET, 209
Stoichiometric information, 196
Stoichiometry, 124, 125, 195, 196, 201, 221
Stokes red-shift, 197
Structural complementarity, 16, 67
Structural
 defects, 133
 information, 69, 209
 model(s), 75, 93
 proteins, 47, 51, 59
 complementarity, 16, 67
Structured network, 28
Structured water, 26, 28, 74
Structure factors, 179, 180
Structure factor phases, 180
Structure refinement, 181
Subcellular organelles, 39, 73
Subcellular particles, 7, 40, 42
Sub-dispersion, 94, 97
Submicroscopic and microscopic particles, 42
Substance transport, 37
 across membranes, 4, 40, 101–122
Substrate, 17, 123, 195
Successive conformational changes, 186
Suction pressure, 111
Sugar, 27, 60, 63, 74, 116, 118, 119, 143
Sugar-phosphate radical, 64
Sun, 189, 192
Superhelices, 66
Suppliers of protons, 18
Supra-cellular architecture, 45, 139
Supraliminal finite-step stimulus, 164

Supraliminal infinite-step stimulus, 164
Supraliminal stimuli, 159, 170
Supramolecular
  assemblies, 7, 57
  level, 28
  structures, 7, 29
Surface
  area, 19, 37, 42, 88, 89, 98, 103, 104, 133, 139, 222
  area of contact, 30
  area per polar head, 35
  charges, 78–88
  chemical groups, 78
  tension, 8, 15, 19, 32
  tension coefficient, 19, 21, 36, 219
Surfactants, 32, 33
Suspension of cells, 1, 76, 98
Suspension of spherules, 96
Svedberg (S), 42
Svedberg theoretical model, 42
Symporter, 185
Symport system, 185
Synapse, 156
Synthetic lipid bilayers, 93

T

Tagged proteins, 197
Taylor series, 11, 82
Telophase, 44
Temporal
  behaviour, 144, 160, 162
  derivatives, 169
  variation, 105, 160, 164, 167, 170
Tensorial order, 188
Tertiary structure, 55, 56, 66
Tetra-coordinated water molecules, 23
Tetrahedral structure, 22
Tetrameric complex, 7
Tetrameric potassium channels, 77, 182
Tetramer, 25, 56, 57, 183, 195
Tetraethylammonium (TEA), 162
Tetra methyl silane (TMS), 212
Tetranucleotides, 63
Tetrapeptides, 50
Tetrodotoxin (TTX), 120, 162
Theoretical model(s), 42, 80, 140
Theory of
  dielectrics, 89
  excitation, 163
  measurements, 138
Thermal
  agitation, 25, 79, 103, 110, 123
  agitation energy, 24, 112
  inertia, 19
  insulator, 20
Thermodynamic stability, 75
Thermodynamic systems, 189
Thermolysis, 20, 28
Thermostability, 196
Thickness of the diffuse double layer, 84
Thickness of the hydrophobic layer, 95, 98
Three-dimensional structure, 2, 40, 54, 56, 64
Three-like structure, 2, 40, 54, 56, 64
3D protein structure, 55, 56, 180
  fibrous, 56
  globular, 56
  membranar, 56
Threshold, 157, 159
Threshold potential, 160
Thylakoid membranes, 189
Thylakoid sacks, 43
Thymine, 60
Time constant, 157
Tissue, 2, 3, 7, 18, 28, 45, 47, 73, 84, 114, 115, 137, 147, 206
Torsion stress, 66
Total current, 162, 163
Total fluorescence emission, 202
Total interaction energy, 15
Transcytosis, 101, 102
Transfer RNA, 41
Transfer of excitation, 198, 202
Transient currents, 160
Transient dipole, 16
Transition dipoles, 199, 200
Transition state ensemble, 58
Transitory structure, 23
Translational diffusion, 75, 76
Transmembrane
  current, 161, 162, 164, 167
  current density, 168
  domains, 185, 207
  ionic currents, 160
  potential, 147–155, 160, 161, 163–170, 182
  resistance, 158
Transport
  across membranes, 4, 40, 42, 47, 62, 101–122, 137, 140
  medium, 28
  vesicles, 209
Transversal ionic currents, 167
Trinucleotides, 63
Tripeptide, 50
Triplet state, 197, 198
Tritiated water, 21
Tritium, 21
Truncated cones, 37

Turgor pressure, 115
Twisted superhelix, 66
Two dimensional crystals, 186

U
Ultracentrifuges, 42
Unconventional geometry, 139
Unexcited acceptor, 198
Unexcited molecule, 198
Uni-cellular organisms, 196, 204
Uniform magnetic field, 209
Unimolecular reaction, 165
Uniporter, 117, 185, 188
Unit cell, 179, 180
Unit surface area, 104
Unmyelinated axon conductances, 158
Unmyelinated vertebrate axon, 158
Unpaired electrons, 42, 211
Unprogrammed DNA synthesis, 68
Unregulated secretion, 101
Uracil, 60

V
Vacuoles, 43, 115
van der Waals forces, 15, 17
van der Waals interactions, 10, 15, 17, 23, 65, 75
van der Waals spheres, 55
van't Hoff laws, 111–113
Variable fields, 89
Variation of entropy, 27, 46, 120
Vascular and bronchial tree, 138
Vectorial equation, 104, 224
Vectorial process, 188
Very short-range interactions, 17
Vibrational relaxation, 197
Virial equations, 113, 114
Viruses, 39, 73, 84, 86, 101
Voltage clamp method, 163
Voltage-clamp amplifier, 174
Voltage-clamp technique, 160, 161, 174
Voltage clamp setup, 166
Voltage dependent channels, 155
Voltage-dependent potassium channels, 181
Voltage-gated, 183
Voltage-gated K^+ channels, 181, 182
Voltage-sensitive gate, 182
Voltage sensor, 182

Volume, 4, 11, 25, 35, 81, 87, 95, 103, 110-112, 114, 115, 123, 133, 137, 139, 141, 179, 191, 206, 222
Volume fraction, 96, 98
Viscosity coefficient, 19, 85, 104, 195

W
Water
 ascension, 111
 channels, 115, 182
 content, 18
 fluidity, 25
 hating, 20, 30
 hydrocarbon interface, 35
 molecules, 7, 10, 18, 20-30, 50, 54, 55, 74, 78, 107, 110-112, 115, 124, 126, 182, 219
 molecule association, 24
 molecule cluster, 24
 oxygen, 20
 percentage, 18
 shells, 28
 volume, 25
Wave equation, 170
Weak interactions, 16
Wetting, 15, 28
Wheel representation, 52, 53
Whole-cell recording, 175
Wire electrodes, 160

X
X monophosphate nucleotide (XMP), 61
X-ray beam, 69
X-ray crystallography, 68, 177, 178, 181, 182, 216
X-ray diffraction, 24, 56, 68, 69, 177–181, 209
X-ray diffraction diagram, 65

Y
Yeasts, 40, 43, 44, 98, 141, 196, 207, 208
Yellow fluorescent protein (YFP), 196, 207, 208

Z
Zeta potential (electokinetic potential), 85, 87
Zwitterion, 48
Zwitterionic, 32
Zygote, 45, 46